Springer Series in

OPTICAL SCIENCES 146

founded by H.K.V. Lotsch

Springer Series in

OPTICAL SCIENCES

The Springer Series in Optical Sciences, under the leadership of Editor-in-Chief *William T. Rhodes*, Georgia Institute of Technology, USA, provides an expanding selection of research monographs in all major areas of optics: lasers and quantum optics, ultrafast phenomena, optical spectroscopy techniques, optoelectronics, quantum information, information optics, applied laser technology, industrial applications, and other topics of contemporary interest.

With this broad coverage of topics, the series is of use to all research scientists and engineers who need up-to-date reference books.

The editors encourage prospective authors to correspond with them in advance of submitting a manuscript. Submission of manuscripts should be made to the Editor-in-Chief or one of the Editors. See also www.springer.com/series/624

Please view available titles in *Springer Series in Optical Sciences* on series homepage http://www.springer.com/series/624

Jesper Glückstad
Darwin Palima

Generalized Phase Contrast

Applications in Optics and Photonics

Springer

Professor Jesper Glückstad, PhD, DSc
DTU Fotonik, Department of Photonics Engineering
Technical University of Denmark
DK-2800 Kgs. Lyngby, Denmark

Darwin Palima, Assistant Professor, PhD
DTU Fotonik, Department of Photonics Engineering
Technical University of Denmark
DK-2800 Kgs. Lyngby, Denmark

Programmable Phase Optics: www.ppo.dk

Published by Springer,
P.O. Box 17, 3300 AA Dordrecht, The Netherlands
In association with
Canopus Academic Publishing Limited,
15 Nelson Parade, Bedminster, Bristol, BS3 4HY, UK

Springer Series in Optical Sciences ISSN 0342-4111 e-ISSN 1556-1534

ISBN 978-90-481-2838-9 ISBN 978-90-481-2839-6 (eBook)

DOI 10.1007/978-90-481-2839-6

Springer Dordrecht Heidelberg London New York

Library of Congress Control Number: 2009931245

Printed on acid-free paper

9 8 7 6 5 4 3 2 1

Springer is part of Springer Science+Business Media (www.springer.com)

Preface

This book is based on the authors' work for more than a decade, initiated by a generic patent application for the generalized phase contrast (GPC) method in the mid-nineties. In short, the GPC invention propounds a generalization of Nobel Laureate Fritz Zernike's original phase contrast method not only in terms of a wider domain of theoretical operation but, in particular, also opening up new and exciting applications beyond optical microscopy. After the issuing of this key patent for GPC, a number of associated application patents have been filed, in addition to more than 150 papers and conference presentations on the theoretical and experimental aspects of GPC and its applications. A culmination came in early 2005 when one of the authors (J. Glückstad) defended his dissertation on the GPC method, for which he obtained a higher doctorate degree (Doctor of Science) from the Technical University of Denmark. It was at this point that the idea was originally fostered to write a monograph and explain to a wider audience about GPC and its applications in contemporary optics and photonics. The present book is strongly supported by a rich portfolio of research work, both of theoretical and experimental nature, which have been undertaken in collaboration with a number of scientists around the world whom we would like to explicitly acknowledge for their key contributions: L. Lading, H. Toyoda, T. Hara, Y. Suzuki, N. Yoshida, P. C. Mogensen, R. L. Eriksen, V. R. Daria, S. Sinzinger, P. J. Rodrigo, C. A. Alonzo, N. Arneborg, I. Perch-Nielsen, P. Bøggild, J. Jahns, P. Ormos and L. Kelemen.

Copenhagen, Denmark, 1 July 2009

Jesper Glückstad Darwin Palima

"With the phase-contrast method still in the first somewhat primitive stage, I went in to the Zeiss Works in Jena to demonstrate. It was not received with such enthusiasm as I had expected. Worst of all was one of the oldest scientific associates, who said: 'If this had any practical value, we would ourselves have invented it long ago'. Long ago, indeed!"

Fritz Zernike

Contents

Chapter 1
Introduction

The term "phase contrast" was originally coined in allusion to the conventional practice of using contrast agents in microscopy to view details in biological samples. Biological specimens are essentially transparent, owing to minimal absorption heightened by the typically thin sample preparations. Thus, they generate poor images with insufficient contrast under a standard microscope. Instead of relying on external contrast agents to improve absorption, the Dutch physicist Frits Zernike invented a phase contrast microscope that uses the phase alterations imparted by a transparent sample onto an incident illumination as the source of contrast in the microscope image to render vivid details of the specimen. By eliminating extraneous chemical agents, Zernike's phase contrast microscope is able to show clear images of living samples, which led to a breakthrough in medicine and biology and earned him the 1953 Nobel Prize in Physics.

The study of biological specimens illustrates one among numerous possible uses of phase visualization. The imaging and visualization of optical phase, such as wavefront modulation, disturbances or aberrations, is a challenging yet often vital requirement in optics. A number of techniques can be applied in fields ranging from optical component testing through to wavefront sensing whenever a qualitative or quantitative analysis of an optical phase disturbance is required. In general, a phase disturbance cannot be directly viewed and a suitable method must therefore be sought to extract information about the wavefront from an indirect measurement. An example of this is the generation of fringe patterns in an interferometer, which gives information about the flatness of an optical component without requiring a physical measurement of the component surface. In this book, we discuss a powerful phase contrast technique coined "generalized phase contrast" [1] that we have developed for the visualization of phase disturbances, outlining the considerable improvements this method offers over previous schemes that lead to a variety of powerful applications in optics and photonics.

Phase visualization is typically achieved through interferometry and a number of interferometric techniques can be classed as common-path interferometry. In a common-path interferometer (CPI) the signal and reference beams travel along the same optical axis and interfere at the output of the optical system. Put simply, this means that we perturb a portion of the wavefront we wish to measure to create a reference wavefront, and it is the interference between the unperturbed wavefront information and this synthesized reference wave that allows the visualization of the phase information in the original wavefront. The Zernike phase contrast method [2] is possibly the most widely known implementation of CPI. However, CPIs exist in many different forms, such as the point diffraction interferometer, dark central ground filtering, field absorption and phase contrast methods. Although special cases such as the Zernike method have been previously treated, a comprehensive approach to the analysis of a generic CPI has been lacking.

1.1 The Generalized Phase Contrast Method

One of the aims of this book is to set down a rigorous analytical framework describing the operation, design and optimization of common-path interferometers. The approach we use is based on a generalization of Zernike's phase contrast approach [2, 3], which breaks away from the restrictions of the so-called "small-scale" phase approximation that limits Zernike's original method. The generalized phase contrast (GPC) approach provides a straightforward methodology for the optimization of a broad range of different CPI types and supports and enhances the more empirical methods for CPI optimization that are generally employed. Since the GPC method is not limited by the operational constraints of Zernike's original method, it is possible to convert phase information into high-contrast intensity distributions by a careful choice of the spatial filtering parameters to match the phase disturbance. Hence, the theoretical framework of the GPC method conveniently provides for designing CPI systems that address a range of applications and potentially achieves optimal performance in terms of accuracy, visibility and peak irradiance.

We use Zernike's original method as the starting point in our explanation of the requirements for a generalized description and clarify the limiting conditions for its application and explain the requirements for a generalized description. Subsequently, we discuss how our approach, based on the GPC method, can be used for the optimization of the whole class of generic CPI systems. We provide a detailed description of the mathematical approach that we have adopted and analyse the importance of the so-called synthetic reference wave (SRW) profile for obtaining optimal performance from a CPI. We then consider the parameters that can best quantify the performance of a generic CPI system and use the simultaneous optimization of the visibility and peak irradiance as the two key parameters describing the output of any CPI. We demonstrate that our analysis can be used to treat any one of a number of different CPI types with different Fourier plane filtering parameters. Through the introduction of a combined filter parameter, we develop graphical methods for the optimization of the visibility and peak irradiance for a given phase distribution at the input of the CPI. We apply our analysis to some well-known CPI systems and consider how they might be further optimized.

Of particular importance in the development of CPI systems is the desire to produce a linear mapping between the input phase of a wavefront and the output intensity. The work of Henning [4, 5] is an extension of the linear phase-to-intensity mapping range for the Zernike method. Using the analysis presented in this book we can demonstrate that it is possible to generalize Henning's phase contrast method to have a wider functionality. We take this opportunity to look in detail at the difference between ambiguous and unambiguous phase-intensity mapping and discuss the possible improvements to some currently applied CPI systems.

1.2 From Phase Visualization to Wavefront Engineering

The generalized treatment of common-path interferometry offers possible enhancements to phase contrast microscope systems that allows for accurate quantitative phase microscopy and wavefront sensing in general. Most proposals for a CPI-based quantitative phase microscopy use phase-shifting interferometry to retrieve the input phase using multiple intensity measurements but fall back to Zernike's plane wave SRW model for analysis. Thus, they can run into increasing errors for the part of the phase structures modulated away from the centre of the observation field. The SRW description in GPC reduces this error and allows accurate quantitative phase microscopy over a wider field of view.

The GPC framework affords an improved understanding of the intricacies of the phase contrast mechanism and paves the way for deploying GPC-based CPI systems across diverse applications beyond phase microscopy. Aside from accurately visualizing unknown phase inputs, GPC allows us to exploit contemporary spatial light-modulation technologies to directly control the spatial phase of an input wavefront and generate desired intensity patterns. Equations derived from the generalized framework can be used for optimising visibility and light efficiency of arbitrary analogue phase-encoded wavefronts. Synthesizing high-efficiency intensity patterns is made possible by using practically lossless phase-only modulation at the input plane and filter planes of a GPC system. For experimental expediency, one may choose to adopt ternary-phase encoding, or even binary-only phase encoding, which may also be optimized within the GPC framework to generate intensity patterns with high efficiency.

Equipped with a well-characterized phase-to-intensity mapping system, we can envision a converse system that synthesizes desired phase patterns from an input intensity pattern. This intensity-to-phase conversion technique, realized based on a reversal of the GPC method, is referred to as reverse phase contrast (RPC). The fundamental idea behind the RPC method is the desire to be able to obtain spatial phase modulation by use of a simple and robust amplitude modulator inserted in a CPI configuration. As a potential application, one can use an amplitude modulator to generate a phase hologram and benefit from the advantages of a phase hologram over an amplitude hologram.

1.3 GPC – an Enabling Technology

The parameter optimization prescribed by the GPC method and its reversed mode, the RPC method, lends itself to interfacing with powerful applications that can exploit and benefit from the versatility of the technique. A direct application of the generalized phase contrast analysis is in performing more accurate measurements that allow for quantitative analysis of phase objects, such as biological specimens, or in other wavefront-sensing applications where accurate phase information is needed. Using contemporary spatial modulator technologies, user-programmed phase information can be used with GPC to generate high-efficiency intensity projections for image display and beam-shaping. Accurate phase-to-intensity conversion may also be exploited in optical encryption and decryption systems where vital information can be phase-encoded and scrambled for increased security and then descrambled with the proper key and the information retrieved from its phase-encoded form with a generalized phase contrast technique.

Coupled to microscope objectives, GPC-generated intensity patterns can be used for spatially controlled light–matter interaction, such as in materials processing. These microstructured light patterns are also effective in manipulating microscopic particles such as colloidal suspensions and biological cells, since optical forces exert a significant effect on particle dynamics at these length scales. Thus, phase contrast, which started as a tool for visualizing biological samples, can be exploited in its generalized form to generate light for controlled manipulation of the samples under observation. GPC's capacity for energy-efficient laser pattern projection makes it an essential element in the optical toolbox for the expanding set of photonics applications.

The core framework of the generalized phase contrast method is discussed in Chapters 2 and 3, supplemented with a phasor-based analysis outlined in Chapter 4. Wavefront sensing and analysis using GPC is developed in Chapter 5, where we show how GPC can prescribe optimal parameters for these applications. We also demonstrate how GPC can reduce the error in quantitative phase microscopy. Optimization for wavefront engineering is treated in Chapter 6, where we outline various encoding possibilities for binary projections and advance into the more general aspects of grey-level encoding. These optimizations are utilized in Chapter 7, where we present applications of GPC-based light projections. The highly successful application of GPC in interactive real-time multi-particle optical trapping and micromanipulation is discussed in Chapter 8. We present alternative GPC implementations in Chapter 9, such as the planar integrated micro-optics platform. We re-examine the common-path interferometer in Chapter 10, this time considering an amplitude-modulated input, and optimize this system to develop the reverse phase contrast (RPC) method. Operating effectively in reverse as the phase contrast effect, RPC can convert an amplitude input into a phase-only modulated light. We present several experimental demonstrations, which include high-speed phase modulation using a digital micromirror device. Finally, we apply generalized phase contrast principles in phase-only optical cryptography in Chapter 11. Exploiting the robustness of the micro-optics platform, a miniaturized GPC implementation can be an attractive option for applications in optical encryption and decryption. The concept of miniaturization is also combined with elements of GPC and RPC in a novel optical decryption system proposed in this chapter.

1.4 GPC as Information Processor

The plurality and diversity of GPC-powered applications are unified in GPC's character as an information processor. In these applications, one may consider GPC as an information-processing channel where input phase information is processed and communicated as output intensity.

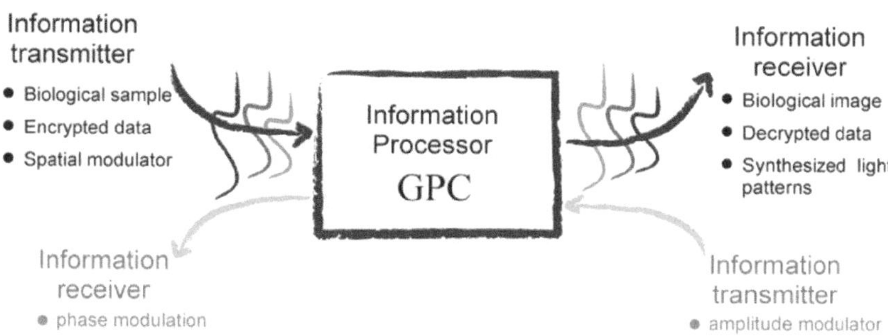

In sensing applications, the output visualizes the otherwise invisible phase data to gain vital information about an object under study. In wavefront synthesis, the user defines the input data to specify the information about intensity output characteristics. The reverse phase contrast shows that information may also be sent along the opposite direction through this information processor.

References

1. J. Glückstad, "The Generalized Phase Contrast Method", 322 pp. (Doctor of Science Dissertation, Technical University of Denmark, 2004).
2. F. Zernike, "How I discovered phase contrast", Science **121,** 345-349 (1955).
3. J. W. Goodman, *Introduction to Fourier Optics* (McGraw-Hill, San Francisco, 2nd ed., 1996).
4. H. B. Henning, "A new scheme for viewing phase contrast images," Electro-optical Systems Design **6,** 30-34 (1974).
5. G. O. Reynolds, J. B. Develis, G. B. Parrent, Jr., B. J. Thompson, *The New Physical Optics Notebook: Tutorials in Fourier Optics,* (SPIE Optical Engineering Press, New York 1989) Chap. 35.

Chapter 2
Generalized Phase Contrast

Light intensity is easily quantified by using calibrated detectors that can directly exploit the energy flux from an incident light. Spatial intensity variations can be imaged using an array of such energy detectors, as in a camera, for instance. On the other hand, light phase is invisible to energy detectors and is usually detected indirectly by exploiting phase-dependent phenomena that affect intensity. For example, intercepting light with a lenslet array would generate an array of spots at the common focal plane of the lenslets and any phase perturbations could be deduced from observed changes in the configuration of the intensity spots. When using coherent illumination, a common method consists of introducing a reference beam and then analyzing the phase-dependent interference pattern to determine the phase perturbation. Working without the benefit

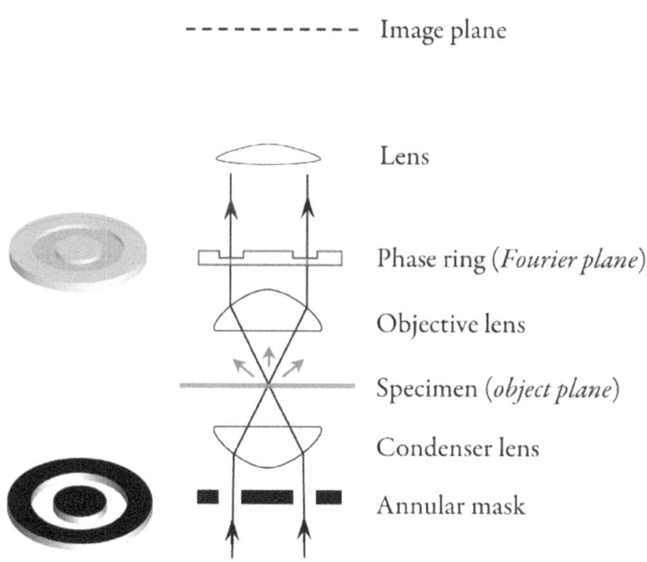

Fig. 2.1 A simplified schematic of a typical Zernike phase contrast microscope.

of coherent laser sources, Gabor invented the first holograms capable of interferometrically recording phase information by adapting Zernike's phase contrast configuration, where the reference and the object beams propagate along a common path to ensure coherence [1]. Aside from coherence, common-path interferometry also surmounts typical experimental hurdles that tend to smear out the interference pattern with its relative tolerance to vibrations and fluctuations in the ambient conditions, which becomes a major problem when the reference beam travels along a different path.

The accuracy of the extracted phase information from the output of an interferometer is dependent on assumptions about the reference wave, and this is no different for a common-path interferometer. Thus it is vital to examine how a phase contrast method models the reference wave in order to understand its limitations. In this chapter, we examine the assumptions employed in Zernike's phase contrast method. Although sufficient for very thin phase objects like biological samples, its limited range of validity necessitates a generalized formulation to encompass a wider range and broaden the horizon for possible applications.

2.1 Zernike Phase Contrast

The Dutch physicist Fritz Zernike received a Nobel Prize in 1953 for demonstrating the phase contrast method and inventing the phase contrast microscope. Zernike's invention paved the way for breakthroughs in medicine and biology by making living biological samples, like cells or bacteria, clearly visible under a microscope. Being generally colourless and transparent, biological samples are essentially invisible under a regular microscope unless one employs contrast dyes that can potentially harm the cells and prevent the observation of natural biological processes.

Zernike's phase contrast method [2–5] renders vivid details of transparent objects by converting the phase perturbations introduced by the object into observable intensity fluctuations by the use of a phase shifting filter at the spatial Fourier plane that imparts a relative phase shift on the undiffracted light components. A simplified schematic is shown in Fig. 2.1, which is based on the eventual implementation that uses conical sample illumination and a phase ring filter (Zernike also considered different combinations of illuminations and filters, as shown in ref. [4]).

Thin biological specimens are typically weak phase objects that introduce minimal phase perturbations, $\phi(x,y)$, onto an incident light. Thus it is sufficient to describe the incoming phase distribution by a "small-scale" phase approximation where the largest phase deviation is typically taken to be significantly less than $\pi/3$ [5]. When the input phase distribution is confined to this limited range, a Taylor expansion to first order is sufficient for the mathematical treatment so that the input wavefront can be written as

$$\exp\left[j\phi(x,y)\right] \approx 1 + j\phi(x,y). \tag{2.1}$$

For this first order approximation, the constant term represents the undeflected light while the spatially varying second term represents scattered light. The light corresponding to the two terms in this "small-scale" phase approximation can be spatially separated by placing the input phase distribution at the front focal plane of a lens to generate the corresponding spatial Fourier transformation at the back focal plane. In this geometry, light represented by the constant term is focused on-axis while the varying term is scattered off-axis, assuming an on-axis plane wave illumination of the phase object[†]. Owing to the weak-phase approximation, any unscattered component from the varying term may be reasonably neglected. Zernike realized that it is possible to generate interference between the two phase-quadrature terms in Eq. (2.1) by introducing a small quarter-wave-shifting plate to act on the focused light. As a result, the output intensity becomes

$$I(x',y') \approx \left| j + j\phi(x',y') \right|^2 \approx 1 + 2\phi(x',y'), \tag{2.2}$$

which enables phase visualization characterized by a linear phase-to-intensity transformation within the valid regime of the small-scale approximation. An approximately linear phase-to-intensity conversion is therefore achieved by phase contrast microscopes when studying thin and transparent biological specimens.

It should be noted that a three-quarter waveplate works equally well to produce phase contrast, but the plus sign in Eq. (2.2) is negated, leading to so-called negative phase contrast. Although linear, the phase-to-intensity mapping only applies to weak phase objects, which makes the second term in Eq. (2.2) significantly smaller than the constant term. This results in a very restricted intensity modulation depth. A substantial improvement in the visibility can be achieved in a Zernike phase contrast visualization, at the expense of light efficiency, by strongly dampening the focused light in addition to the phase shift required to generate the contrast [5].

2.2 Towards a Generalized Phase Contrast Method

In the general case, the input phase modulation is not limited to a small-scale perturbation and, hence, a first-order series expansion, as in the Zernike approximation will insufficiently represent the phase-modulated input. In this generalized regime, higher-order terms in the expansion need to be taken into account, so the expansion takes the form:

$$\exp\left[j\phi(x,y) \right] \approx 1 + j\phi(x,y) - \tfrac{1}{2}\phi^2(x,y) - \tfrac{1}{6} j\phi^3(x,y) + \tfrac{1}{24}\phi^4(x,y) + \ldots \tag{2.3}$$

[†] Most modern phase contrast microscopes do not use on-axis plane wave illumination but a superposition of plane waves incident at a cone of illumination angles, similar to the schematic in Fig. 2.1. In this case, the lens focuses the undiffracted light, the constant term in Eq. (1), into a ring at the back focal plane where a phase ring introduces a quarter-wave phase shift to produce interference and phase visualization at the output.

The spatially varying terms can potentially contribute to the undiffracted light even for weak phase objects, $\phi(x,y) \le \pi/3$, but they are much smaller compared to the constant first term in the Taylor series expansion and can, thus, be reasonably neglected without serious errors. However, the spatially varying terms can significantly contribute to the on-axis light for inputs with larger modulation depths. In this case, they can no longer be neglected or considered as separate from the focused light, as is customary in the Zernike approximation. The contributions from the spatially varying terms can result in a significant modulation of the focal spot amplitude on the back focal plane of the lens. These terms can, in fact, result in either constructive or destructive interference with the on-axis light, although the net result, based on conservation of energy, will be an attenuation of the focused light amplitude which can only achieve a maximum value for a perfect, unperturbed plane wave at the input.

However, examples abound in the literature where the contribution of the higher-order terms in the Taylor expansion is neglected and it is assumed that only the first term in the Taylor series expansion contributes to the strength of the focused light [8–15]. In particular, some frequently cited derivations of phase contrast [12, 14], whilst correct within the small-scale Zernike approximation, generate significant errors if extended to cover larger-scale phase perturbations. However, the fact that certain results derived for small-scale phase contrast imaging (including for example the dark-field method) are expressed by use of a general phasor notation, $\exp\left[j\phi(x,y)\right]$, may explain why some flawed analyses continue to propagate in the phase contrast literature. For phase contrast or dark field imaging of large-scale phase objects, the use of a first-order Taylor expansion based analysis, which is actually only valid within the regime of the small-scale phase approximation, is unacceptable.

For phase objects breaking the first-order Zernike approximation we must identify an alternative mathematical approach to that of the Taylor expansion given by Eq. (2.3). A Fourier analysis of the phase object provides for a more suitable technique for completely separating the on-axis and higher spatial frequency components. This gives the following form for $\exp\left[j\phi(x,y)\right]$, where $(x,y)\in\Omega$ defines the spatial extent of the phase object:

$$\exp\left[j\phi(x,y)\right]=\left(\iint_{\Omega}\mathrm{d}x\mathrm{d}y\right)^{-1}\iint_{\Omega}\exp\left[j\phi(x,y)\right]\mathrm{d}x\mathrm{d}y+\text{“higher frequency terms”} \quad (2.4)$$

In this Fourier decomposition the first term is a complex valued constant linked to the on-axis focused light from a phase object defined within the spatial region, Ω, and the second term describes light scattered by spatially varying structures in the phase object. Comparing Eq. (2.3) and Eq. (2.4), it is apparent that the first term of Eq. (2.3) is a poor approximation to the first term of Eq. (2.4) when operating beyond the Zernike small-scale phase regime.

A key issue to keep in mind when analysing the effect of spatial filtering on the incoming light diffracted by phase perturbations is the definition of what spatially constitutes focused and scattered light. In the previous description of Zernike phase contrast it was assumed that the focused light is spatially confined to a somewhat unphysical delta function, which is evident when taking the Fourier transform of Eq. (2.1):

$$\Im\{\exp[j\phi(x,y)]\}\approx\delta(f_x,f_y)+j\Im\{\phi(x,y)\} \tag{2.5}$$

where (f_x,f_y) indicates coordinates in the spatial frequency domain. Ensuring that only the focused light is subjected to the quarter-wave phase delay requires a filter with unphysical delta function dimensions. As we know, any aperture truncation that typically occurs within a practical optical system will lead to a corresponding spatial broadening of the focused light. It is therefore essential that we define the terms "focused light" and "scattered light" explicitly for such a system. Furthermore, the finite aperture effects from a physical phase-shifting filter must be accounted for so as to accurately describe its influence on the observed output intensity. We will also need to re-examine the required phase shift for the filter since the quarter-wave delay was derived based on the "small-scale" phase approximation under the unphysical assumptions of delta functions for the focused light and apertures.

In this context it is necessary to look more carefully at the sequence of apertures confining the light wave propagation through a typical optical set-up – one that we shall describe in detail in the succeeding chapters. After properly accounting for the aperture effects, we carry the analysis to a level that allows us to determine the appropriate filter parameters. This combined analysis forms the core of the generalized phase contrast method.

References

1. D. Gabor, "Holography, 1948-1971," from Nobel Lectures, Physics 1971-1980, Editor Stig Lundqvist, World Scientific Publishing Co., Singapore, 1992
2. F. Zernike, "How I discovered phase contrast", Science **121,** 345-349 (1955).
3. F. Zernike, "Phase contrast, a new method for the microscopic observation of transparent objects. Part I," Physica **9**, 686-698 (1942).
4. F. Zernike, "Phase contrast, a new method for the microscopic observation of transparent objects. Part II," Physica **9**, 974-986(1942).
5. J. W. Goodman, *Introduction to Fourier Optics* (McGraw-Hill, San Francisco, 2nd ed., 1996).
6. H. B. Henning, "A new scheme for viewing phase contrast images", Electro-optical Systems Design **6,** 30-34 (1974).
7. G. O. Reynolds, J. B. Develis, G. B. Parrent, Jr., B. J. Thompson, *The New Physical Optics Notebook: Tutorials in Fourier Optics*, (SPIE Optical Engineering Press, New York 1989) Chap. 35.

8. H. H. Hopkins, "A note on the theory of phase-contrast images", Proc. Phys. Soc. B., **66**, 331-333 (1953).
9. S. F. Paul, "Dark-ground illumination as a quantitative diagnostic for plasma density", Appl. Opt., **21**, 2531-2537 (1982).
10. R. C. Anderson and S. Lewis, "Flow visualization by dark central ground interferometry", Appl. Opt. **24**, 3687 (1985).
11. M. P. Loomis, M. Holt, G. T. Chapman and M. Coon, "Applications of dark central ground interferometry", Proc. of the 29th Aerospace Sciences Meeting, AIAA **91-0565**, 1-8 (1991).
12. D. Malacara, *Optical shop Testing, 302-305* (John Wiley & Sons, New York 2nd ed., 1992).
13. A. K. Aggarwal and S. K. Kaura, "Further applications of point diffraction interferometer", J. Optics (Paris) **17**, 135-138 (1986).
14. M. Born and E. Wolf, *Principles of Optics*, 426-427 (Pergamon Press, 6th ed., 1980).
15. C. A. Mack, "Phase contrast lithography", Proc. SPIE **1927**, 512-520 (1993).
16. Y. Arieli, N. Eisenberg and A. Lewis, "Pattern generation by inverse phase contrast", Opt. Comm. **138**, 284-286 (1997).

Chapter 3
Foundation of Generalized Phase Contrast:
Mathematical Analysis of Common-Path Interferometers

A well-focused standard microscope forms a transparent image of a transparent object, and hence will not be suitable for visualizing the features of such an object. However, although the features remain invisible, the transparent image replicates the characteristic phase variations of the object. Thus the spatial features of the transparent object can be visualized by subjecting its transparent image to interferometric measurements. Rather than introducing an external reference beam, a common-path interferometer synthesizes the reference wave using the undiffracted light from the object. A phase contrast microscope, such as that of Zernike, can therefore be built around a common-path interferometer. Unfortunately, the assumptions in a standard Zernike phase contrast analysis limit its domain of applicability to objects satisfying the small-scale phase approximation. Moreover, the use of a plane-wave approximation for the illumination implies a similar description of the undiffracted beam, which then mathematically focuses to a Dirac delta function and requires a matching but unphysical Dirac-delta filter.

In this chapter, we use a common-path interferometer as a model system for developing the generalized phase contrast (GPC) method. The resulting generalized formulation encompasses input wavefronts with an arbitrarily wide phase modulation range and realistically incorporates physical effects arising from intrinsic system apertures. Generalized phase contrast not only allows us to prescribe optimal parameters in phase contrast microscopy but also enables us to expand phase contrast applications into novel contexts beyond microscopy that will be treated in succeeding chapters.

3.1 Common-Path Interferometer: a Generic Phase Contrast Optical System

A commonly applied architecture for common-path interferometry is illustrated in Fig. 3.1. This architecture is based on the so-called $4f$ optical processing configuration and provides an efficient platform for spatial filtering. An output interferogram of an unknown phase object or phase disturbance is obtained by applying a truncated on-axis filtering operation at the spatial frequency domain between two Fourier transforming lenses (L1 and L2). The first lens performs a spatial Fourier transform so that directly propagating, undiffracted light is focused to the on-axis filtering region whereas light representing the spatially varying object information is scattered to locations outside this central region.

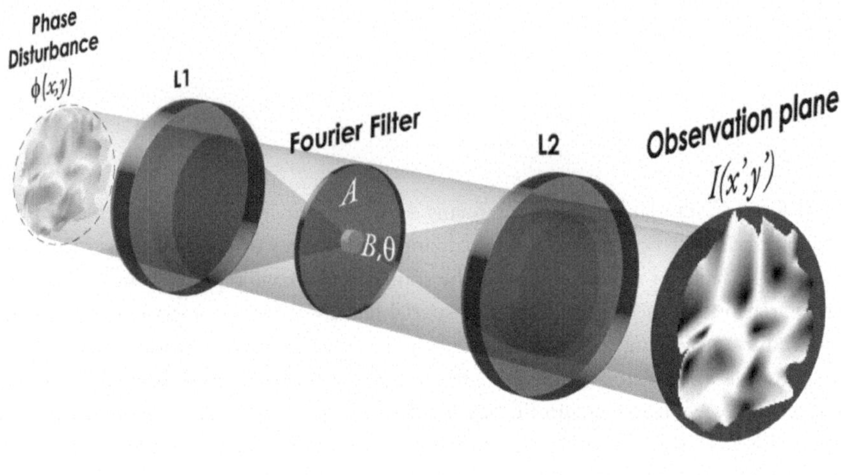

Fig. 3.1 Generic CPI based on a $4f$ optical system (lenses L1 and L2). The input phase disturbance is shown as an aperture-truncated phase function, $\phi(x,y)$, which generates an intensity distribution, $I(x',y')$, in the observation plane by an on-axis filtering operation in the spatial frequency plane. The values of the filter parameters (A, B, θ) determine the type of filtering operation.

We can describe a general Fourier filter in which different phase shifts and amplitude damping factors are applied to the "focused" and "scattered" light. In Fig. 3.1, we show a circularly symmetric Fourier filter described by the amplitude transmission factors, A and B, as well as phase shifts, θ_A and θ_B, for the "scattered" and "focused" light, respectively. For simplicity, all of the succeeding discussions will simply characterize the filter using the relative phase shift, $\theta = \theta_B - \theta_A$, since, after all, it is the relative phase, θ, that affects the output and not the actual values of θ_A and θ_B. The parameters A, B and θ provides a generalized filter specification and properly choosing their values can replicate any one of a large number of commonly used spatial filtering types (i.e. phase contrast, dark central ground, point diffraction and field-absorption filtering). Without a Fourier filter, the second Fourier lens will simply perform an inverse transform, albeit with reversed coordinates, to form an inverted image of the input phase variations at the observation plane. By applying a Fourier filter, the second Fourier lens transforms the phase-shifted, focused on-axis light to act as a synthetic reference wave (SRW) at the output plane. The SRW interferes with the scattered light to generate an output interference pattern that reveals features of the input phase modulation. In the following section we discuss the importance of the SRW and show how it influences, among other things, the choice of the Fourier filter parameters.

3.2 Field Distribution at the Image Plane of a CPI

Having described the generic optical system that makes up the CPI, we now turn to a detailed analytical treatment of the important elements in this system. Let us assume a circular input aperture with radius Δr truncating the phase disturbance modulated onto a collimated, unit amplitude, monochromatic field of wavelength λ. We can describe the complex amplitude, $a(x,y)$, of the light emanating from the entrance plane of the optical system shown in Fig. 3.1 by,

$$a(x,y)=\mathrm{circ}(r/\Delta r)\exp\left[j\phi(x,y)\right], \tag{3.1}$$

using the definition that the circ-function is unity within the region $r=\sqrt{x^2+y^2}\leq\Delta r$ and zero elsewhere.

Similarly, we assume a circular on-axis centred spatial filter of the form:

$$H(f_x,f_y)=A\left[1+\left(BA^{-1}\exp(j\theta)-1\right)\mathrm{circ}\left(f_r/\Delta f_r\right)\right] \tag{3.2}$$

where $B\in[0;\ 1]$ is the chosen filter transmittance of the focused light, $\theta\in[0;\ 2\pi]$ is the applied phase shift to the focused light and $A\in[0;\ 1]$ is a filter parameter describing field transmittance for off-axis scattered light as indicated in Fig. 3.1. The spatial frequency coordinates are related to spatial coordinates in the filter plane such that:

$$(f_x,f_y)=(\lambda f)^{-1}(x_f,y_f) \text{ and } f_r=\sqrt{f_x^2+f_y^2} \text{ [1].}$$

Assuming, for simplicity, a unity-magnification imaging, the output field is obtained by performing an optical Fourier transform (denoted by $\Im\{\ \}$ operator) of the input field from Eq. (3.1) followed by a multiplication of the filter parameters in Eq. (3.2) and a second optical Fourier transformation (Note: from here on, the second Fourier operation is replaced by the inverse Fourier operation $\Im^{-1}\{\ \}$ since their only difference is a negation of coordinates, which is an image inversion of the image in the output plane). In mathematical form, the sequence of operations is shown below:

$$\begin{aligned}\Im^{-1}\left\{H(f_x,f_y)\Im\{a(x,y)\}\right\}=A\Big[&\exp\left(j\phi(x',y')\right)\mathrm{circ}\left(r'/\Delta r\right)\\&+\left(BA^{-1}\exp(j\theta)-1\right)\Im^{-1}\left\{\mathrm{circ}\left(f_r/\Delta f_r\right)\Im\left\{\mathrm{circ}\left(r/\Delta r\right)\exp(j\phi(x,y))\right\}\right\}\Big]\end{aligned} \tag{3.3}$$

3.2.1 Assumption on the Phase Object's Spatial Frequency Components

The task of performing the inverse Fourier transform operation, $\Im^{-1}\{\ \}$, at the right-hand side of Eq. (3.3) is facilitated by a proper understanding of the different quantities that factor into the operand given by $\mathrm{circ}(f_r/\Delta f_r)\Im\{\mathrm{circ}(r/\Delta r)\exp(j\phi(x,y))\}$. From the properties of the Fourier transform, this can be rewritten as $\mathrm{circ}(f_r/\Delta f_r)\left(\Im\{\mathrm{circ}(r/\Delta r)\}\otimes\Im\{\exp(j\phi(x,y))\}\right)$, where $\otimes$ denotes the convolution operation. The appearance of these three quantities, $\mathrm{circ}(f_r/\Delta f_r)$, $\Im\{\mathrm{circ}(r/\Delta r)\}$ and $\Im\{\exp(j\phi(x,y))\}$, at the Fourier or spatial frequency plane are considered. The first two terms are both circularly symmetric (or have no azimuthal dependence) and thus can both be illustrated in a 1D radial plot with spatial frequency f_r as the horizontal axis, as shown in Fig. 3.2.

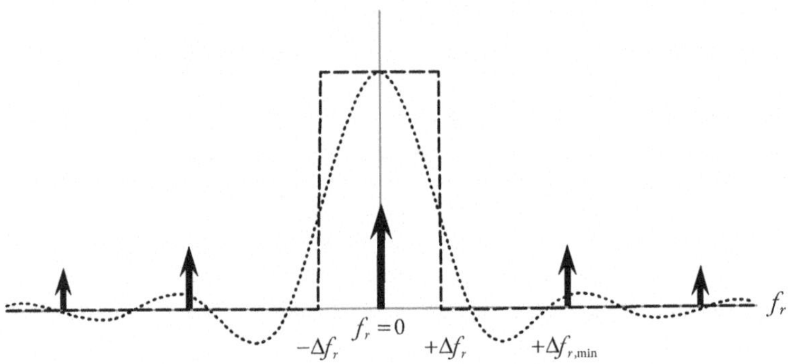

Fig. 3.2 Relevant Fourier plane quantities: phase-shifting region, $\mathrm{circ}(f_r/\Delta f_r)$ (dashed); Airy function $\Im\{\mathrm{circ}(r/\Delta r)\}$ (dotted), and object frequency components $\Im\{\exp(j\phi(x, y))\}$ (arrows). $f_{r\min}$ indicates the smallest radial spatial frequency component of a given phase object.

However, the third term, $\Im\{\exp(j\phi(x,y))\}$, is in general not circularly symmetric but may be characterized by diffraction orders or a sum of weighted Dirac-delta functions centered at various locations at the spatial frequency plane. To one of these delta functions, specifically the zero-order or on-axis centered $\delta(f_r)$, we assign the generally complex-valued weight $\bar{\alpha}$ derived from the Fourier analysis of a given input phase by Eq. (2.3). Among the remaining higher-orders that exist as off-axis delta functions, at least one (first-order) would be closest to the origin and, based on its frequency components, its location on the frequency plane can be denoted as $(f_{x\min}, f_{y\min})$. The f_r-axis of the radial plot shown in Fig. 3.2 has been chosen to pass through $(f_{x\min}, f_{y\min})$, which enables us to plot the diffraction orders along the line defined by the origin and the point $\left(f_{x\min}, f_{y\min}\right)$ at the frequency plane. Thus, the higher-order that appears closest to the origin intersects the f_r-axis at a radial frequency $f_{r\min} = \sqrt{f_{x\min}^2 + f_{y\min}^2}$.

The third term appears as part of the convolution $\Im\left\{\mathrm{circ}\left(r/\Delta r\right)\right\} \otimes \Im\left\{\exp(j\phi\left(x, y\right)\right\}$, which generates a sum of weighted Jinc functions ($\propto \Im\left\{\mathrm{circ}\left(r/\Delta r\right)\right\}$), each centered at the location of the corresponding delta function. If the minimum non-zero frequency, $f_{r\min}$, is set to satisfy the condition

$$\left|f_{r\min}\right| >> 1/\Delta r, \tag{3.4}$$

then there will be a negligible overlap (refer to the arrows and the dotted line in Fig. 3.2) between the on-axis centered Jinc function and the closest neighboring Jinc function centered at $f_{r\min}$. Adding the cautious condition

$$\Delta f_r < \left|f_{r\min}\right|/2, \tag{3.5}$$

we find that the term $\mathrm{circ}\left(f_r/\Delta f_r\right)\left(\Im\left\{\mathrm{circ}\left(r/\Delta r\right)\right\} \otimes \Im\left\{\exp(j\phi\left(x, y\right)\right\}\right)$, which evaluates the convolution within the phase shifting region of the Fourier filter, may be approximated by $\Im\left\{\mathrm{circ}\left(r/\Delta r\right)\right\}$ itself multiplied by the complex-valued term $\bar{\alpha}$. Recalling that $\bar{\alpha}$ is the coefficient of $\delta\left(f_r\right)$ in Fig. 3.2, we can calculate it as the average of the phase object over the input area (i.e. Eq. (2.3) as written below:

$$\bar{\alpha} = \left[\pi(\Delta r)^2\right]^{-1} \iint_{\sqrt{x^2+y^2} \leq \Delta r} \exp\left[j\phi\left(x, y\right)\right] \mathrm{d}x\mathrm{d}y = \left|\bar{\alpha}\right| \exp\left(j\phi_{\bar{\alpha}}\right). \tag{3.6}$$

Upon incorporating all these considerations, the inverse Fourier transform in Eq. (3.3) can then be evaluated as

$$\begin{aligned} &\Im^{-1}\left\{H\left(f_x, f_y\right)\Im\left\{a\left(x, y\right)\right\}\right\} \cong \\ &\qquad A\left[\exp\left(j\phi\left(x', y'\right)\right)\mathrm{circ}\left(r'/\Delta r\right) + \bar{\alpha}\left(BA^{-1}\exp\left(j\theta\right) - 1\right)g\left(r'\right)\right], \end{aligned} \tag{3.7}$$

where

$$g\left(r'\right) = \Im^{-1}\left\{\mathrm{circ}\left(f_r/\Delta f_r\right)\Im\left\{\mathrm{circ}\left(r/\Delta r\right)\right\}\right\} = 2\pi\Delta r \int_0^{\Delta f_r} J_1\left(2\pi\Delta r f_r\right) J_0\left(2\pi r' f_r\right) \mathrm{d}f_r \tag{3.8}$$

can be considered as the generating function for the synthetic reference wave (SRW). Finally, the output intensity (or irradiance) is obtained by taking the squared modulus of the field given in Eq. (3.7). The generally complex-valued and object-dependent term, $\bar{\alpha}$, corresponding to the amplitude of the focused light plays a significant role in the expression for the interference pattern described by Eq. (3.7). Referring to the discussion in Chapter 2, we are now able to confirm that the frequent assumption, that the amplitude of the focused light is approximately equal to the first term of the Taylor expansion in Eq. (2.3), can generally result in misleading interpretations of the interferograms generated at the CPI output when unwittingly applied beyond its limited domain of validity.

3.2.2 The SRW Generating Function

To summarise, we performed an optical Fourier transform of the input field from Eq. (3.1), filtered the components by a multiplication with the filter transfer function in Eq. (3.2) and then subjected the product to a second optical Fourier transform (corresponding to an inverse Fourier transform with inverted coordinates). As a result, we obtained an expression for the intensity $I(x',y')$ describing the interferogram at the observation plane of the *4f* CPI set-up:

$$I(x',y') = A^2 \left|\exp\left[j\phi(x',y')\right]\mathrm{circ}\left(r'/\Delta r\right) + \bar{\alpha}\left[BA^{-1}\exp(j\theta)-1\right]g(r')\right|^2 \tag{3.9}$$

To proceed, we need to find an accurate working expression for the SRW to complement the derived output intensity expression. Earlier, we used the zero-order Hankel transform [1] to describe the SRW generating function, $g(r')$. For an applied circular input aperture with radius, Δr, and a Fourier filter whose central phase shifting region corresponds to a spatial frequency radius, Δf_r, we obtained the following expression for the SRW generating function by use of the zero-order Hankel transform (c.f. Eq. (3.8)):

$$g(r') = 2\pi\Delta r \int_0^{\Delta f_r} J_1(2\pi\Delta r f_r)\, J_0(2\pi r' f_r)\,\mathrm{d}f_r \tag{3.10}$$

As is evident from its origin in Eq. (3.8), the SRW generating function incorporates effects due to the finite extent of the input aperture and the phase-shifting central region of the filter. This suggests that, due to its influence on the SRW, proper matching of these apertures will significantly impact the performance of the common-path interferometer. To simplify the analysis, and yet maintain validity over different choices of the input aperture size, we will introduce a dimensionless term, η, to specify the size of the central filtering region. This requires a "length scale" reference in the spatial Fourier domain, which we will take to be the radius of the main lobe of the Airy function generated by the Fourier transform of the circular input aperture alone. Thus, denoting the Airy disc radius as R_2 and the radius of the central filtering region as R_1, we can formally define the dimensionless filter parameter size as

$$\eta = R_1/R_2 = (0.61)^{-1}\,\Delta r \Delta f_r\,, \tag{3.11}$$

where Δr is the radius of the input aperture and Δf_r is the (spectral) radius of the central filtering region. The 0.61 factor arises from the radial distance to the first zero crossing of the Airy function corresponding to half of the Airy mainlobe factor, of 1.22 [1].

Applying the dimensionless filter size into Eq. (3.10) and then subsequently performing a series expansion in r', we obtain the following expression for the SRW generating function:

$$g(r')=1-J_0(1.22\pi\eta)-(0.61\pi\eta)^2 J_2(1.22\pi\eta)(r'/\Delta r)^2 \\ +\left\{\frac{1}{4}(0.61\pi\eta)^3\left[2J_3(1.22\pi\eta)-0.61\pi\eta J_4(1.22\pi\eta)\right]\right\}(r'/\Delta r)^4+\ldots \tag{3.12}$$

In this expansion, the SRW has been expressed in radial coordinates normalized to the radius of the imaged input aperture to maintain applicability regardless of the choice of input aperture size. Moreover, this provides for convenient scaling to account for any magnification within the imaging system although, for simplicity, a direct imaging operation of unity magnification is assumed for the remainder of our analysis. It is apparent from Eq. (3.12) that the generated SRW will change as a function of the dimensionless parameter expressing the radius of the central filtering region. Additionally, it is clear that the SRW spatial profile will not necessarily be flat over the system output aperture. This is an important, yet often neglected, factor in determining the performance of a CPI.

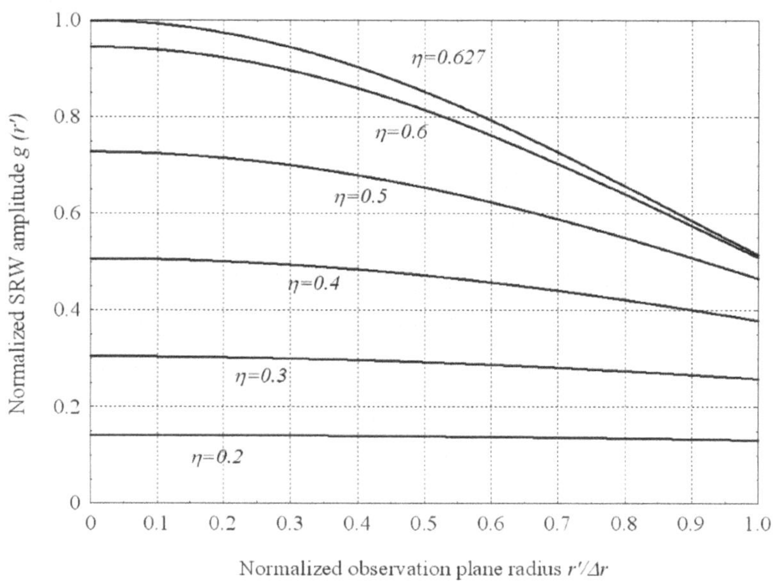

Fig. 3.3 Plot of the spatial variation of the normalized synthetic reference wave (SRW) amplitude $g(r')$ as a function of the normalized CPI observation plane radius for a range of η values from 0.2 to 0.627. This plot shows that a large value of η produces significant curvature in the SRW across the aperture, which will cause a distortion of the output interference pattern. In contrast, a low value of η generates a flat SRW, but at the cost of a reduction in the SRW amplitude.

Figure 3.3, shows the input-normalized amplitude of the SRW generating function for different η values, each plot displayed as a function of the output radius coordinate, r', normalized to the system aperture radius, Δr. It can be seen from the plots that as η increases so does the strength of the SRW, as can be expected from qualitative arguments. However, the curvature also increases with increasing η, thus distorting the wavefront profile of the SRW. From the point of view of optimising a CPI, it is desirable to properly select the filter size so that the curvature of $g(r')$ is negligible over a sufficiently large spatial region of the system aperture centred around $r'=0$. Firstly, we can choose to limit the range of η so that $g(r')$ never exceeds 1. This identifies an upper limit determined by the first zero crossing of the Bessel function $J_0(1.22\pi\eta)$ where $\eta \approx 0.627$. Secondly, it is very important to keep η as small as possible to make sure that the scattered object light is not propagated through the zero-order filtering region. Finding a minimum applicable η-value is less apparent, but obviously choosing a very small value will reduce the strength of the SRW to an unacceptably low level compromising the fringe visibility in the interference with the diffracted light. The term η can thus have a significant impact on the resulting interferometric performance and is of the same importance as the filter parameters A, B and θ when designing a CPI.

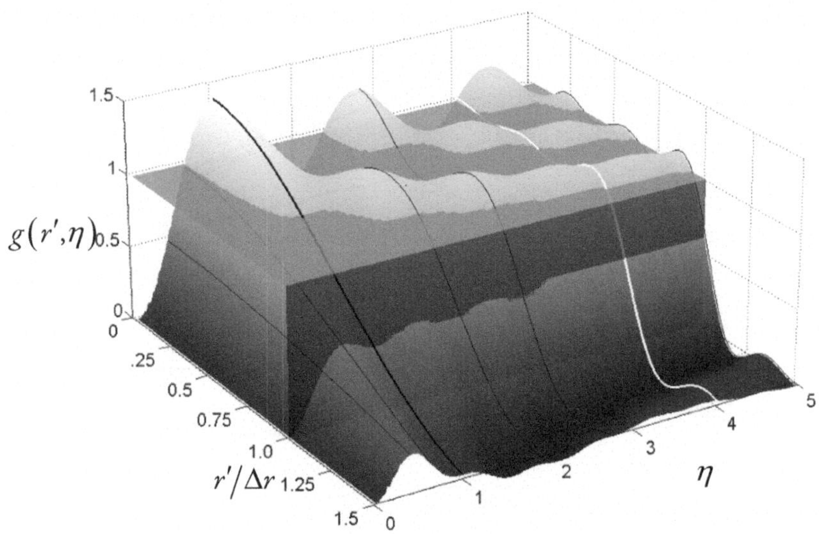

Fig. 3.4 Plot showing the evolution of the SRW generating function from $\eta=0$ to $\eta=5$. The radial profile of the generating function approaches $\mathrm{circ}(r'/\Delta r)$ as η is increased. The ideal top-hat scenario is indicated by the reference planes drawn at g=1 and at $r'/\Delta r=1$. Representative profiles are traced at $\eta=0.40, 0.627, 1.0$ (thick black trace), 2.0, 2.75, 4.0 (thick white line), and $\eta=5$.

Figure 3.4 illustrates the input-normalized amplitude of the SRW generating function as a function of the output radius coordinate, $r'/\Delta r$, but this time for a wider range of η values. We can deduce from this plot that, assuming we can allow $g(r')$ to occasionally exceed the value 1, then we can define a second regime for choosing the value of η (or filter radius). We can denote this alternate operating regime as a so-called large-η range, as opposed to the small-η range defined by the previously found interval range, as opposed to the small-η range defined by the previously found interval $0 \leq \eta \leq 0.627$. Phase objects having relatively small spatial frequency content $f_{r\min}$ need to operate within the previously defined small-η regime while those that have a relatively large $f_{r\min}$ may do so within the large-η regime, to make the representation of the output intensity in Eq. (3.9) as spatially accurate as possible.

We saw that depending on the accuracy needed for the description of the interferograms one can choose to include a number of spatial higher-order terms from the expansion in Eq. (3.12). The influence of the higher-order terms has the largest impact along the boundaries of the imaged aperture. For η-values smaller than 0.627 and when operating within the central region of the image plane, spatial higher-order terms are of much less significance and we can safely approximate the synthetic reference wave with the first and space invariant term in Eq. (3.12):

$$g\left(r' \in central\ region\right) \approx 1 - J_0\left(1.22\pi\eta\right) \tag{3.13}$$

so that we can simplify Eq. (3.9) to give:

$$I\left(x',y'\right) \approx A^2\left|\exp\left(j\phi\left(x',y'\right)\right) + K\bar{\alpha}\left(BA^{-1}\exp\left(j\theta\right) - 1\right)\right|^2 \tag{3.14}$$

where $K = 1 - J_0\left(1.22\pi\eta\right)$. The influence of the finite on-axis filtering radius on the focused light, incorporated in the K parameter, is thus effectively included as an extra "filtering parameter" so that the four-parameter filter set $\{A, B, \theta, K(\eta)\}$ together with the complex object-dependent term, $\bar{\alpha}$, effectively defines the type of filtering scheme we are applying.

3.2.3 The Combined Filter Parameter

Having determined a suitable operating range for the CPI in terms of the production of a good SRW as determined by the generating function $g(r')$, we must now examine the role that the remaining filter parameters play in the optimization of a CPI.

Looking at Eq. (3.14), we see that the different filter parameters (A, B, θ) can be combined to form a single complex-valued term, C, the combined filter term, such that:

$$C = |C|\exp\left(j\psi_C\right) = BA^{-1}\exp\left(j\theta\right) - 1. \tag{3.15}$$

Therefore, Eq. (3.14) can be simplified to give:

$$I\left(x',y'\right) = A^2\left|\exp\left(j\phi\left(x',y'\right) - j\psi_C\right) + K\bar{\alpha}|C|\right|^2 \tag{3.16}$$

where the usual filter parameters can be retrieved from the combined filter parameter using

$$\begin{cases} BA^{-1} = \sqrt{1+2|C|\cos\psi_C + |C|^2} \\ \sin\theta = \left(BA^{-1}\right)^{-1}|C|\sin\psi_C \end{cases} \tag{3.17}$$

Since it is a complex variable, the combined filter term C, which effectively describes the complex filter space, can be considered to consist of a vector of phase ψ_C and length $|C|$ as expressed by Eq. (3.15). Thus in order to obtain an overview of the operating space covered by all the possible combinations of three independent filter parameters (A, B, θ) we can now, instead, choose to consider a given filter in terms of the two combined parameters ψ_C and $|C|$. However, referring to Eq. (3.16), it can be seen that the filter parameter, A, also appears independently of the combined filter term, C. Fortunately, this issue can be resolved by considering that the term BA^{-1} from Eq. (3.15) must be constrained in the following way:

$$\begin{cases} BA^{-1} < 1 \Rightarrow A = 1, \quad B = |C+1| \\ BA^{-1} = 1 \Rightarrow A = 1, \quad B = 1 \\ BA^{-1} > 1 \Rightarrow B = 1, \quad A = |C+1|^{-1} \end{cases} \tag{3.18}$$

These constraints arise from the adoption of a maximum irradiance criterion minimising unnecessary absorption of light in the Fourier filter, which reduces both irradiance and the signal-to-noise ratio in the CPI output.

Any given filter can be explicitly defined by a given value of the two parameters ψ_C and $|C|$ therefore we can use a single plot to display the location of a given filter graphically within the complex filter space. Such a plot is shown in Fig. 3.5 where we plot the magnitude of the combined filter parameter $|C|$ against its phase ψ_C.

There exist different families of operating curves in this complex filter space, each of which can be traced out by keeping a term such as BA^{-1} constant while θ is varied or vice versa (these form the fine grid like structure in Fig. 3.5). Plotting the operating curves for C this way makes it relatively simple to identify particular operating regimes for different classes of filters. For example, we are particularly interested in the operating curve for a lossless Fourier filter, a filter in which $BA^{-1} = 1$, since this corresponds to a class of filters for which optical throughput is maximized. The lossless operating curve is shown as the bold line in Fig. 3.5. We can derive the expression for the shape of the lossless operating curve by using the following identity:

$$\exp(j\theta) - 1 = 2\left|\sin(\theta/2)\right|\exp\left(j(\theta+\pi)/2\right) \tag{3.19}$$

and combining Eq. (3.15) and Eq. (3.19) we obtain an expression for the lossless operating curve, for which $|C|$ is defined for two distinct regions as:

$$\begin{cases} |C| = 2\sin\left(\psi_C - \pi/2\right) \quad \forall \quad \psi_C \in \left[\pi/2;\ 3\pi/2\right] \\ |C| = 0 \quad \forall \quad \psi_C \notin \left[\pi/2;\ 3\pi/2\right] \end{cases} \tag{3.20}$$

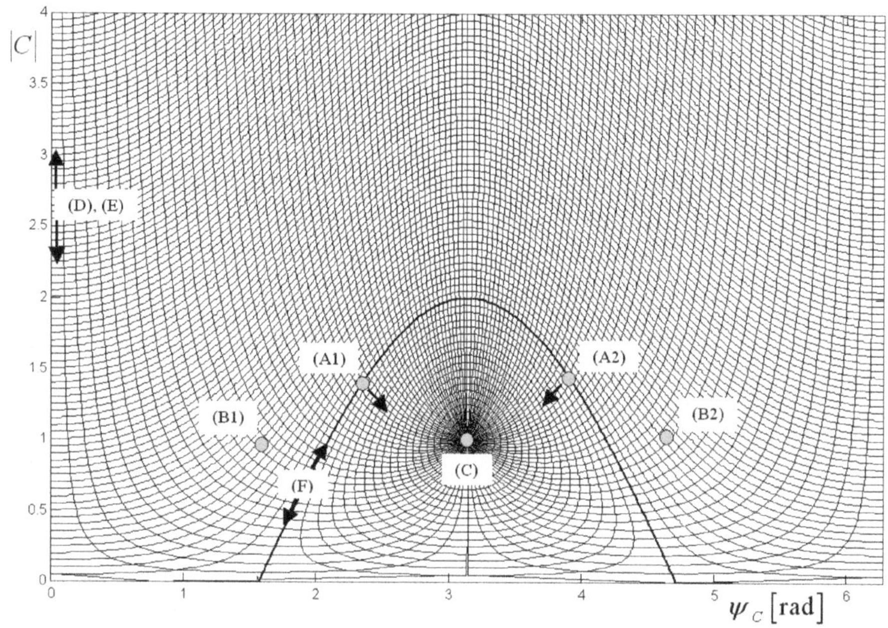

Fig. 3.5 Complex filter space plot of the modulus of the combined filter parameter, $|C|$, against the phase ψ_C over the complete 2π phase region. The use of these combined parameters (defined in Eqs. (3.15) and (3.17)) allows us to simultaneously visualise all the available combinations of the terms A, B and θ. The bold curve is the operating curve for a phase-only (lossless) filter, whilst the fine grid represents operating curves for differing values of the filter terms A, B and θ. We have marked operating regimes for a number of CPI architectures including: (A) Zernike, (B) Henning, (C) dark central ground and (D) field absorption filtering and (E,F) point diffraction interferometers. Full filter details for these techniques are summarised in Table 1.

Table 1 Comparison of filter parameters for the different CPI types highlighted in Fig. 3.5.

Method	**Filter parameters**	**References**	**Fig. 3.5 label**
Zernike phase contrast	$A=1, B\in[0.05;1], \theta=\pm\pi/2$	[1, 2]	(A1), (A2)
Henning phase contrast	$A=2^{-1/2}, B=1, \theta=\pm\pi/4$	[3, 4]	(B1), (B2)
Dark central ground filter	$A=1, B=0, \theta=0$	[5–7]	(C)
Field absorption filter	$A<1, B=1, \theta=0$	[8]	(D)
PDI*/Smartt interferometer	$A<1, B=1, \theta=0, K\ll 1$	[9–12]	(E)
Phase-shifting PDI	$A=1, B=1, \theta\in[0;2\pi], K\ll 1$	[13, 14]	(F)

* PDI indicates: point diffraction interferometry.

The development of this complex filter space plot makes it extremely simple to compare different filter parameter selections that have been proposed and independently dealt with in the scientific literature. In Fig. 3.5, we have superimposed the operating points for some of the different CPI filter types that are commonly used. The filter type, parameters and the corresponding labels and references are summarised in Table 1.

The filter space plot not only provides a convenient method of comparing different CPI architectures, but also forms a useful basis for exploring the effectiveness of new CPI configurations. Referring to Fig. 3.5, it can be seen that many of the filter types used in CPIs lie on or close to the lossless operating curve defined by Eq. (3.20). The arrows on the plot indicate the general regime in the complex filter space, in which a given filter type can be operated. To our knowledge, this identification and labelling of known cases from the literature in a single general filter space representation has not been previously demonstrated. We believe that it provides a very useful basis for understanding the inherent differences or similarities between existing filtering methods. Later, we will use the complex filter space plot as a framework on which to optimize the visibility and the irradiance of a CPI operated as a wavefront sensing system.

3.3 Summary and Links

In this chapter we developed the fundamental GPC framework by analyzing common path interferometers (CPI). We considered systems based on spatial filtering around the optical axis in the spatial Fourier domain, where we have abandoned the simplifying assumptions in typical treatments. We have derived conditions for high interferometric accuracy taking account of the so-called synthetic reference wave. We also identified a combined filter parameter that places all spatial filters of different CPI schemes in the same phase space domain, drastically simplifying the comparative analysis of different filtering techniques. The Zernike filter is one among many schemes that are subsumed by this generalized treatment. We will continue to build upon this basic framework throughout the book as we adopt refinements that address the varying demands and design freedoms of different applications. Practical illustrations will be discussed in Chapter 5, where we characterise wavefront-sensing CPI systems using a combined filter parameter. We will show that the generalized mathematical treatment here can be successfully applied to the interpretation and possible modification for the optimization of a number of different but commonly used CPI architectures. In Chapter 6, we will adapt the basic framework to incorporate design freedoms available at the input plane when using GPC for wavefront engineering. There we will discuss how filter parameters can be fully optimized to match known inputs. The framework developed here is also used as the starting point for discussing the different alternative implementations in Chapter 9. Finally, we will again build upon the basic formulation developed here when we consider CPIs with amplitude-modulated inputs when we discuss the reverse phase contrast method in Chapter 10. Meanwhile, we will develop a phasor chart method in the next chapter, which will come in handy in understanding and optimizing generalized phase contrast systems.

References

1. J. W. Goodman, Introduction to Fourier Optics (McGraw-Hill, San Francisco, 2nd ed., 1996).
2. F. Zernike, "How I discovered phase contrast", Science **121**, 345-349 (1955).
3. H. B. Henning, "A new scheme for viewing phase contrast images", Electro-optical Systems Design **6**, 30-34 (1974).
4. G. O. Reynolds, J. B. Develis, G. B. Parrent, Jr., B. J. Thompson, *The New Physical Optics Notebook: Tutorials in Fourier Optics*, (SPIE Optical Engineering Press, New York 1989) Chap. 35.
5. S. F. Paul, "Dark-ground illumination as a quantitative diagnostic for plasma density", Appl. Opt., **21**, 2531-2537 (1982).
6. R. C. Anderson and S. Lewis, "Flow visualization by dark central ground interferometry", Appl. Opt. **24**, 3687 (1985).
7. M. P. Loomis, M. Holt, G. T. Chapman and M. Coon, "Applications of dark central ground interferometry", Proc. of the 29th Aerospace Sciences Meeting, AIAA **91-0565**, 1-8 (1991).
8. C. S. Anderson, "Fringe visibility, irradiance, and accuracy in common path interferometers for visualization of phase disturbances", Appl. Opt. **34**, 7474-7485 (1995).
9. C. Koliopoulus, O. Kwon, R. Shagam, J. C. Wyant and C. R. Hayslett, "Infrared point-diffraction interferometer", Opt. Lett. **3**, 118-120 (1978).
10. R. N. Smartt, W. H. Steel, "Theory and application of point-diffraction interferometers", Japan J. Appl. Phys. **14**, 351-356 (1975).
11. W. P. Linnik, "Ein einfaches interferometer zur prüfung von optischen systemen", Proc. Acad. Sci. USSR **1**, 208-211 (1933).
12. P. M. Birch, J. Gourlay, G. D. Love, and A. Purvis, "Real-time optical aberration correction with a ferroelectric liquid-crystal spatial light modulator", Appl. Opt. **37**, 2164-2169 (1998).
13. C. R. Mercer, K. Creath, "Liquid-crystal point-diffraction interferometer for wavefront measurements", Appl. Opt. **35**, 1633-1642 (1996).
14. C. R. Mercer, K. Creath, "Liquid-crystal point-diffraction interferometer", Opt. Lett. **19**, 916-918 (1994).

Chapter 4
Phasor Chart for CPI-Analysis

As we saw in the previous discussion it is usually a challenging task to obtain an overview of the output from a common-path interferometer (CPI) for a given input phase distribution and the set of filter parameters (A, B, θ) and $K(\eta)$. We therefore include a brief description of a graphical phasor-chart method for CPI-analysis that we have developed and will discuss how to apply it. We explain how this original phasor chart can be reconfigured for use with the combined filter parameter, C, that we introduced in the preceding section and show how both types of phasor charts can be used as powerful visualization tools in the analysis of any given CPI. The primary role of the graphical phasor chart is to illustrate the relationship between the intensity at the output of the CPI and the size of the input phase perturbation, in other words, the phase-to-intensity mapping for a given set of filter parameters.

4.1 Input Phase to Output Intensity Mapping

The results from the mathematical analysis of a CPI in Chapter 3 describe the intensity of the output interferogram as (c.f. Eq. (3.14)):

$$I(x',y') \approx A^2 \left|\exp\left(j\phi(x',y')\right) + K\bar{\alpha}\left(BA^{-1}\exp\left(j\theta\right)-1\right)\right|^2 . \tag{4.1}$$

Since the output intensity is independent of a uniform phase offset, we can rewrite the output intensity equation as

$$I(x',y') \approx A^2 \left|\exp\left(j\tilde{\phi}(x',y')\right) + K\left|\bar{\alpha}\right|\left(BA^{-1}\exp\left(j\theta\right)-1\right)\right|^2 , \tag{4.2}$$

where we have omitted a uniform phase offset, $\exp(i\phi_{\bar{\alpha}})$, and the image of the input phase modulation has been replaced by the relative phase, $\tilde{\phi}(x',y') = \phi(x',y') - \phi_{\bar{\alpha}}$.

By including the implied temporal field oscillations, which were not explicitly shown earlier for simplicity, we can regard this output as a superposition of three phasors:

$$(1)\ \exp[j\tilde{\phi}(x',y') + j\omega t]$$
$$(2)\ (BA^{-1})K|\bar{\alpha}|\exp(j\theta + j\omega t)$$
$$(3)\ K|\bar{\alpha}|\exp(j\pi + j\omega t),$$

where the result is further scaled by A. Two equivalent ways of illustrating the phasor addition are shown in Fig. 4.1. Figure 4.1(a) shows a conventional phasor layout, while Fig. 4.1(b) shows an alternative layout that we will use to develop a framework for a phasor chart that is convenient for assessing the effect of different filter parameters on CPI performance.

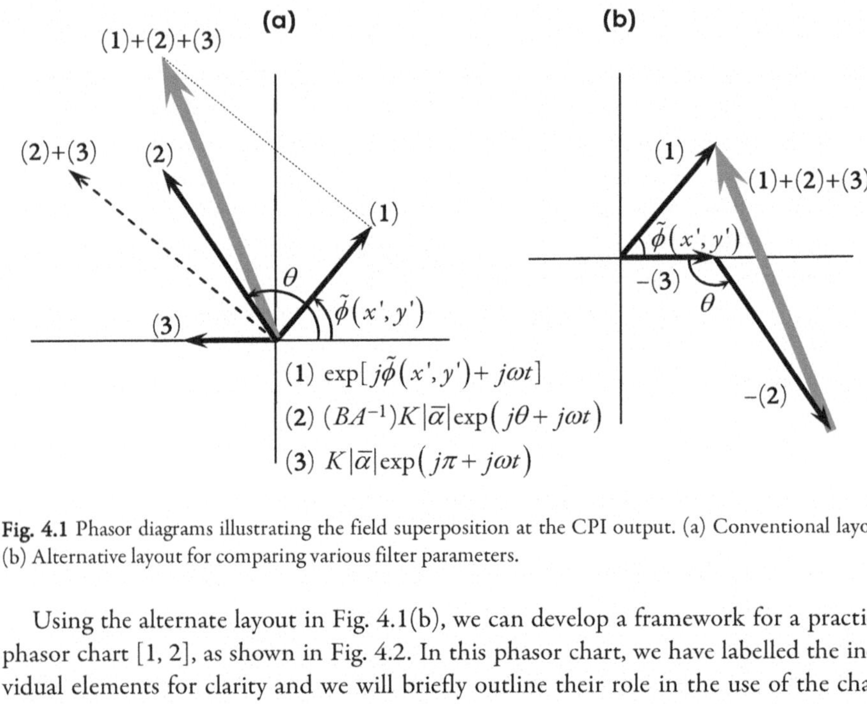

Fig. 4.1 Phasor diagrams illustrating the field superposition at the CPI output. (a) Conventional layout; (b) Alternative layout for comparing various filter parameters.

Using the alternate layout in Fig. 4.1(b), we can develop a framework for a practical phasor chart [1, 2], as shown in Fig. 4.2. In this phasor chart, we have labelled the individual elements for clarity and we will briefly outline their role in the use of the chart. The input phase deviation, $\tilde{\phi}$, within the range 0 to 2π is given on the unity phase circle. The set of all possible filter parameters that can alter the −(3) phasor in Fig. 4.1(b) can be drawn as a set of circles having a common tangent point with the imaginary axis at the origin and whose centres are located along the positive real axis. The example shown in Fig. 4.2 shows eight such circles, showing a range of $K|\bar{\alpha}|$ values from 1/8 to 1. As previously explained, the term $K|\bar{\alpha}|$ represents a combination of properties relating to the input phase distribution and the relationship between the radius of the Fourier filter and that of the main lobe of the Airy function in the Fourier plane. In short, it describes the amplitude which when focused in the Fourier plane is incident on the central filter region generating the synthetic reference wave at the output.

For each circle, we define an angular coordinate, θ, measured anticlockwise from the negative real axis, to describe possible orientations for the –(**2**) phasor in Fig. 4.1(b). Finally, the quadratic intensity scale (shown at the bottom of Fig. 4.2) is the method by which we map the input phase value to the intensity at the output. The lower end of the scale is fixed in place on the chart at a position determined by the three filter parameters (A, B, θ) and the intersection of the intensity scale with the unity phase circle gives the phase-to-intensity mapping. The effect of the amplitude transmission filter parameters A and B can be understood in terms of their effect on the position at which the zero point of the quadratic intensity scale is fixed (labelled 'fix point' in Fig. 4.3). Adjusting these parameters has the effect of shifting the zero point of the intensity scale radially by a factor, BA^{-1}, with respect to the corresponding $K|\bar{\alpha}|$ circle, along a line, the angular position of which is given by the third filter parameter, θ.

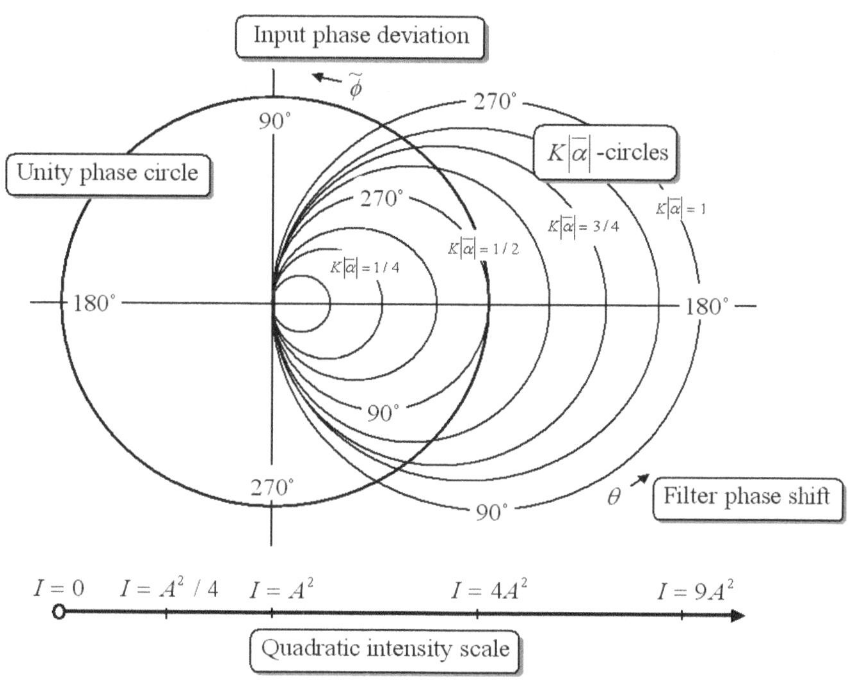

Fig. 4.2 The basic layout of the graphical phasor chart for mapping an input phase, $\tilde{\phi}$, to an output intensity, I, for a CPI with the Fourier filter parameters (A, B, θ). The key elements of the chart are labelled for clarity.

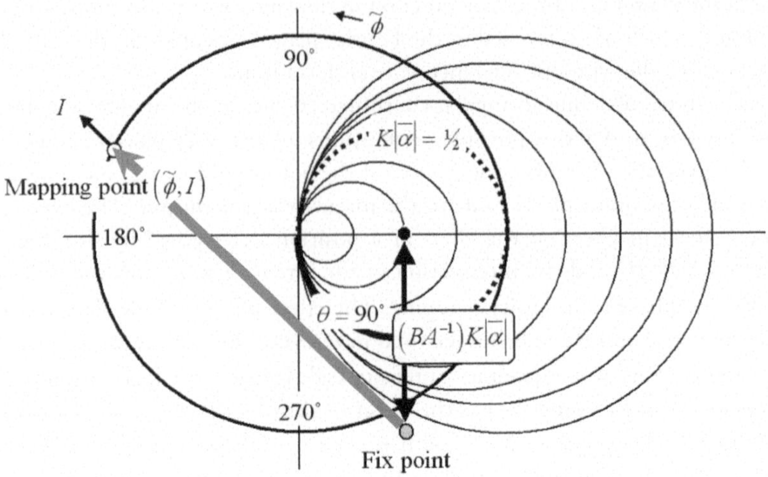

Fig. 4.3 Application of the phasor chart for $K|\bar{\alpha}|=1/2$ and $\theta=\pi/2$, with a radial scaling for the zero point of the quadratic intensity scale determined by the amplitude transmission terms: $A=1/2$ and $B=1$. Having fixed the zero point of the intensity scale at the desired point, the phase to intensity mapping is determined from the intersection of the quadratic intensity scale with the unity phase circle.

A simple example of the application of the phasor chart serves to demonstrate how it is used. The first step is to identify the appropriate value of $K|\bar{\alpha}|$ for the situation we wish to evaluate.

The example shown in Fig. 4.3 illustrates the resultant phasor for $K|\bar{\alpha}|=1/2$ and $\theta=\pi/2$. The centre point of this $K|\bar{\alpha}|$-circle is marked on the chart and the angle θ is measured about this point. The amplitude transmission terms A and B have the values 1/2 and 1 respectively, which effectively scales the radius of the $K|\bar{\alpha}|$-circle by a factor of $BA^{-1}=2$. This scaling serves to fix the zero point of the quadratic intensity scale at the required position. Finally, the intersection of the quadratic intensity scale with the unity phase circle occurs at the points $(\tilde{\phi}, I)$.

4.2 Modified Phasor Chart Based on Complex Filter Parameter

For the majority of cases we are working with the combined filter parameter, $C=BA^{-1}\exp(j\theta)-1$, obtained from Eq. (3.14) and we would ideally like to modify the graphical phasor chart to work with this parameter [3]. The original phasor chart can therefore be modified to operate directly with this combined and complex filter parameter, C, instead of the basic filter parameters (A, B, θ). This modified phasor chart is illustrated in Fig. 4.4.

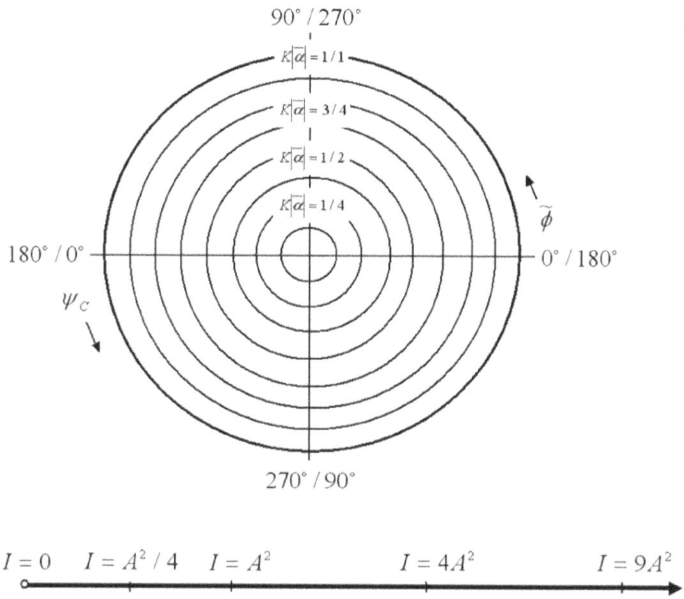

Fig. 4.4 The modified phasor chart which has been adapted to work directly with the two components of the combined filter parameter: $|C|$ and ψ_C.

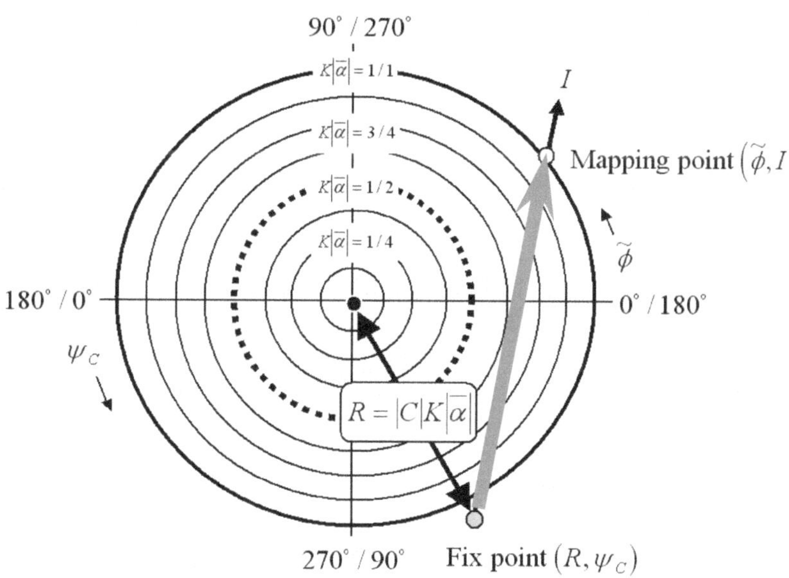

Fig. 4.5 Application of the modified phasor chart to a CPI with the same zero intensity point as that described by the phasor chart in Fig. 4.4.

The most noticeable feature in this modified chart is that all the $K|\bar{\alpha}|$-circles are now arranged concentrically within the unity phase circle. Referring to Fig. 4.5 it can be seen that the scale of the unity phase circle has the same orientation as in the original chart. The phase of the combined filter parameter, ψ_C, is measured about the same centre point as $\tilde{\phi}$, but the scales are shifted by 180° (the axis labels on the modified chart give both phase values with $\tilde{\phi}$ first). The same quadratic intensity scale is used and is given in terms of the basic filter parameter, *A*, as in the original chart. If required, this parameter can be extracted from the combined filter parameter, *C*, by use of the relationships described in Eq. (3.17) and the constraints in Eq. (3.18).

In Fig. 4.5, we show how the modified phasor chart can be applied to graphically analyze the same system as used in the example illustrated by Fig. 4.3. In this case, the parameter, ψ_C, gives the phase describing the position for the zero point of the intensity scale on the modified chart. Whilst the radial scaling for the position of the zero intensity point simply depends on the factor $|C|$, resulting in a radial parameter, *R*, in the coordinate system given by $R = K|C||\bar{\alpha}|$. Thus we find that $|C| = \sqrt{5}$ and $\psi_C \approx 117^\circ$ gives the same fix point location as in Fig. 4.3, which can be described by the polar coordinate set $(R,\ \psi_C) = (\sqrt{5}/2,\ 117^\circ)$. Since the fix point location is the same in both phasor chart types, it can therefore be directly transferred between them. In most cases we find it useful to first locate the intensity scale zero point by use of a modified phasor chart as shown in Fig. 4.4 and then transfer the coordinates directly onto the original phasor chart shown in Fig. 4.2. It should also be noted that finding the phase-intensity mapping as the intersection between the quadratic intensity scale and the $\tilde{\phi}$ unit-circle follows exactly the same procedure in both phasor charts.

4.3 Summary and Links

In this chapter we developed graphical phasor chart methods for the analysis of common path interferometers. As we will see in the next chapter, this allows convenient and intuitive optimization of the visibility and peak irradiance, which improves the resulting output phase contrast for a given phase distribution at the input of a common path interferometer. In particular we could take advantage of the combined filter parameter, which places all common path interferometers in the same phase space domain as shown in Chapter 3. We will explore this further in the next chapter as we apply our phasor charts to some concrete example systems and consider how they can be further optimised using the phasor chart scheme. We also apply phasor charts to find a phase-to-intensity mapping that matches device constraints when we consider image projection in Chapter 7. The phasor chart scheme developed here is again utilized in Chapter 9 to study CPIs subjected to amplitude-modulated inputs to help visualize the required optimization for achieving the reverse phase contrast effect.

References

1. J. Glückstad, "Graphic method for analyzing common path interferometers," Appl. Opt. **37**, 8151-8152 (1998).
2. J. Glückstad, "Graphic method for analyzing common path interferometers," Optics & Photonics News **9**, 2 (1998).
3. J. Glückstad and P. C. Mogensen, "Optimal phase contrast in common-path interferometry," Appl. Opt. **40** , 268-282 (2001).

Chapter 5
Wavefront Sensing and Analysis Using GPC

The need to perceive invisible phase perturbations can arise in a wide range of imaging systems, from phase microscopes that non-invasively image transparent biological samples *in vivo* to adaptive optical systems that measure and compensate for phase aberrations to improve quality in more general imaging situations. The common-path interferometer (CPI) is an appealing tool for extracting and quantifying phase perturbations. A CPI applies the same principle of phase-to-intensity transformation as amplitude-dividing interferometers (e.g. Michelson or Mach-Zender) but with the advantage of heightened robustness to mechanical vibrations and ambient air fluctuations owing to the internally synthesized reference wave. Having laid out the framework for the generalized phase contrast, a natural extension is to exploit GPC to accurately measure unknown phase, which we will explore in this chapter. We will apply GPC analysis to determine how filter parameters can be adapted in order to tailor a CPI for accurate wavefront sensing. This involves optimizing the fringe visibility and peak irradiance, as well as extending the linear phase-to-intensity mapping regime. GPC can prescribe the appropriate filter size to prevent scattered light at the input from propagating through the central filtering region and thereby prevent an erroneous interpretation of the output interference patterns. GPC can also be used to tailor the other filter parameters to extend the linear regime of the phase-to-intensity mapping. Moreover, one may take note of the actual filter parameters and apply the GPC model to extract accurate information. The generalized phase contrast framework allows us to properly describe the synthetic reference wave (SRW), which enables proper interpretation of the CPI output intensity over an arbitrarily wide phase range. Thus, one does not always have to minimize the SRW curvature since the GPC description of the SRW can be exploited to accurately determine the unknown phase input. Whether using GPC to expand the linear regime of the phase-to-intensity mapping, or to account for the SRW-based inhomogeneity, the method paves the way for shifting from qualitative to quantitative phase imaging that can be utilized both for microscopy and other generalized wavefront-sensing applications.

5.1 GPC Mapping for Wavefront Measurement

Extracting meaningful information about wavefront phase perturbations using a common-path interferometer is facilitated when the CPI generates output patterns that possess good contrast. In practical implementations, one may encounter input constraints such as low light levels. When this happens, an output that would, ideally, exhibit high contrast could suffer from poor signal-to-noise ratio (SNR) when swamped with optics and detector noise. Thus the output peak irradiance becomes essential to guarantee a decent SNR.

Denoting the maximum and minimum intensity levels at the output as I_{max} and I_{min} respectively, the contrast is described by the visibility, V, which can be expressed as:

$$V = \frac{I_{max} - I_{min}}{I_{max} + I_{min}} \tag{5.1}$$

We have previously determined that for a CPI input field with mean value, $\bar{\alpha}$, which is characterized by an unknown wavefront phase perturbation, $\phi(x, y)$, having a dynamic range of $\Delta\tilde{\phi}$, the output intensity is (c.f. Eq. (3.16))

$$I(x',y') = A^2 \left| \exp\left(j\tilde{\phi}(x',y') - j\psi_C \right) + K|\bar{\alpha}||C| \right|^2 \tag{5.2}$$

where (x',y') are the output plane coordinates, $C = |C|\exp[j\psi_C]$ is the combined filter parameter, and we have replaced the phase image, $\phi(x',y')$, with a relative phase $\tilde{\phi}(x',y') = \phi(x',y') - \phi_{\bar{\alpha}}$ upon omitting a uniform phase offset, $\exp(i\phi_{\bar{\alpha}})$. This result assumes a homogeneous SRW, which is a suitable description when studying the central region of the phase perturbation. The spatial variation becomes crucial for accurate measurements in the periphery and this will be thoroughly analyzed in section 5.6.

The phase-to-intensity mapping described by Eq. (5.2) enables us to determine the visibility and peak irradiance of the CPI output, which we can write, respectively, as

$$\begin{cases} V = (maxcos - mincos)\left[(K|\bar{\alpha}||C|)^{-1} + K|\bar{\alpha}||C| + maxcos + mincos \right]^{-1} \\ I_{max} = A^2\left[1 + (K|\bar{\alpha}||C|)^2 + 2K|\bar{\alpha}||C|\, maxcos \right] \end{cases} \tag{5.3}$$

where

$$\begin{cases} mincos = \min\left[\cos\left(\tilde{\phi} - \psi_C \right) \right] \\ maxcos = \max\left[\cos\left(\tilde{\phi} - \psi_C \right) \right], \quad \forall \tilde{\phi} \in \left[-\Delta\tilde{\phi}/2;\ \Delta\tilde{\phi}/2 \right] \end{cases} \tag{5.4}$$

These expressions for the visibility and peak irradiance can be simplified in some special cases, which we will treat later. We will first consider an analysis of the general case as given in the full versions of Eqs. (5.2)–(5.4). The variable K is held fixed in this analysis, for high fringe accuracy, as was previously discussed.

It is apparent that a general analysis over the entire domain covered by the complex filter space plot of Fig. 3.5 is complicated by the appearance of the input-dependent variables, $|\bar{\alpha}|$ and $\Delta\tilde{\phi}$, in the expressions for the visibility and the peak irradiance. Instead of arbitrarily choosing combinations of $|\bar{\alpha}|$ and $\Delta\tilde{\phi}$ prior to an optimization analysis for the filter parameters, we model the input phase perturbations by a random distribution of phasors over the phase range $\Delta\tilde{\phi}$ generating a spatial average or wavefront Strehl ratio [1] corresponding to $|\bar{\alpha}|$. For example, we can choose to analyse phase perturbations modelled by a *uniform* random distribution of phasors. In this case, the two terms $|\bar{\alpha}|$ and $\Delta\tilde{\phi}$ are related by the absolute value of a sinc-function in the following way [2]:

$$|\bar{\alpha}|_{uniform} = \left|\mathrm{sinc}\left(\Delta\tilde{\phi}/2\right)\right| \tag{5.5}$$

Using this expression to substitute for $|\bar{\alpha}|$ in Eq. (5.2) and Eq. (5.3) we can produce expressions for the visibility and peak irradiance, which depend only on the input phase modulation depth and the combined filter parameter. Having derived expressions for the visibility and peak irradiance it is instructive to compare their variation as a function of the phase modulation depth, so we are able to systematically compare the performance of a generic CPI system for different ranges of phase perturbations at the input. Using the complex filter space plot as a framework we are able to display the variation in the visibility and peak irradiance at the CPI output as a function of the complex components, ψ_C and $|C|$, of the combined filter parameter. By using a pair of complex filter space plots, we can thus show the optimum filter type for a given phase perturbation to maximize the visibility or the peak irradiance.

In Fig. 5.1 we show a set of contour plot pairs for the visibility and peak irradiance with a range of input phase distribution depths $\Delta\tilde{\phi}$ from (a) $\pi/8$ through to (d) $3\pi/2$. These are shown in the form of normalized greyscale contour plots in which the maximum value is represented by the darkest region. All four sets of complex filter space plots cover the same range for the parameters ψ_C and $|C|$, from 0 to 2π radians and 0 to 4 respectively. The operating curve for a lossless filter is shown on all the plots. Referring to Fig. 5.1 it is evident that a small $\Delta\tilde{\phi}$ (see Fig. 5.1(a)) does not provide much overlap between regions of high visibility and strong peak irradiance. For the larger values of $\Delta\tilde{\phi}$, like those in Fig. 5.1(b) and Fig. 5.1(c), an overlap can be found between the regions of high visibility and peak irradiance. However there is no overlap between regions in which these values are maximized illustrating that there will always be a trade-off between the absolute optimization of these two terms. It is of interest to note that as the value of $\Delta\tilde{\phi}$ increases further to $3\pi/2$ and beyond (see Fig. 5.1(d)) then the overlap between high values of the visibility and the peak irradiance diminishes.

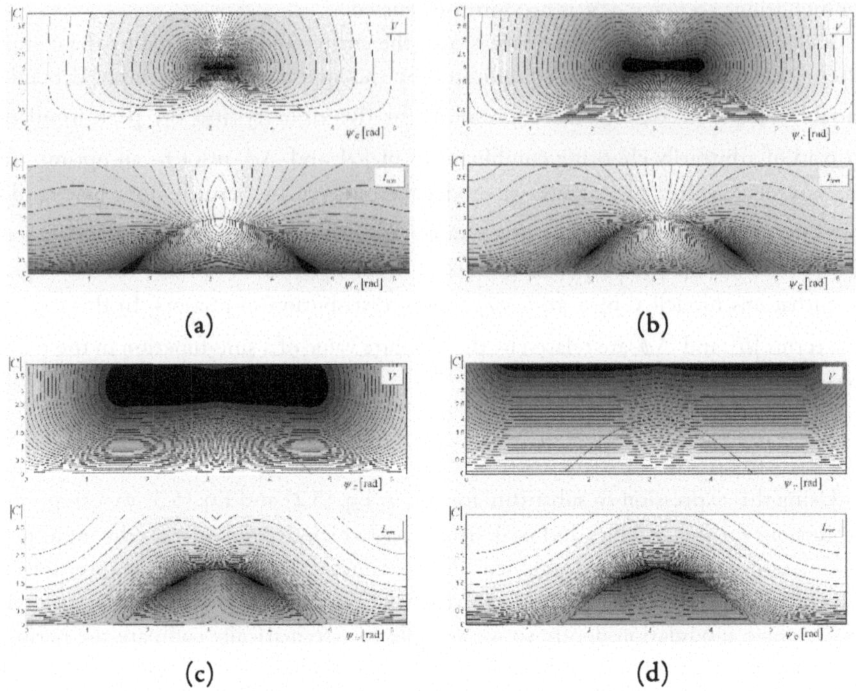

Fig. 5.1 Complex filter space plots for the optimization of the visibility, V, and peak irradiance, I_{max}, with different input phase distributions. We show four pairs of plots illustrating the variation in the visibility and peak irradiance at the output of a CPI for a uniform random phase distribution of dynamic range (a) $\pi/8$ (b) $\pi/3$ (c) π and (d) $3\pi/2$ for a constant value of $K = 1/2$. The greyscale of the contour plots defines the relative visibility or peak irradiance (with the darkest regions representing the maximum value) that can be achieved by varying the Fourier filter parameters. The operating curve for a lossless filter is shown on all the plots.

The 'upward shift' towards higher values of $|C|$, and broadening of the maximum visibility region in the filter space domain as $\Delta\tilde{\phi}$ increases can be understood by deriving the expression for C in case of optimal visibility:

$$C_{V=1} = \left(K|\bar{\alpha}|\right)^{-1} \exp\left(j\pi + j\tilde{\phi}_{V=1}\right), \quad \forall\ \tilde{\phi}_{V=1} \in \left[-\Delta\tilde{\phi}/2;\ \Delta\tilde{\phi}/2\right]. \tag{5.6}$$

In this case, we use the fact that the minimum intensity will have to be zero within the interval $\Delta\tilde{\phi}$ in order to have optimal visibility. The maximum visibility region in Fig. 5.1 thus broadens as a direct consequence of increasing $\Delta\tilde{\phi}$. The upward shift also becomes clear from Eq. (5.6). To have optimal visibility, the combined filter parameter must have a magnitude $|C_{V=1}| = \left(K|\bar{\alpha}|\right)^{-1}$. For a fixed K, increasing the dynamic range,

$\Delta\tilde{\phi}$, (but keeping $\Delta\tilde{\phi}<2\pi$) lowers $|\bar{\alpha}|$ (e.g. see Eq. (5.5) for a uniform distribution). Therefore, the optimal $|C|$ shifts upwards when $\Delta\tilde{\phi}$ increases.

Another interesting observation related to the contour plots in Fig. 5.1 is that $C=2\exp(j\pi)=-2$ provides a very good compromise filter in most cases both for obtaining high visibility and strong peak irradiance. Moreover, this corresponds to a lossless phase-only filter with $\theta=\pi$ and $A=B=1$ as is evident from our previous definition of C.

5.2 Optimal Unambiguous Intensity-to-Phase Mapping

A CPI output pattern with good visibility and peak irradiance may have little value if it does not lend itself to accurate retrieval of the unknown phase perturbation. This occurs, for instance, when there are phase ambiguities and multiple phase values correspond to a particular observed intensity level. For phase perturbations in the range $\Delta\tilde{\phi}\leq\pi$ we have the possibility of optimising both the visibility and irradiance whilst simultaneously obtaining a monotonically increasing mapping of the input phase to the output intensity. Unfortunately, this excludes relatively simple filters like the lossless π-phase filter that was convenient in the previous analysis where the unambiguous mapping of phase to intensity was not the primary concern.

In order to obtain the required monotonic increase in the phase-to-intensity mapping we must map the respective minima and maxima of the input phase disturbance $\Delta\tilde{\phi}$ to the corresponding limits in the output intensity distribution. We therefore get the following expression for the visibility:

$$\begin{aligned} V &= \left[I\left(\Delta\tilde{\phi}/2\right)-I\left(-\Delta\tilde{\phi}/2\right)\right]\left[I\left(\Delta\tilde{\phi}/2\right)+I\left(-\Delta\tilde{\phi}/2\right)\right]^{-1} \\ &= 2\sin\left(\Delta\tilde{\phi}/2\right)\sin\psi_C\left[\left(K|\bar{\alpha}||C|\right)^{-1}+K|\bar{\alpha}||C|+2\cos\left(\Delta\tilde{\phi}/2\right)\cos\psi_C\right]^{-1} \end{aligned} \tag{5.7}$$

where the corresponding peak irradiance is given by:

$$I_{\max}=I\left(\Delta\tilde{\phi}/2\right)=A^2\left|\exp\left(j\Delta\tilde{\phi}/2-j\psi_C\right)+K|\bar{\alpha}||C|\right|^2 \tag{5.8}$$

As in the analysis of the previous section, we can again derive an analytic expression for the combined single filter parameter, C, that generates optimal visibility for the unambiguous intensity-to-phase mapping over the range $\Delta\tilde{\phi}$:

$$C_{V=1}=\left(K|\bar{\alpha}|\right)^{-1}\exp\left(j\pi-j\Delta\tilde{\phi}/2\right) \tag{5.9}$$

This corresponds to a filter consisting of the following amplitude and phase values:

$$\begin{cases} BA^{-1} = \sqrt{1-2\left(K|\bar{\alpha}|\right)^{-1}\cos\left(\Delta\tilde{\phi}/2\right)+\left(K|\bar{\alpha}|\right)^{-2}} \\ \sin\theta = \left(BA^{-1}K|\bar{\alpha}|\right)^{-1}\sin\left(\Delta\tilde{\phi}/2\right) \end{cases} \tag{5.10}$$

Another interesting constraint enforced on the filter to ensure unambiguous intensity-to-phase mapping is that the available phase space for C is reduced as $\Delta\tilde{\phi}$ increases, approaching its maximum value of π. In fact it can be shown that the range of ψ_C we can use for a given $\Delta\tilde{\phi}$, is:

$$\psi_C \in \left[\Delta\tilde{\phi}/2;\ \left(\pi-\Delta\tilde{\phi}/2\right)\right] \tag{5.11}$$

where the single value $\psi_C = \pi/2$ is obtained for the largest possible unambiguous range $\Delta\tilde{\phi} = \pi$.

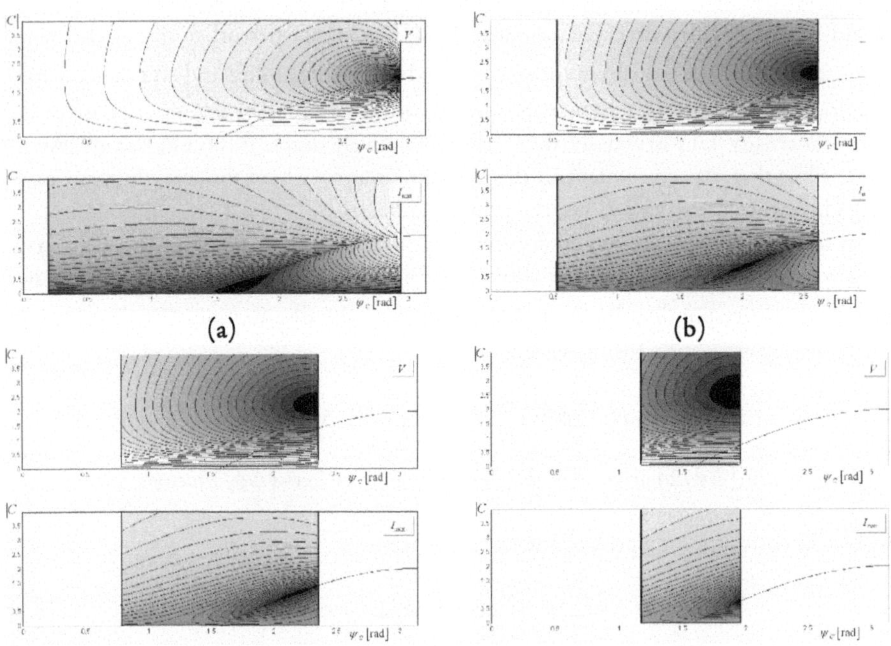

Fig. 5.2 Optimization of the visibility and peak irradiance for an unambiguous direct mapping of input phase to output intensity for a range of uniform random phase distributions of dynamic range (a) π/8, (b) π/3, (c) π/2 and (d) 3π/4. A truncation of the allowed values for ψ_C arises from the unambiguous phase-to-intensity mapping requirement. As the depth of the input phase distribution increases, the allowed values of the combined filter parameter are further constrained.

The constraints on the values that ψ_C can take given by Eq. (5.11) can be easily identified on the complex filter space plots for the visibility and irradiance. In Fig. 5.2, we show sets of such plots in which the visibility and peak irradiance are maximized for different input phase depths $\Delta\tilde{\phi}$ from $\pi/8$ to $3\pi/4$, with the additional constraint of an unambiguous phase-to-intensity mapping. As in the preceding section, we model the phase perturbation using the relationship given in Eq. (5.5).

In Fig. 5.2(a) we show a pair of plots of the visibility and peak irradiance for a phase perturbation $\Delta\tilde{\phi} = \pi/8$. The constraints on the values that ψ_C can take are shown by the sharp vertical boundaries on both plots for ψ_C values of $\pi/16$ and $15\pi/16$ radians. In all the examples shown in Fig. 5.2, we are limited to working within the range of $\psi_C \leq \pi$, therefore a restricted ψ_C axis is used to improve clarity. The appropriate portion of the operating curve for lossless filtering is shown on all the plots. As the value of $\Delta\tilde{\phi}$ increases to $\pi/3$ and $\pi/2$ as shown in Fig. 5.2(b) and Fig. 5.2(c) respectively, it can be seen that the available portion of the phase space domain in which a filter matching our criteria can be placed is further constrained. In the extreme case where $\Delta\tilde{\phi} = \pi$, only a single phase value, $\psi_C = \pi/2$, gives an unambiguous mapping. Referring to the plots from Fig. 5.2(a) to Fig. 5.2(d) it can be seen that the same trends in peak irradiance and maximum visibility are observed as those exhibited for the general optimization of the filter in Fig. 5.1. However, it should be noted that there are differences in the position of the peak irradiance regions and the extent of the maximum visibilities in the contour plots of Fig. 5.2, when compared to those of Fig. 5.1.

5.3 Optimising the Linearity of the Intensity-to-Phase Mapping

Thus far we have developed a rigorous approach for the optimization of the visibility and peak irradiance in a generic CPI and extended this optimization to the removal of ambiguity from the intensity-to-phase mapping. In this section we consider improving the linearity of the unambiguous phase-to-intensity mapping described in the preceding section. Inspired in part by Henning's phase contrast method [3, 4], we look for ways of improving the linearity of the phase-to-intensity mapping, since the user of a CPI is primarily interested in the correlation of a given output intensity to a particular phase perturbation at the input.

The intention of the method invented by Henning was to extend the linearity of the Zernike phase contrast method by introducing field absorption to the Fourier filter, in this case setting the filter parameter $A = 1/\sqrt{2}$ (A is the transmittance outside the phase-shifting region, see Fig. 3.1) The phase shift of the filter was also modified from $\theta = \pm\pi/2$ to $\theta = \pm\pi/4$. Using our analysis, we can confirm that this change of filter parameters extends the linearity as intended. Starting from our derived dependence of the CPI output intensity on the input phase distribution and filter parameters, we obtain the following expression for the Zernike case:

$$\begin{aligned} I_{Zernike}(x',y') &= \left|\exp\left(j\tilde{\phi}(x',y')\right)+(j-1)\right|^2 \\ &= 3+2\left[\sin\left(\tilde{\phi}(x',y')\right)-\cos\left(\tilde{\phi}(x',y')\right)\right] \end{aligned} \tag{5.12}$$

Here we have taken $K|\bar{\alpha}|=1$ and use a positive phase contrast filter with $A=B=1$ and $\theta=\pi/2$ (the alternative negative phase simply generates contrast reversal). This is a basic assumption in the Zernike case where the phase perturbation $\tilde{\phi}(x',y')$ is small around the zero-value (corresponding to the average phase $\phi_{\bar{\alpha}}$). Thus for small phase perturbations Eq. (5.12) simplifies to:

$$I_{Zernike} = 3+2\left(\sin\tilde{\phi}-\cos\tilde{\phi}\right)\rightarrow 1+2\tilde{\phi}+\tilde{\phi}^2 \text{ for } \tilde{\phi}\rightarrow 0 \tag{5.13}$$

This expression has the same limiting value as the well-known form found in many optics textbooks. The quadratic term in Eq. (5.13) quickly becomes important and enforces a natural limitation to the range of linearity of the Zernike phase contrast scheme.

The elegance of the Henning method for extending the linearity can be demonstrated by using the following filter parameters: $A=1/\sqrt{2}$, $B=1$, $\theta=\pi/4$ and $K|\bar{\alpha}|=1$ (positive Henning phase contrast filter):

$$I_{Henning}(x',y') = \tfrac{1}{2}\left|\exp\left(j\tilde{\phi}(x',y')\right)+\left(\sqrt{2}\exp\left(j\pi/4\right)-1\right)\right|^2 = 1+\sin\left(\tilde{\phi}(x',y')\right) \tag{5.14}$$

For small phase perturbations around zero, Eq. (5.14) simplifies to:

$$I_{Henning} = 1+\sin\tilde{\phi}\rightarrow 1+\tilde{\phi} \text{ for } \tilde{\phi}\rightarrow 0 \tag{5.15}$$

Comparing this expression with that given for the Zernike method in Eq.(5.13) we see that linearity has been significantly extended.

Both Zernike's and Henning's phase contrast methods inherently assume that we apply the constraint $K|\bar{\alpha}|=1$ to arrive at the expressions derived in Eqs. (5.13) and (5.15) and found in many optics texts, However, as we have seen throughout this and previous sections, having $K|\bar{\alpha}|=1$ is, in the majority of cases, not the best choice for a realistic operating condition (e.g. $K=½$ is more suitable). Additionally, $|\bar{\alpha}|$ is implicitly less than unity for even small-scale phase perturbations as can be verified by the phasor integration leading to this Strehl factor.

5.4 Generalising Henning's Phase Contrast Method

In light of the analysis from the previous sections, we suggest modifications to Henning's method so that we can guarantee high linearity as derived in Eq. (5.15) as well as high fringe accuracy in practice. The first essential modification to Henning's method is to veer away from the unrealistic and diffractionless Dirac-delta assumption regarding the zero-order filtering, which yields K=1. Taking into consideration the diffraction-broadening of the zero-order beam, a very small filter (K<<1) generates poor contrast while a large filter (K→ 1) may result in an inhomogeneous and highly aberrated synthetic reference wave that can lead to poor accuracy. A practical balance between contrast and accuracy can be achieved by choosing $K = ½$.

As previously described for a general filter in a CPI, we have a spatial intensity distribution given by:

$$I(x',y') = A^2\left[1 + \left(K|\bar{\alpha}||C|\right)^2 + 2K|\bar{\alpha}||C|\cos\left(\tilde{\phi}(x',y') - \psi_C\right)\right], \tag{5.16}$$

and we can, except for a trivial overall intensity scaling, now identify the combined filter parameter, C, that best solves the problem for us using the conditions

$$K|\bar{\alpha}| = ½ \quad \Rightarrow \quad C = 2\exp\left(j\pi/2\right) \tag{5.17}$$

This corresponds to the following choice of the basic filter parameters (c.f. Eq. (3.17))

$$\begin{cases} BA^{-1} = \sqrt{5} \quad \Rightarrow \quad A = 1/\sqrt{5},\ B = 1 \\ \theta = \sin^{-1}\left(2/\sqrt{5}\right) \end{cases} \tag{5.18}$$

and the resulting intensity distribution is

$$I_{\substack{Henning \\ K=½}}(x',y') = \tfrac{2}{5}\left[1 + \sin\left(\tilde{\phi}(x',y')\right)\right] \tag{5.19}$$

One must be cautioned against naïve comparisons between Eq. (5.15) and Eq. (5.19) since the former, although seemingly yielding better irradiance, is based on unrealistic assumptions and hence cannot generally be achieved in practical implementations.

The second modification concerns the $|\bar{\alpha}| = 1$ assumption, which is suitable only for very weak phase perturbations. The result can be improved if we also take the correct, input-dependent average value of $|\bar{\alpha}| \leq 1$ into consideration and use a combined filter parameter given by

$$C = \left(K|\bar{\alpha}|\right)^{-1}\exp\left(j\pi/2\right). \tag{5.20}$$

The corresponding values of the common filter parameters becomes

$$\begin{cases} BA^{-1} = \sqrt{1+(K|\bar{\alpha}|)^{-2}} & \Rightarrow \quad A = 1/\sqrt{1+(K|\bar{\alpha}|)^{-2}},\ B=1 \\ \theta = \sin^{-1}\left(2/\sqrt{1+(K|\bar{\alpha}|)^{-2}}\right) \end{cases}, \tag{5.21}$$

which generates the resulting intensity distribution

$$I_{\substack{Generalized \\ Henning}}(x',y') = 2\left[1+(K|\bar{\alpha}|)^{-2}\right]^{-1}\left[1+\sin\left(\tilde{\phi}(x',y')\right)\right]. \tag{5.22}$$

We have chosen to call this technique with improved linearity the generalized Henning method. In practice it would require an adaptive Fourier filter controlled by a sensing unit that continuously measures the strength of the focused light intensity, $\propto|\bar{\alpha}|^2$, and provides this as feedback to determine the selection of the parameters in the adaptive filter. With such a filter in place, we would be guaranteed optimal linearity for any given phase perturbation covering a range up to around $\pm\pi/3$ which is the approximately linear region of the term $\left(1+\sin\tilde{\phi}\right)$.

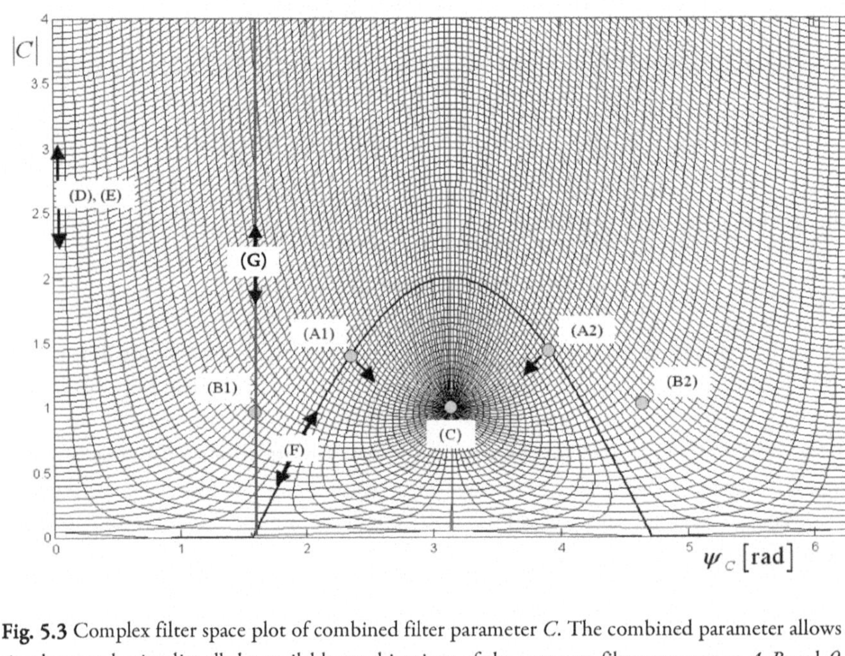

Fig. 5.3 Complex filter space plot of combined filter parameter C. The combined parameter allows us to simultaneously visualise all the available combinations of the common filter parameters A, B and θ. The bold curve is the operating curve for a phase-only (lossless) filter, whilst the fine grid represents operating curves for differing values of the filter terms A, B and θ. The operating regime for common CPI architectures are marked: (A) Zernike, (B) Henning, (C) dark central ground, (D) field absorption filtering, (E,F) point diffraction interferometers, and (G) generalized Henning method.

From Eq. (5.20) it can be seen that including the C-parameter solutions for the generalized Henning method would correspond graphically to tracing out a vertical line parallel to the $|C|$-axis and fixed at $\psi_C = \pi/2$ in the filter phase space plot, as shown in Fig. 5.3 . It should be noted that the filter operating points forming the grid in the filter phase space plot have an irregular distribution, which explains why an apparently simple C-solution generates the need for such an adaptive filter in which one component is fixed and the other two are functions of $|\bar{\alpha}|$ as shown in Eq. (5.21).

The generalized Henning method and its operating regime can be illustrated and understood using the phasor chart we introduced in Chapter 4.

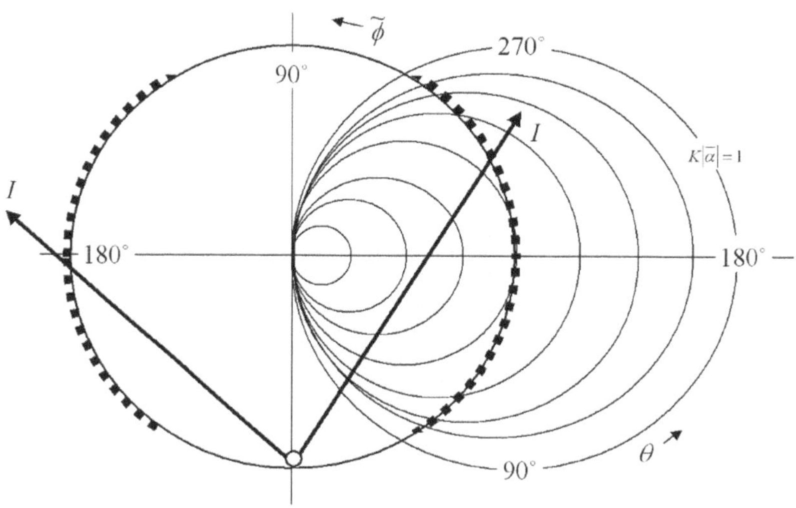

Fig. 5.4 Graphical phasor chart analysis for the generalized Henning method. The zero point of the quadratic intensity scale, I, is fixed at the point where $\tilde{\phi} = 3\pi/2$ on the $\tilde{\phi}$ unit phase circle. The intersection of I with the $\tilde{\phi}$ unit-circle directly gives the output intensity value, which corresponds to the input phase of the CPI. This mapping is approximately linear for the range $\pm\pi/3$ about the points $\tilde{\phi} = 0$ and (indicated by the dashed lines).

In Fig. 5.4, it can be seen that for the phasor chart representation of the generalized Henning method, we fix the zero point for the quadratic intensity scale, I, at the point $\tilde{\phi} = 3\pi/2$ on the $\tilde{\phi}$ unit-circle corresponding to filter parameters given by Eqs. (5.20) and (5.21). It is then possible to directly read out corresponding pairs of values for the phase-to-intensity mapping in the generalized Henning case as is illustrated in Fig. 5.4. Using the phasor chart, we can graphically verify that the range for linear mapping is approximately $\pm\pi/3$ around, indicated by a dashed line on the chart. An additional linear mapping region of $\pm\pi/3$ around $\tilde{\phi} = \pi$ is also indicated. The linearity obtained in the phase-to-intensity mapping is confirmed in the phasor chart by considering the mapping as the graphical intersection points of the quadratic intensity scale with the dashed circular arc. By simple trigonometry we find that the squared length, L^2, measured along the intensity scale intersecting the $\tilde{\phi}$ unit-circle at any given point, is given by:

$$L^2 = \cos^2 \tilde{\phi} + \left(1 + \sin \tilde{\phi}\right)^2 \propto 1 + \sin \tilde{\phi} \tag{5.23}$$

This expression has exactly the same functional form as obtained in Eq. (5.22).

5.5 Linear Phase-to-Intensity Mapping over the Entire Phase Unity Circle

Having demonstrated that it is theoretically possible to improve the linear range of the phase-to-intensity mapping for the CPI through a generalization of the Henning method as described in the preceding section, it is interesting to contemplate how linearity might be extended to generate a mapping for a larger range of input phase values than $\pm\pi/3$. One possible solution for the extension of a linear phase-to-intensity mapping regime would be the addition of a second CPI working in conjunction with the first in the same optical system and thus viewing the same input phase disturbance.

Looking at the phasor chart mapping of the generalized Henning method in Fig. 5.4, we can see that it is indeed possible to cover the whole unity phase circle with an additional interferometer with its intensity scale zero-point fixed at the point corresponding to $\tilde{\phi} = 0$ as shown in Fig. 5.5. For this common-path interferometer we have the combined filter parameter:

$$C = \left(K|\bar{\alpha}|\right)^{-1} \exp\left(j\pi\right) \tag{5.24}$$

corresponding to the filter parameters:

$$\begin{cases} \theta = \pi \\ BA^{-1} = \sqrt{1 - 2\left(K|\bar{\alpha}|\right)^{-1} + \left(K|\bar{\alpha}|\right)^{-2}} \end{cases} \tag{5.25}$$

and the resulting intensity distribution:

$$I(x',y') = 2A^2\left[1 - \cos\left(\tilde{\phi}(x',y')\right)\right] \tag{5.26}$$

The reason we do not explicitly calculate the individual filter values A and B for this interferometer can be understood by looking at the phasor chart in Fig. 5.5. It can be seen that, depending on which $K|\bar{\alpha}|$ -circle we are operating on, we need the possibility for both field absorption (to scale outwards by a factor A^{-1}) or dark ground absorption (to scale inwards by a factor B) to arrive at $\tilde{\phi} = 0$ for the zero point on the intensity scale. Choosing $K = ½$ for high fringe accuracy we can, however, omit these considerations and keep $B = 1$, only varying a single filter parameter, the field absorption, according to Eq. (5.25).

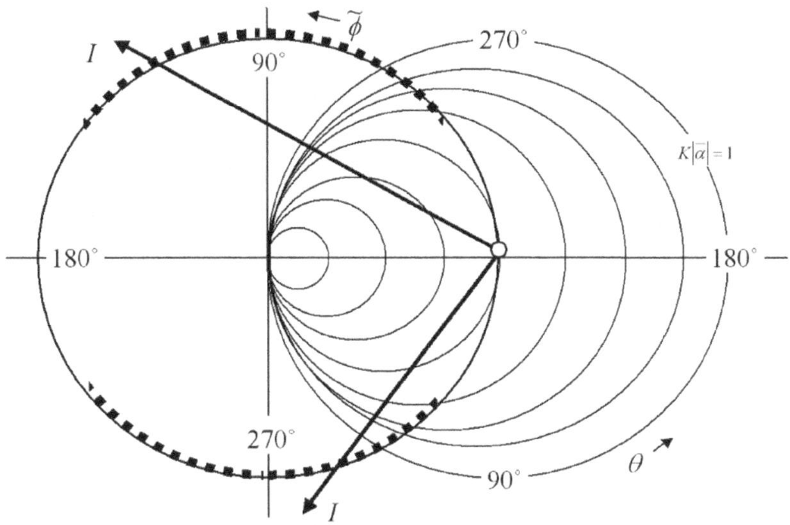

Fig 5.5 Graphical phasor chart for the CPI with filter parameters given by Eqs. (5.24) and (5.25). The zero-point of the quadratic intensity scale, I, is fixed at the point $\tilde{\phi}=0$ on the unit circle. As in Fig.5.4, the intersection of I with the unit circle gives the output intensity for a given input phase. This phase-to-intensity mapping is approximately linear for the range $\pm\pi/3$ about the points $\tilde{\phi}=\pi/2$ and $\tilde{\phi}=3\pi/2$ (indicated by the dashed lines).

Combining a common-path interferometer based on the generalized Henning method with a second common-path interferometer as shown in Fig. 5.5, we show that the whole $\tilde{\phi}$ unit-circle can be covered by an almost linear phase-to-intensity mapping. In Figs. 5.4 and 5.5 we have used dashed lines to indicate how each common-path interferometer covers two ranges extending over $\pm\pi/4$ centred at different $\tilde{\phi}$-points. Denoting the two common-path interferometers, CPI-1 (corresponding to Fig. 5.4) and CPI-2 (corresponding to Fig. 5.5) we can now explicitly identify four linear phase detection intervals over the complete phase circle and these intervals and the corresponding operation modes of both CPIs are indicated in Table 5.1.

This particular configuration of the two CPIs is not the only possibility for establishing linear phase-to-intensity mapping over the whole unit circle. As we can easily confirm by use of the phasor chart any pair of CPIs with intensity scale zero-points displaced by $\pi/2$ on the $\tilde{\phi}$ unit-circle is capable of doing the task. Therefore, the CPI designer can, within the optical constraints given, choose a dual CPI configuration from this range of options that could optimize additional tasks such as the energy throughput or simplicity of filter fabrication or operation.

Table 5.1 Table illustrating the use of a dual CPI configuration for linear phase-to-intensity mapping over a full phase circle.

Phase segment	CPI-1	CPI-2
$\tilde{\phi} \in [-\pi/4 : \pi/4]$	Linear detection	Low intensity mode
$\tilde{\phi} \in [\pi/4 : 3\pi/4]$	High intensity mode	Linear detection
$\tilde{\phi} \in [3\pi/4 : 5\pi/4]$	Linear detection	High intensity mode
$\tilde{\phi} \in [5\pi/4 : 7\pi/4]$	Low intensity mode	Linear detection

We conclude this section by considering an example of a CPI pair configuration that is both relatively simple to implement and energy efficient for each of the CPIs whilst still using $K = ½$ for high fringe accuracy. The phasor chart for this set-up is shown in Fig. 5.6.

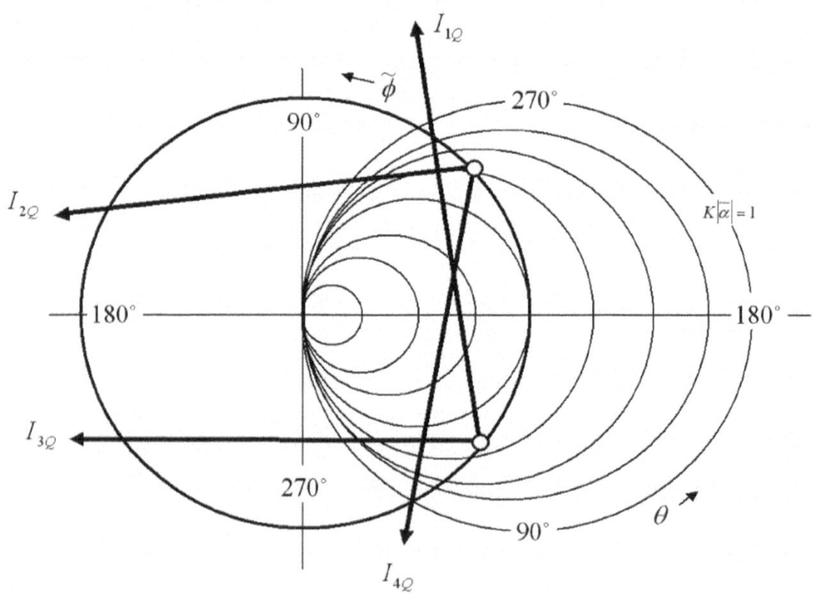

Fig. 5.6 Graphical phasor chart for a symmetric dual CPI configuration with both CPIs shown on the same chart. The zero-point of the quadratic intensity scale, I, for each CPI is fixed at the point $\tilde{\phi} = \pi/4$ and $\tilde{\phi} = 7\pi/4$ on the unit circle, respectively. The labels on the intensity scales indicate the region for which a linear mapping is provided (i.e. 1Q corresponds to the first quadrant; 2Q is the second quadrant and so forth).

The linear mapping regions are explicitly labelled on the respective intensity scales for each CPI where, "1*Q*" corresponds to the first quadrant and "2*Q*" is for the second quadrant and so forth. There are some interesting points that arise from an examination of this dual CPI configuration. First, the zero points for the intensity scales for both CPIs must be placed $\pi/2$ apart on the $\tilde{\phi}$ unit circle, as already described, if we wish to obtain a linear phase-to-intensity mapping over a complete phase circle. Second, since the zero-intensity points for both CPIs are symmetrically positioned about the x-axis and therefore equally displaced from the $\tilde{\phi} = 0°$, the same degree of field absorption is required for both CPI Fourier filters; however, the phase shifts of the two filters are different. We have selected $K = ½$ to fulfil high fringe accuracy requirements, and since $|\bar{\alpha}| \leq 1$, it is apparent that the CPI input can lie anywhere within the confines of the $K|\bar{\alpha}| = ½$ circle. The proximity of the zero-intensity point to the $K|\bar{\alpha}|$ value describing the input phase disturbance determines the extent of the field absorption required. Referring to Fig. 5.6, we can see that the distance of both zero-intensity points from the region enclosed by $K|\bar{\alpha}| = ½$ circle is minimized for this dual CPI configuration, hence reducing the amount of field absorption required, whilst the symmetry ensures the same light throughput and signal-to-noise ratio for the two CPIs.

The final point to note, from the phasor chart in Fig. 5.6, is that the filter term, *B*, is not actually required for the operation of the CPI whilst we operate using the high fringe accuracy requirement of $K = ½$. With $B = 1$ there is no attenuation in the central filtering region which need only act as a phase shifting element. The phasor chart illustrates this point clearly showing that whilst working within the confines of the $K|\bar{\alpha}| = ½$ mapping the zero intensity point of the quadratic intensity scale anywhere on the $\tilde{\phi}$ unit circle can be achieved solely by varying the two filter parameters *A* and θ. This is generally applicable for any linear phase-to-intensity mapping in a CPI.

5.6 Accurate Quantitative Phase Imaging Using Generalized Phase Contrast

We have been using Eq. (3.16) as the phase-to-intensity mapping, which is a reasonable description for the output within the central region. However, applying this mapping in the peripheral region is tantamount to assuming a homogeneous and planar SRW, which is not always appropriate and can reduce the accuracy as we discussed in Chapter 3. In the general case, the spatial variation of the SRW becomes vital to correctly interpret the output. Figure 5.7 illustrates a simple modification to the CPI optical setup that, in principle, allows for the measurement of the generally spatially varying synthetic reference wave (SRW) associated with a phase object. However, in many cases, the GPC model for the SRW suffices and allows for a reasonable interpretation of the output intensity pattern for accurate widefield phase imaging, which, although illustrated below for quantitative phase microscopy, can be extended to more general phase-imaging situations.

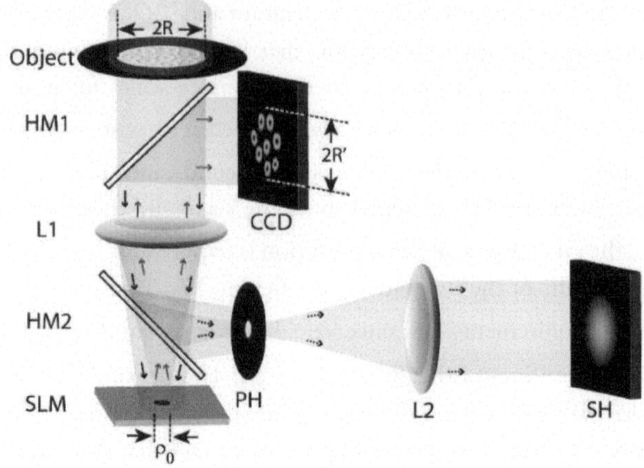

Fig. 5.7 Modified common-path interferometer with a dynamic PCF implemented with a spatial light modulator (SLM). Additional half mirror (HM2), pinhole (PH) and lens (L2) enable the measurement of the complex synthetic reference wave by a Shack-Hartmann sensor (SH).

5.6.1 The Synthetic Reference Wave in Quantitative Phase Microscopy

Microscopy typically involves optics with high numerical aperture (NA). Using a conventional microscope, we can have quantitative phase imaging by relaying the microscope output into the input plane of a common-path interferometer (CPI), such as was done experimentally in ref. [5]. The relatively lower NA in a CPI can be suitably described using a paraxial propagation model, which we adopted for analyzing a GPC-based CPI. Using this model, the GPC output for a lossless filter, normalized with respect to the illumination intensity, is (c.f. Eq. (3.9))

$$I(x',y')=\left|u(x',y')+|\bar{\alpha}|\left[\exp(j\theta)-1\right]g(r')\right|^2 \tag{5.27}$$

where $u(x',y')$ is the image of the phase object and $r'=\sqrt{x'+y'}$. Instead of the using the central value of the SRW profile, K, we now consider the spatial variations, which are in general, object-dependent:

$$g(r')=u(x',y')\otimes\left[\rho_0 J_1(2\pi\rho_0 r')/r'\right] \tag{5.28}$$

where and ρ_0 describes the size of the phase-shifting region in the Fourier plane, as illustrated in Fig. 5.7. The dashed beam path in Fig. 5.7 illustrates a novel optical scheme for extracting the actual SRW of a particular complex object field as. As in the work by Kadono et. al. [6], a half-mirror (HM2) is inserted between lens (L1) and the dynamic

PCF to deflect a measurable fraction of the focused beam to a pinhole of the same radius R_0 as that of the PCF's phase-shifting centre. Along this secondary optical axis, we can place an identical Fourier lens (L2) at a distance *f* from the pinhole, which results in an additional 4*f* setup whose output field offers an accurate measure of the SRW profile. For a large class of objects, this is a slowly varying complex field and can be accurately measured by a Shack-Hartmann wavefront sensor.

In a practical implementation, $u(x',y')$ vanishes outside a region in the image plane defined by the magnified finite extent of the incident beam or a circular input aperture of radius R. Within this region, $J_1(2\pi\rho_0 r')/r'$ is slowly varying in our implementation with a small PCF and approaches a constant as the PCF size is made to shrink. The convolution term in Eq. (5.28) corresponds to a multiplication in the Fourier plane, which represents truncation of the field within a small PCF aperture centred on the zero-frequency component. For PCF aperture sizes that are smaller than the central lobe of the Airy or sinc function generated by the input aperture on the Fourier plane, it is sufficient to approximate the field within the PCF aperture as having an Airy variation that is scaled by the complex zero-frequency value. Using this, the SRW profile is suitably described as

$$g(r') = 2\pi R \int_0^{\rho_0} J_1(2\pi\rho R) J_0(2\pi\rho r') \mathrm{d}\rho \tag{5.29}$$

5.6.2 Limitations of the Plane Wave Model of the SRW

Previous analyses of CPI-based phase-imaging instruments [5–8] have assumed a planar reference wave. In general, the SRW profile described here clearly indicates a radial dependence by noting Eq. (5.29). Using $\theta = \theta_2 = \pi$, Fig. 5.8 compares the SRW profiles obtained by FFT-based simulations and by the GPC-model. These plots are for similar PCF sizes, R_1, and for the case without an input object, $u(x',y') = 1$. Here we treat different aperture radii R or different values of the dimensionless parameter $\eta = RR_1 / 0.61\lambda f$, which serves as a measure of the PCF size relative to that of the Airy disc. The assumption of a planar SRW is only valid if the aperture size (dashed line) is greatly reduced to a limited field of view (see Fig. 5.8(a)). If a larger field of view needs to be utilized, the radial variation can no longer be neglected (see Figs. 5.8(b) and (c)). Note that the GPC-model of the SRW (solid curve) is in good agreement with the FFT-based result (open squares). A careful inspection of the minimal error shows that the GPC-model slightly overestimates the SRW produced by the numerical experiments, which is an artefact of the discrete grid in the FFT simulation. In real experiments, such quantization errors are eliminated by fabricating high-quality input and PCF apertures. Intensity linescans of the corresponding interferograms are also shown in Figs. 5.8 (d)–(f) to illustrate the effect of the spatially varying reference wave. Further reduction of the aperture size makes the SRW flatter at the expense of diminished imaging field of view. Another major drawback associated with the flattening is the weakening of the SRW, which decreases the interferogram contrast, i.e. decrease in signal-to-noise ratio (SNR), which can consequently degrade phase measurement accuracy.

In the GPC-QPI scheme, calculation of the unknown phase distribution becomes more exact over the entire field of view as it accounts for the SRW profile for a range of η values. To maintain high SNR, it is also advisable to make the SRW amplitude-matched with the input object. This suggests the use of $\eta \geq 0.41$ (see Fig. 5.8(b)) with preference for the smaller value for weaker phase objects.

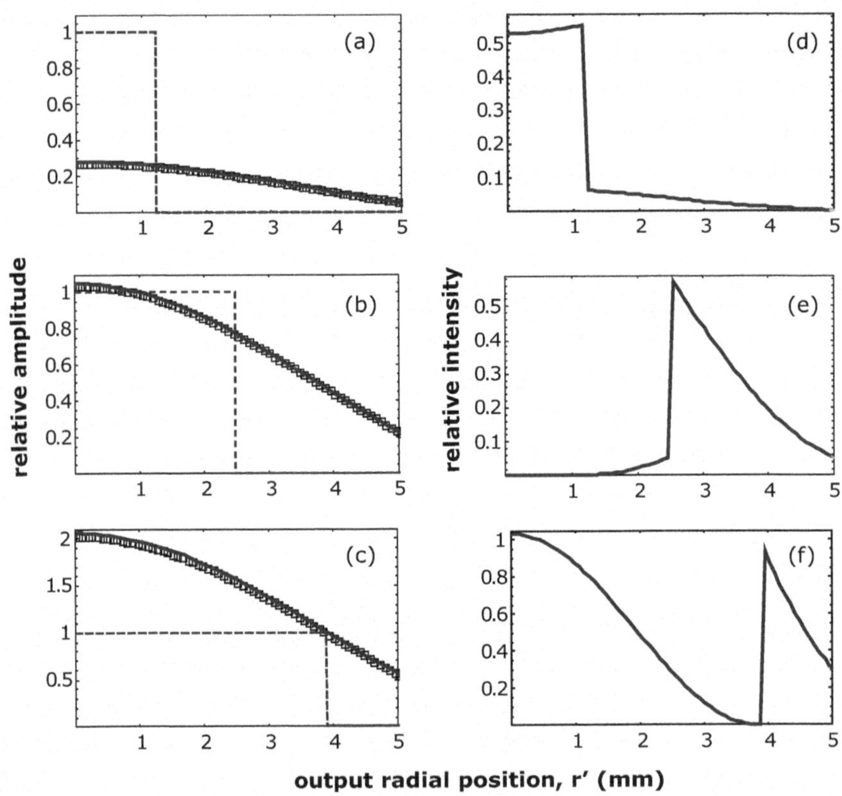

Fig. 5.8 Amplitude profiles of the SRW for $\theta = \pi$ obtained by FFT-based simulation (squares) and by GPC model (solid lines) for aperture sizes corresponding to (a) $\eta = 0.20$, (b) $\eta = 0.41$ and (c) $\eta = 0.64$. The corresponding images of the truncating aperture are also plotted for reference (dashed lines). FFT-calculated output interferograms for (d) $\eta = 0.20$, (e) $\eta = 0.41$ and (f) $\eta = 0.64$.

5.6.3 GPC-Based Phase-Shifting Interferometry

The intensity distribution outside the image aperture (or the "halo intensity") is approximated by

$$I(r'>R') \approx |\bar{\alpha}|^2 \left|\exp(j\theta)-1\right|^2 \left[g(r'>R')\right]^2 \tag{5.30}$$

We note from Eq. (5.30) that the halo intensity is simply a scaled version of the squared modulus of the SRW in the region $r'>R'$. Experimentally, this suggests that using the same detector array already in place, an accurate and real-time averaged measurement of the zero-frequency amplitude may be obtained by considering the energy ratio

$$|\bar{\alpha}|^2 = \frac{1}{4\sin(\theta/2)} \frac{\sum_{m,n} I\left(\sqrt{x'^2_m + y'^2_n} > R'\right)}{\sum_{m,n}\left[g\left(\sqrt{x'^2_m + y'^2_n} > R'\right)\right]^2}. \tag{5.31}$$

The calculation involved in Eq. (5.31) can be performed rapidly since data points of $g\left(\sqrt{x'_m + y'_n} > R'\right)$ obtained from Eq. (5.29) can be stored in a lookup table. It is interesting to note that the halo intensity, habitually ignored in CPI measurements, now becomes a significant part of the detected interferogram. The manner in which $|\bar{\alpha}|$ is determined above is a better alternative to the currently known approach [5, 6], which requires a cumbersome insertion of additional beam splitter and point detector. The output intensity can now be rewritten as

$$\begin{aligned} I(x',y') \approx & \left|u(x',y')\right|^2 + 4|\bar{\alpha}|^2 \sin^2(\theta/2)\left[g(r')\right]^2 \\ & +4|\bar{\alpha}|\sin(\theta/2)\left|u(x',y')\right| g(r')\cos\left[\phi(x',y') - \phi_{\bar{\alpha}} - (\theta+\pi)/2\right]. \end{aligned} \tag{5.32}$$

Unambiguous measurement of the relevant phase, $\tilde{\phi}(x',y') = \phi(x',y') - \phi_{\bar{\alpha}}$, with an unconstrained dynamic range is possible with the use of three interferograms I_0, I_1 and I_2 corresponding to $\theta = \theta_0 = 0$, $\theta_1 = \pi/2$ and $\theta_2 = \pi$, respectively. From Eq. (5.32), these three intensity distributions are given by

$$I_0(x',y') \approx \left|u(x',y')\right|^2, \tag{5.33}$$

$$\begin{aligned} I_1(x',y') \approx & \left|u(x',y')\right|^2 + 2|\bar{\alpha}|^2\left[g(r')\right]^2 \\ & + 2|\bar{\alpha}|\left|u(x',y')\right| g(r')\left\{\sin\left[\tilde{\phi}(x',y')\right] - \cos\left[\tilde{\phi}(x',y')\right]\right\}, \end{aligned} \tag{5.34}$$

and

$$I_2(x',y') \approx \left|u(x',y')\right|^2 + 4|\bar{\alpha}|^2\left[g(r')\right]^2 - 4|\bar{\alpha}|\left|u(x',y')\right| g(r')\cos\left[\tilde{\phi}(x',y')\right], \tag{5.35}$$

where the relevant phase, $\tilde{\phi}(x',y')$, can then be extracted from

$$\tan\left[\tilde{\phi}(x',y')\right]=\frac{2I_1(x',y')-I_2(x',y')-I_0(x',y')}{I_0(x',y')-I_2(x',y')+4|\bar{\alpha}|^2\left[g(r')\right]^2}. \tag{5.36}$$

In practical experiments, determination of $|\bar{\alpha}|$ by Eq. (5.31) is best carried out using I_2 due to the strong halo light that accompanies this interferogram. Note that the phase image of a generally complex-valued object (i.e. one that modulates both amplitude and phase of the incident field) is obtained which can be combined with Eq. (5.33) to completely map the complex object. In the next section, we investigate the robustness of the GPC-based SRW description to determine categories of input objects in which the model is considered adequate.

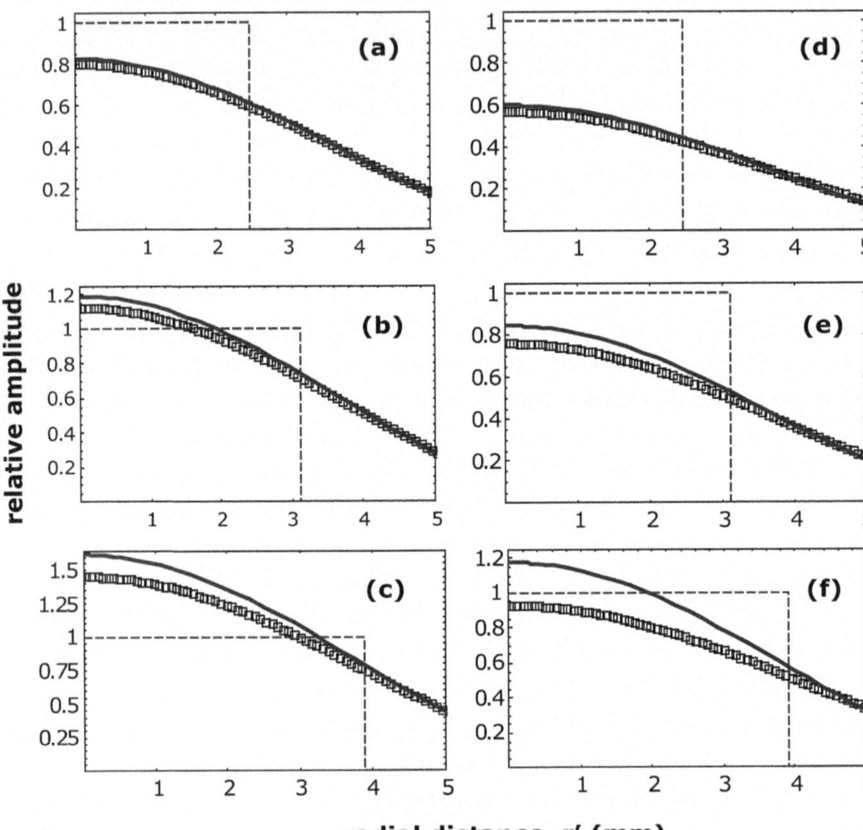

Fig. 5.9 SRW amplitude profiles for $\theta=\pi$ obtained by FFT-based simulation (squares) and by GPC model (solid lines) for an input π-phase disc of different fill factor and aperture size combinations (a) $\eta=0.41$, $F=0.1$, (b) $\eta=0.51$, $F=0.1$, (c) $\eta=0.64$, $F=0.1$, (d) $\eta=0.41$, $F=0.2$, (e) $\eta=0.51$, $F=0.2$, and (f) $\eta=0.64$, $F=0.2$. The corresponding images of the truncating aperture are also plotted for reference (dashed lines).

5.6.4 Robustness of the GPC Model of the SRW

A simple test object we have considered is a circular π-phase disc concentric with the aperture. Figure 5.9 shows the effect of changing the aperture size, proportional to η, and fill-factor *F* of the π-phase disc on the SRW. An increase in *F* causes a decrease in the SRW strength, which is due to the decrease in $|\overline{\alpha}|$. We note that as η increases, the GPC-representation of the SRW becomes less accurate, especially for larger *F*. A good compromise between interferometric contrast and SRW accuracy is therefore found by choosing $\eta \approx 0.41$.

5.6.5 GPC-Based Quantitative Phase Imaging

Let us now illustrate the reconstruction procedure for different 2D phase objects. We consider a centred π-phase disc of fill factor $F = 0.1$ as a test object and use $\eta = 0.41$ in numerical simulations. We generate interferograms for three PCF phase shifts $\theta = \theta_0 = 0$, $\theta_1 = \pi/2$ and $\theta_2 = \pi$, which are needed to reconstruct the phase information according to Eq. (5.36). For a planar model of the SRW, we replace $|\overline{\alpha}|^2 \left[g(r')\right]^2$ in Eq. (5.36) by a constant, $t_{dc}^2 I_{in}$, representing the fraction of incident light transmitted sthrough the PCF aperture. This is a numerical equivalent of the measurement by a pinhole-detector tandem described by Kadono et al. [6].

To illustrate the large errors in the measured phase when using a plane wave model of the SRW, we consider a test object comprised of alternating $\pi/2$ and $-\pi/2$ phase discs with total fill factor F = 0.2 positioned close to the edge of the aperture in a circular arrangement. The three interferograms for this particular case are given in Fig. 5.10(a)–(c). When assuming a planar SRW, a considerable error is found in the calculated phase as evidently seen in Fig. 5.10(d). In terms of percent deviation from the true phase value of $\pm\pi/2$, a maximum phase error of about 37.4% is observed, or about λ/12. On the other hand, the GPC-based QPI scheme considerably suppresses the error to a minimal amount of 1.3%, or λ/306.

Another interesting phase object we have considered is a helical phase front with a circular obstruction at its centre. Motivations for this choice of test object include the fact that a helical or vortex phase contains phase values over the full phasor cycle. Secondly, with this object's tractable Fourier transform, which consists of a zero-order beam surrounded by an annular beam [9] whose radius increases with the topological charge ℓ, we are able to perform a test on the accuracy of the GPC-QPI method as a function of the object's spatial frequency content. Figures 5.11(a)–(c) depict the output intensities corresponding to the three interferometric measurements. Following the procedures we have outlined above, we obtain the residual error measurements corresponding to the planar SRW assumption and the GPC-QPI method as shown in

Figs. 5.11(d) and (e) for the case where the unobstructed annular region of the helical phase with $\ell = 10$ covers a fill factor $F = 0.2$. In this particular case, we noted maximum peripheral errors of $\lambda/9$ and $\lambda/505$ for the planar and the GPC model, respectively. This corresponds to an impressive accuracy improvement in excess of a factor of 50. Furthermore, we investigated the effect of varying the topological charge ℓ on the observed accuracy, represented by ε. We noted that both the planar and the GPC-based SRW representation produce larger values of ε for smaller ℓ as shown in Fig. 5.12. This is anticipated from the fact that the tails from the higher spatial frequency component (in this case, from the annular beam) get closer to the central region of the PCF as ℓ decreases but remain unaccounted for by either model. Nevertheless, the GPC-model still results in about an order of magnitude improvement even for $\ell = 1$.

Our observation that phase errors for the plane wave SRW model can reach up to several tens of nanometres for optical wavelengths substantiate those described in previous experimental and numerical investigations [10, 11]. In those works, experimental verification of the accuracy of QPI schemes was carried out using a well-known topometry measurement technique with an atomic force microscope (AFM). In Ref. [9], a new type of CPI was used called the spiral phase contrast microscopy (SPCM) that employs a spiral phase filter. Like in the conventional Zernike phase contrast approximation, SPCM assumes that the reference wave emanating from the

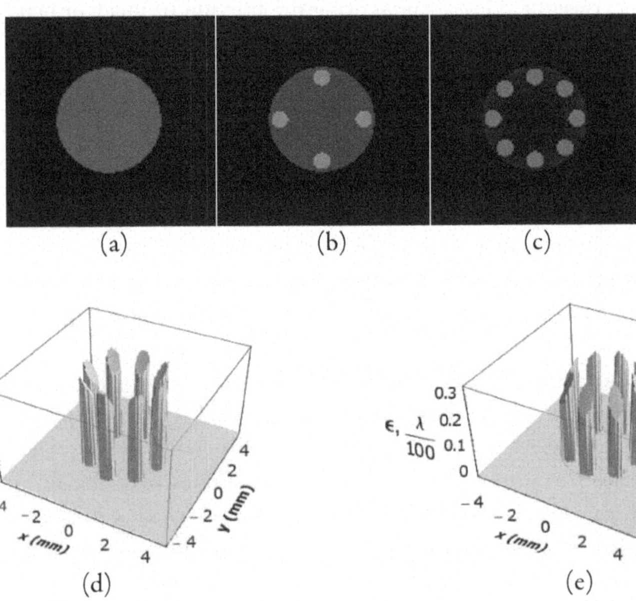

Fig. 5.10 Interferograms for an object consisting of alternating $\pi/2$ and $-\pi/2$ phase discs obtained with PCF shifts (a) $\theta_0 = 0$, (b) $\theta_1 = \pi/2$, (c) $\theta_2 = \pi$ and plots comparing the residual phase error obtained when the (d) planar and (e) GPC model of SRW are assumed. $\eta = 0.4$ is used.

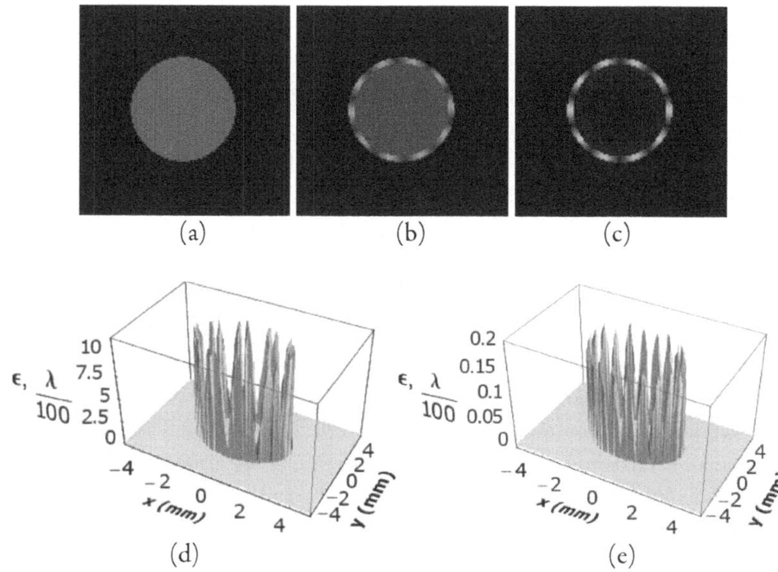

Fig. 5.11 Interferograms for an obstructed helical phase of charge $\ell=10$ obtained with PCF shifts (a) $\theta_0=0$, (b) $\theta_1=\pi/2$, (c) $\theta_2=\pi$ and plots comparing the residual phase error obtained when the (d) planar and (e) GPC model of SRW are assumed. $\eta=0.4$ is used.

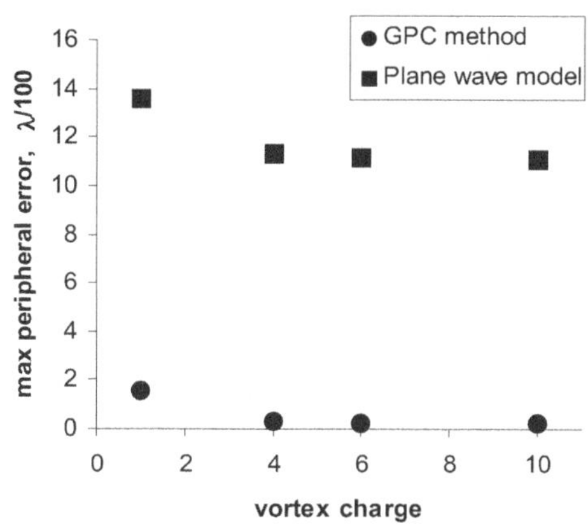

Fig. 5.12 Maximum peripheral phase error as a function of the topological charge of a centrally obstructed vortex phase object.

transmissive centre of the spiral filter is a plane wave. It is interesting to note that phase values measured experimentally under this approximation underestimate those from AFM measurements by up to 40%. This underestimation in the measured phase is also observed with the numerical results for planar SRW we have presented above. The work of Wofling, et. al. [11] also showed that errors can be significantly suppressed if the object-dependent and spatially varying SRW profile can be somehow extracted from the interferometric information. At the expense of computational cost, it was shown that the SRW profile can be better approximated as an n-th degree polynomial through nonlinear optimization algorithms. Higher integer n results in better accuracy but longer computational time. In contrast, the proposed GPC-QPI scheme prescribes a semi-analytic representation of the SRW that thereby enables rapid extraction of the phase image through Eq. (5.36), especially for most objects that weakly perturb the phase, such as micro-organisms [7] and erythrocytes [5, 8]. Our numerical results have also revealed that the GPC model of the SRW can produce QPI accuracy down to ~1 nm level without resorting to computationally intensive iterative methods, for objects with sufficient separation between low and high spatial frequency components.

The foregoing analysis and the supporting numerical results point to bounds on the achievable accuracy that are imposed by the theoretical models used as basis for processing the phase contrast output to achieve quantitative phase imaging. We have shown that the GPC-based SRW model can lead to a superior phase measurement accuracy of ~1 nm without the need for iterative calculations. Thus, GPC-based QPI extends the nanometre-accuracy over the entire aperture while maintaining rapid acquisition rates [5] – making it possible to accurately and simultaneously probe dynamics of multiple biological specimens in a colony over the entire field of view. The widely used planar SRW model tends to limit the effective field of view for accurate QPI using CPIs as it can lead to considerable amounts of error in phase depth measurements near the aperture periphery. In an experimental implementation, this model-based error will persist and can become a limiting factor even after accounting for and correcting the other sources of error, such as noise and detector constraints (e.g. grey-level quantization and nonlinearities), among others. Other variants of the common-path interferometer recently described in the literature, including SPCM and diffraction phase microscopy (DPM) [1] have also assumed a plane reference wave when reconstructing a phase image. We believe that similar extensions to the model of the reference wave in SPCM and DPM, which both employ a finite-sized region in spatially filtering the zero-order beam, can also result in considerable enhancement in the overall accuracy of these methods.

5.7 Summary and Links

In this chapter we have compared a range of well-known common path interferometers using the filter space plots that we have developed. This allowed us to determine and indicate how their performance might be improved according to a chosen figure of merit. Using the criteria of high fringe accuracy, high visibility and peak irradiance we have shown that it is possible to optimise a CPI system for operation with a given dynamic range of phase distribution at the input. Using the complex filter space plots developed in Chapters 3 and 4, we have shown that the lossless operating curve for a non-absorbing filter provides a very good first choice for a variety of filtering applications. Dwelling within the lossless operating curve is crucial to wavefront detection schemes especially when working in a photon limited regime. However, we also found that field absorption becomes increasingly necessary for large-scale input phase perturbations if the visibility is to be maximised. Owing to its validity over an expanded phase dynamic range, we showed that using GPC analysis can improve the linearity of the phase-to-intensity mapping in currently applied CPI systems. Using the Henning phase contrast method as an illustrative example, we derived a generalization of this method that should offer considerable practical improvements. We also discussed the extension of linear, unambiguous phase-intensity mapping to the full phase circle and demonstrated, through the use of our phasor charts, that this can be achieved using two CPI systems operated in parallel. This approach is related to phase shifting interferometry, which we also examined in detail. We showed that the analytical framework in GPC can be applied in the optimization of CPI-based phase shifting interferometry, thus enabling accurate quantitative phase imaging. The generalized treatment in GPC lends itself to even wider applications beyond traditional phase contrast imaging, which will be explored when we venture into novel phase contrast applications in succeeding chapters.

References

1. A. van den Bos, "Aberration and the Strehl ratio", J. Opt. Soc. Am. A, **17**, 356-358 (2000).
2. G. D. Love, N. Andrews, P. M. Birch, D. Buscher, P. Doel, C. Dunlop, J. Major, R. Myers, A. Purvis, R. Sharples, A. Vick, A. Zadrozny, S. R. Restaino, and A. Glindemann, "Binary adaptive optics: atmospheric wavefront correction with a half-wave phase shifter", Appl. Opt. **34**, 6058-6066 (1995); addenda **35**, 347-350 (1996).
3. H. B. Henning, "A new scheme for viewing phase contrast images", Electro-optical Systems Design 6, 30-34 (1974).
4. G. O. Reynolds, J. B. Develis, G. B. Parrent, Jr., B. J. Thompson, *The New Physical Optics Notebook: Tutorials in Fourier Optics*, (SPIE Optical Engineering Press, New York 1989) Chap. 35.
5. N. Lue, W. Choi, G. Popescu, T. Ikeda, R. R. Dasari, K. Badizadegan and M. S. Feld, "Quantitative phase imaging of live cells using fast Fourier phase microscopy," Appl. Opt. **46**, 1836–1842 (2007).
6. H. Kadono, M. Ogusu and S. Toyooka, "Phase shifting common path interferometer using a liquid-crystal phase modulator," Opt. Commun. **110**, 391–400 (1994).
7. T. Noda and S. Kawata, "Separation of phase and absorption images in phase-contrast microscopy," J. Opt. Soc. Am. A **9**, 924–931 (1992).

8. G. Popescu, L. P. Deflores, J. C. Vaughan, K. Badizadegan, H. Iwai, R. R. Dasari and M. S. Feld, "Fourier phase microscopy for investigation of biological structures and dynamics," Opt. Lett. **21**, 2503–2505 (2004).
9. C. S. Guo, X. Liu, J. L. He and H. T. Wang, "Optimal annulus structures of optical vortices," Opt. Express **12**, 4625–4634 (2004).
10. S. Bernet, A. Jesacher, S. Fürhapter, C. Maurer and M. Ritsch-Marte, "Quantitative imaging of complex samples by spiral phase contrast microscopy," Opt. Express **14**, 3792–3805 (2006).
11. S. Wolfling, E. Lanzmann, M. Israeli, N. Ben-Yosef M and Y Arieli, "Spatial phase-shift interferometry – a wavefront analysis technique for three-dimensional topometry," J. Opt. Soc. Am. A **22**, 2498–2509 (2005).
12. G. Popescu, T. Ikeda, R. R. Dasari and M. S. Feld, "Diffraction phase microscopy for quantifying cell structure and dynamics," Opt. Lett. **31**, 775–777 (2006).

Chapter 6
GPC-Based Wavefront Engineering

The interaction between light and matter, as light traverses through a material, will invariably leave an information imprint on the propagating light. This phenomenon is utilized in *sensing* applications where the perturbed light is used to deduce undetermined characteristics of a sample. Another interesting application of this imprinting is in the *synthesis* of light with designed characteristics using engineered modifications of the material's optical properties. Arising from fundamentally similar phenomena, it is thus not surprising that, over time, techniques originally designed for sensing applications eventually find their way into engineered light synthesis. Holography is a classic example of this cross-over. The term "holography" was coined to reflect the technique's capacity to record the full information from a sensing wavefront's phase and amplitude to preserve information about the perturbing material being studied. Later, computer holography was invented to synthesize light using holograms that are mathematically or iteratively determined, instead of being optically recorded. Far from being a trivial exercise, the synthesis application of a sensing technique is usually faced with a different set of theoretical and experimental hurdles that can potentially offer new degrees of design freedom to enhance performance. Computer holography, for instance, can minimize the typical twin image and spurious zero-order problems of conventional holography to achieve superior diffraction fidelity.

In this chapter, we will explore how the generalized phase contrast method can be used for wavefront synthesis. In the previous chapters we used the GPC framework to optimize output conditions under the constraint of unknown input wavefront phase disturbances. We saw that when a CPI is applied to wavefront sensing or the visualization of unknown phase objects, the GPC method specifies the filter phase and aperture size parameters for achieving optimal performance in extracting and displaying the phase information carried by an incoming wavefront. The capacity to optimize output conditions even when constrained by unknown inputs in sensing applications indicates that we can expect a much-enhanced performance in GPC-based synthesis since we can exploit the additional freedom to modify the input parameters. We will now explore how GPC can be used to find optimal input and filter parameters to synthesize wavefronts possessing desired output characteristics.

6.1 GPC Framework for Light Synthesis

Figure 6.1 shows a schematic illustration of a model optical setup for implementing GPC-based synthesis of patterned light at a designated output plane. With the phase contrast filter (PCF) removed, it functions as a $4f$ imaging setup that replicates object information from the input to the output plane with unit magnification. A phase modulating spatial light modulator (SLM) at the input plane allows phase perturbations to be imprinted onto an incident light and this phase object is imaged at the output. The PCF serves to synthesize a reference wave from the propagating light to generate an interference pattern at the output that mimics the features of the input phase modulation. Phase-only modulation at the input and filter planes minimizes absorption losses, thus enabling energy-efficient synthesis of patterned light when properly optimized according to the GPC framework. The imaging-based geometry allows for straightforward input phase encoding, thus doing away with the computational burden typically encountered when synthesizing light patterns by computer holography. If needed, one may substitute lenses with different focal-length ratios to achieve a desired magnification, or use accessory optics to relay a scaled version of the synthesized light to the final operating region. For notational simplicity, we will use the unity magnification setup and neglect the easily tractable coordinate inversion that accompanies imaging in the succeeding mathematical development that applies the GPC framework to find optimal parameters for light synthesis.

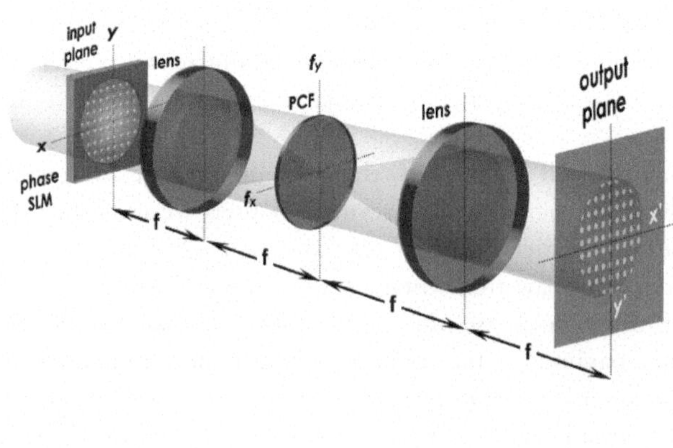

Fig. 6.1 Typical optical $4f$-setup used to synthesize patterned light using the generalized phase contrast method.

Let us consider a field at the input plane that is given by

$$p(x,y)=a(x,y)\exp\left[i\phi(x,y)\right], \tag{6.1}$$

where $a(x,y)$ is an amplitude profile that arises either due to the illumination beam, or from a limiting aperture and $\phi(x,y)$ is the coupled phase modulation. This is formally equivalent to the phase sensing case with the crucial difference that the phase information, $\phi(x,y)$, is now user-defined and, hence, fully configurable. The PCF at the Fourier plane is typically of the form,

$$H(f_x,f_y)=1+\left[\exp(i\theta)-1\right]\mathrm{circ}\left(\frac{f_r}{\Delta f_r}\right), \tag{6.2}$$

where $f_r=\sqrt{f_x^2+f_y^2}$. Note that we have dropped the absorption factors originally present in the wavefront sensing filter to ensure optimal energy throughput. As will be outlined later, we will instead exploit available freedoms at the input for optimization. The associative groupings in Eq. (6.2) are chosen to explicitly model how the filter generates a synthetic reference wave (SRW). The first term in the filter simply transmits all the Fourier components and, hence, projects an image of the input phase variations at the output plane. The second term is a low-pass filter whose radial cutoff frequency, Δf_r, is determined by the circular aperture. At the output plane, the low-pass-filtered image of the input phase variations, scaled by a multiplicative complex factor, $\exp(i\theta)-1$, serves as a reference wave for the directly imaged phase pattern. Thus, the synthesized intensity pattern at the output plane is an interferogram given by

$$I(x',y')=\left|a(x',y')\exp\left[i\phi(x',y')\right]+\left[\exp(i\theta)-1\right]p_l(x',y')\right|^2, \tag{6.3}$$

where $p_l(x',y')$ is the low-pass filtered image of the input, with coordinate inversion effects neglected, as previously mentioned. The filtered image is given by

$$p_l(x',y')=\Im^{-1}\left\{\mathrm{circ}\left[f_r/\Delta f_r\right]\Im\left\{a(x,y)\exp\left[i\phi(x,y)\right]\right\}\right\}, \tag{6.4}$$

where $\Im\{\ldots\}$ and $\Im^{-1}\{\ldots\}$ represent forward and inverse Fourier transforms, respectively and the spatial frequency coordinates, (f_x,f_y), are related to physical spatial coordinates in the filter plane, (x_f,y_f), such that $(f_x,f_y)=(\lambda f)^{-1}(x_f,y_f)$.

To gain insight on the filtered image and the synthesized reference wave, we exploit the fact that the PCF cut-off frequency is typically small and the relevant field at the Fourier plane is contained within a small region centred at the zero-order. A convenient approximation for the field within this limited region is given by

$$\mathrm{circ}\left[f_r/\Delta f_r\right]\Im\left\{a(x,y)\exp\left[i\phi(x,y)\right]\right\}\approx\bar{\alpha}\,\mathrm{circ}\left[f_r/\Delta f_r\right]\Im\left\{a(x,y)\right\}. \tag{6.5}$$

This approximates the spatial profile around the zero-order based on the expected profile from an unmodulated input. The profile is then rescaled by a complex scaling factor,

$$\bar{\alpha} = |\bar{\alpha}|\exp(\mathrm{i}\phi_{\bar{\alpha}}) = \int a(x,y)\exp[\mathrm{i}\phi(x,y)]\mathrm{d}x\mathrm{d}y \Big/ \int a(x,y)\mathrm{d}x\mathrm{d}y\,, \tag{6.6}$$

which correctly sets the magnitude and phase of the zero-order. The scaling factor may be alternately interpreted as a normalized zero-order, where the normalization factor is the zero-order for an unmodulated input. Using the approximate field within the PCF, the synthesized reference wave is then obtained as

$$\begin{aligned} s_r(x',y') &= [\exp(\mathrm{i}\theta)-1]\,p_l(x',y') \\ &= \bar{\alpha}[\exp(\mathrm{i}\theta)-1]\,g(x',y'), \end{aligned} \tag{6.7}$$

where we have isolated the complex scaling factors from the SRW spatial profile,

$$g(x',y') = \Im^{-1}\left\{\mathrm{circ}\left[f_r/\Delta f_r\right]\Im\{a(x,y)\}\right\}. \tag{6.8}$$

The synthesized light is then suitably described as

$$I(x',y') = \left|a(x',y')\exp\left[\mathrm{i}\phi(x',y')\right] + \bar{\alpha}[\exp(\mathrm{i}\theta)-1]\,g(x',y')\right|^2, \tag{6.9}$$

6.2 Optimizing Light Efficiency

To illustrate the optimization procedure for GPC-based synthesis of patterned light, we begin with the case of a uniformly illuminated circular input aperture, $a(x,y)=\mathrm{circ}[(r/\Delta r]$. As done previously, we can introduce a normalized filter size, η, to express the filter cut-off frequency, Δf_r, relative to a cut-off that matches the Airy disc generated by an unmodulated input. Using the formulation developed for this geometry in Chapter 3, we may rewrite the output within the central output region as

$$I(x',y') = \left|\exp\left[\mathrm{i}\tilde{\phi}(x',y')\right] + K|\bar{\alpha}|[\exp(\mathrm{i}\theta)-1]\right|^2, \tag{6.10}$$

where, for notational consistency, we denote output plane coordinates by (x',y'), adopt a relative phase notation, $\tilde{\phi}(x',y')=\phi(x',y')-\phi_{\alpha}$, and use K to denote the central value of the SRW profile.

The freedom to tweak the input phase distribution when synthesizing output intensity patterns relaxes the constraints originally present when sensing and measuring unknown phase disturbances. The parameter η can, in most cases, be chosen to completely encompass the zero-order light with the result that the term, K, tends to unity (see mathematical description in Eq. (3.12) and relevant discussions in Chapter 3). For this particular case, the SRW approaches a top hat profile where we can achieve nearly 100% light efficiency. For smaller and irregular phase patterns fine-tuning of η in the region 0.4–0.6 provides for an efficient operating regime while maintaining minimal losses.

6.2.1 Dark Background Condition for a Lossless Filter

Using a lossless filter and phase-only input encoding reduces absorption losses and ensures optimal light throughput through the optical system. This confines potential losses to improper light distribution at the output plane. The light efficiency of a synthesized light distribution can be optimized if we avoid generating spurious light at designated dark regions. To achieve this, we will set as our principal design criterion that the synthesized illumination satisfies a dark background condition. This criterion may be written as

$$I\left(x_0', y_0'; \tilde{\phi}_0\right) = 0 \tag{6.11}$$

where $\left(x_0', y_0'\right)$ indicates the coordinates of the designated dark background and $\tilde{\phi}_0$ is the relative phase shift generating a zero-intensity level at the observation plane.

Applying this design criterion to the output intensity expression in Eq. (6.10) yields the following expression for a lossless phase-only filter

$$K|\bar{\alpha}|\left[1-\exp(j\theta)\right] = \exp\left(j\tilde{\phi}_0\right) \tag{6.12}$$

A key point arising from Eq. (6.12) is that we gain a simple way to express a new design criterion that relates the spatial average value, $|\bar{\alpha}|$, of any input phase pattern to the zero-order phase shift, θ, of a matched Fourier phase filter. This is highly convenient since it allows us to determine an optimal filter phase shift after choosing an input phase according to any experimental constraint. Moreover, it points towards extra means of optimization by encoding the phase modulation when the filter phase has a restricted dynamic range or is fixed.

A dark background is achieved by complete destructive interference, which requires the interfering terms to have equal magnitudes. Taking the modulus of Eq. (6.12), the requirement for equal magnitudes imposes that the left-hand side of Eq. (6.12) should have unit amplitude. Since K is, by definition, positive, applying trigonometric manipulations on the modulus equation expresses the dark background condition as:

$$K|\bar{\alpha}| = \left|1-\exp(j\theta)\right|^{-1} = \tfrac{1}{2}\left|\sin(\theta/2)\right|^{-1} \tag{6.13}$$

This expresses the design criterion for achieving optimal efficiency for intensity patterns with dark background when using a lossless phase contrast filter.

6.2.2 Optimal Filter Phase Shift

When contemplating using phase contrast for light synthesis, rather than phase sensing, the typical instinct is to adapt Zernike's phase shift of $\theta = \pi/2$. From the preceding analysis, we obtained Eq. (6.13), which is a key result for a fully transmissive wavefront-engineered GPC mapping. This makes it possible to deduce the range of valid phase parameters fulfilling the design criteria specified by Eq. (6.11). Knowing that the largest possible value that the term $K|\bar{\alpha}|$ takes on is unity leads to the following solution interval for the filter phase shift in Eq. (6.13) within a full phase-cycle:

$$\theta_{\text{optimal}} \in [\pi/3,\ 5\pi/3] \tag{6.14}$$

It is remarkable that the Zernike approximation, $K|\bar{\alpha}| = 1$, requires a filter shift of $\theta = \pi/3$ and not the conventional value, $\theta = \pi/2$, used in a Zernike microscope. Moreover, although $\theta = \pi/2$ lies within the optimal solution interval, it requires that $K|\bar{\alpha}| = 1/\sqrt{2}$, which is again incompatible with the approximation. Thus, blindly operating a Zernike microscope in reverse for light synthesis can potentially yield sub-optimal results, unless the filter parameter K and input parameter $\bar{\alpha}$ in the GPC formulation are carefully considered.

From Eq. (6.13) we also observe that $K|\bar{\alpha}|$ can take on a value limited to the interval:

$$K|\bar{\alpha}| = [½;\ 1] \tag{6.15}$$

Equation (6.13), together with the solution intervals described by Eqs. (6.14) and (6.15), specify the design parameters for achieving optimal performance when encoding phase inputs to synthesize desired outputs. It also provides insights on the conditions that allow for optimal performance when extracting and displaying the phase information carried by the incoming wavefront.

Now, assuming that we have a fixed and fully transmissive phase-only filter, the best choice for the filter parameter is one that admits the largest dynamic range of phasor values at the input. Since increasing the input dynamic range generally lowers $|\bar{\alpha}|$, the smallest possible *real* value, $K|\bar{\alpha}| = ½$, is accordingly desirable, which implies $\theta = \pi$. This leads to the output intensity distribution:

$$I(x', y') = 2\left[1 - \cos\left(\phi(x', y')\right)\right] \tag{6.16}$$

6.2.3 Optimal Input Phase Encoding

Having chosen a filter $\theta = \pi$ to allow for the widest input dynamic range, we now proceed to find the input conditions for optimizing the output. We start by applying the condition $K|\bar{\alpha}| = ½$ chosen earlier and then use the definition of $\bar{\alpha}$ in Eq. (6.6).

Upon separating the real and imaginary terms, we obtain the following two requirements for designing an optimal input phase function, $\phi(x,y)$:

$$\begin{aligned} K\Omega^{-1}\int_{\Omega}\cos\left[\tilde{\phi}(x,y)\right]\mathrm{d}x\mathrm{d}y &= ½ \\ K\Omega^{-1}\int_{\Omega}\sin\left[\tilde{\phi}(x,y)\right]\mathrm{d}x\mathrm{d}y &= 0 \end{aligned} \tag{6.17}$$

Comparing Eqs. (6.16) and (6.17), we notice that it is only the first requirement in Eq. (6.17) that is directly related to the output intensity in Eq. (6.16) via the cosine term. Since we can always choose between two phasors that have the same cosine value (with 0 and π as the only exceptions), the second requirement can be fulfilled, independently of the first requirement, by simply complex conjugating an appropriate number of phasor values. This design freedom is a key feature of the GPC-method that makes it possible to solely concentrate on the first requirement when synthesizing a desired and virtually lossless grey level intensity pattern.

The first requirement in Eq. (6.17) can be fulfilled by several means, including: dynamic phase-range adjustment, fill-factor encoding, phase-histogram adjustment, spatial scaling of the phasor pattern and raster encoding, among others. In a histogram adjustment technique one will typically start out with a desired relative intensity distribution $I_{desired}(x',y')$ where the maximum achievable intensity level is unknown but relative intensity levels are known and the lowest intensity level is fixed by the background criterion of Eq. (6.11). The procedure is now to adjust the histogram for $I_{desired}(x',y')$, while maintaining identical relative intensity level ratios, until the first requirement in Eq. (6.17) is fulfilled. Subsequently, the second requirement in Eq. (6.17) is satisfied by complex conjugating an appropriate amount of the phasors. The simplest approach involves complex conjugating every second identical phasor value found, for example, by a simple raster search.

In attempting to satisfy the second requirement in Eq. (6.17), one can turn the "phasor flipping" step into an advantageous tool and thereby gain an extra degree of design freedom. One possible consideration involves taking the spatial distribution of the phasors into account to manipulate the spatial frequency content when flipping phasors. For example, choosing neighbouring phasors to have a maximum difference between them introduces high spatial frequency modulation and optimizes the separation between low and high spatial frequency terms at the Fourier filter plane to facilitate the filtering process.

Replacing a phasor and encoding its conjugate, although preserving the synthesized output intensity, can potentially influence the output phase. Therefore, another potential consideration when flipping phasors is to aim at purposely encoding a spatial phase distribution on the synthesized output. Adding to its energy efficiency, the potential for encoding the output phase in a GPC-based pattern synthesis raises another advantage of the phase-only modulation over, say, directly modulating the intensity with an amplitude modulator. This additional design freedom in GPC can be potentially exploited to achieve desired propagation behaviour since the output phase influences how an intensity pattern propagates.

6.3 Phase Encoding for Binary Output Intensity Patterns

6.3.1 Ternary Input Phase Encoding

In many cases, patterns with equalized output intensity levels on a dark background are sufficient. In the succeeding analysis, we therefore focus on the encoding of the input phase levels to synthesize outputs with binary intensity levels. As will soon become apparent, a derivation based on three input phase levels (i.e. ternary phase encoding) allows for the synthesis of a wide range of binary intensity patterns and subsumes the simpler, yet important, binary phase level encoding as a special case.

Let us consider an input aperture whose total illuminated area is designated as Ω. In ternary phase encoding, we divide the input aperture into three sub-regions that may be discontinuous, with areas Ω_0, Ω_1 and Ω_2, and encoded with phase values ϕ_0, ϕ_1 and ϕ_2, respectively. We aim to derive general relationships between the input phase parameters and the Fourier filter parameters that obey the design criterion we have already identified earlier. We can express the total aperture area and its average phase modulation as the sum of the phase-weighted sub-areas:

$$\Omega_0 \exp(j\phi_0) + \Omega_1 \exp(j\phi_1) + \Omega_2 \exp(j\phi_2) = \Omega|\bar{\alpha}|\exp(j\phi_\alpha) \tag{6.18}$$

Using the shifted phase notation and expressing the sub-areas as fractions of the total area, such that $F_1 = \Omega_1/\Omega$ and $F_2 = \Omega_2/\Omega$, we get

$$(1 - F_1 - F_2)\exp\left(j\tilde{\phi}_0\right) + F_1 \exp\left(j\tilde{\phi}_1\right) + F_2 \exp\left(j\tilde{\phi}_2\right) = |\bar{\alpha}| \tag{6.19}$$

We are interested in synthesizing binary intensity patterns with levels determined by the input phase values. In this case, we map the dark background region with $\left(\Omega_0,\, \tilde{\phi}_0\right)$ and associate the output intensity level, I, to both $\left(\Omega_1,\, \tilde{\phi}_1\right)$ and $\left(\Omega_2,\, \tilde{\phi}_2\right)$ at the input plane. It follows that:

$$I\left(\tilde{\phi}_1\right) = I\left(\tilde{\phi}_2\right) = I \tag{6.20}$$

Two unique phase values correspond to the same intensity in a symmetric condition, as can be verified from the phasor chart analysis in Chapter 4. To simplify the analysis, we use this symmetry to introduce

$$\Delta\phi = \tilde{\phi}_1 - \tilde{\phi}_0 = \tilde{\phi}_0 - \tilde{\phi}_2 . \tag{6.21}$$

Incorporating this into Eq. (6.19) and applying Eq. (6.13) to substitute for $|\alpha|$, we obtain

$$F_1\left[\exp(j\Delta\phi) - 1\right] + F_2\left[\exp(-j\Delta\phi) - 1\right] = K^{-1}\left[1 - \exp(j\theta)\right]^{-1} - 1 \tag{6.22}$$

Isolating the real and imaginary parts of Eq. (6.22), we respectively obtain the following set of equations:

$$\begin{cases} F_1 + F_2 = (2K-1)\left[2K(1-\cos\Delta\phi)\right]^{-1} \\ F_1 - F_2 = \sin\theta\left[2K\sin\Delta\phi(1-\cos\theta)\right]^{-1} \end{cases} \tag{6.23}$$

This may also be expressed in terms of the fractional areas, such that:

$$\begin{cases} F_1 = (4K)^{-1}\left[(2K-1)(1-\cos\Delta\phi)^{-1} + \sin\theta\left(\sin\Delta\phi(1-\cos\theta)\right)^{-1}\right] \\ F_2 = (4K)^{-1}\left[(2K-1)(1-\cos\Delta\phi)^{-1} - \sin\theta\left(\sin\Delta\phi(1-\cos\theta)\right)^{-1}\right] \end{cases} \tag{6.24}$$

Since we have focused on solutions where identical intensity levels are obtained in both the F_1-region and the F_2-region we can define the resulting illumination compression factor, σ, in the following way:

$$\sigma = (F_1 + F_2)^{-1} = \left[1-(2K)^{-1}\right]^{-1}(1-\cos\Delta\phi) \tag{6.25}$$

The minimum compression factor, occurring for $F_1 + F_2 = 1$, corresponds to uniform illumination at the output, whereas the maximum compression factor is found for $K \to ½$.

When using K=1, it may be convenient to rewrite Eq. (6.24) as

$$\begin{cases} F_1 = 1/8\left[\left(\sin(\Delta\phi/2)\right)^{-2} + 2\left(\sin\Delta\phi\tan(\theta/2)\right)^{-1}\right] \\ F_2 = 1/8\left[\left(\sin(\Delta\phi/2)\right)^{-1} - 2\left(\sin\Delta\phi\tan(\theta/2)\right)^{-1}\right] \end{cases}, \tag{6.26}$$

which can be solved to relate $\Delta\phi$ and the compression factor:

$$\Delta\phi = 2\sin^{-1}\left(\sqrt{\sigma/4}\right) \tag{6.27}$$

6.3.2 Binary Input Phase Encoding

An interesting special case can be deduced from Eq. (6.22) by setting $F_2 = 0$, where we find that:

$$F = F_1 = \left[K\left(1-\exp(j\theta)\right)-1\right]\left[K\left(1-\exp(j\Delta\phi)\right)\left(1-\exp(j\theta)\right)\right]^{-1} \tag{6.28}$$

implying that, for the binary phase modulation case, we must have:

$$\begin{cases} K = \tfrac{1}{2}\left[\tan(\Delta\phi/2)/\tan(\theta/2)+1\right], & \forall\ (\Delta\phi \neq \pi,\ \theta \neq \pi) \\ K = [2(1-2F)]^{-1}, & \forall\ \Delta\phi = \theta = \pi \end{cases} \tag{6.29}$$

in order for the fill factor, F, to be real-valued.

This result turns out to be the special case that corresponds to the set of solutions where a binary phase pattern serves as the input. We will now determine how this set of solutions provides for efficient encoding without requiring analogue phase levels at the input.

In a practical reconfigurable and binary modulated system it should be possible to add and remove illuminated regions depending on the specific application without affecting the illumination strength significantly. Changing the number and/or the size of each illumination region, results in a change in the fraction of modulated light with the fill factor. By calculating the visibility it is possible to have a rough measure of how much the illumination efficiency varies as a function of the fill factor.

Denoting the maximum and minimum intensity levels at the output as I_{max} and I_{min} the visibility, V, can be expressed in terms of the output intensity minima and maxima such that:

$$V = \frac{I_{max} - I_{min}}{I_{max} + I_{min}} \tag{6.30}$$

When encoding a binary input phase with the largest dynamic range (i.e. $0/\pi$-modulated input used with a matching $\theta = \pi$ filter shift) we can use the output intensity equation, Eq. (6.10), to obtain the visibility as

$$V = 4K\left|1 - 2F_\pi\right|\left[1 + 4K^2\left(1 - 2F_\pi\right)^2\right]^{-1} \tag{6.31}$$

where we have expressed $|\bar{\alpha}|$ in terms of the fill factor, F_π, which denotes the fractional area of π-modulation at the input and becomes the eventual fill factor of the output pattern.

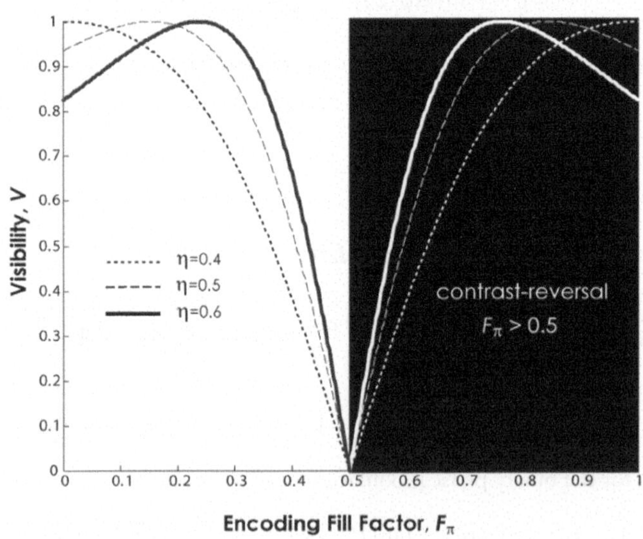

Fig. 6.2 Plot of the visibility as a function of the fractional area of π modulation, F_π, for different filter sizes ($\eta = 0.4$, $\eta = 0.5$ and $\eta = 0.6$) when using 0-π binary input phase modulation with a π-shifting PCF. Encoding with fill factors greater than 0.5 leads to contrast-reversed outputs.

In Fig. 6.2, we show a plot of the visibility as a function of the encoding fill factor, F_π, denoting the fractional area of π-modulation within the input aperture. We have plotted three curves for $\eta = 0.4$, $\eta = 0.5$ and $\eta = 0.6$, corresponding to filters having phase-shifting regions of different sizes. In binary phase-encoded applications such as optical micromanipulation the illumination would most likely occupy a fractional area less than 0.25 because there is an upper limit for how close and dense the light can be directed in order to give a sufficient separation throughout the plane of operation. We can see from Fig. 6.2 that a filter with $\eta = 0.5$ ensures a nearly constant visibility for $F_\pi < 0.25$. The visibility deteriorates for encoding fill factors approaching $F_\pi = 0.5$ as they result in a weaker SRW (falling to zero at $F_\pi = 0.5$). The visibility curve shows reflection symmetry about $F_\pi = 0.5$ and improves for $F_\pi > 0.5$. However, encoding with $F_\pi > 0.5$ generate contrast-reversed patterns (with π–modulated regions appearing dark while 0-modulated regions are bright). Hence, good visibility is only obtained for intensity patterns whose bright regions occupy a relatively small fraction of the imaged aperture region. This constraint applies only when encoding 0-π binary phase inputs. In general, GPC-based light patterning can generate high-visibility patterns with arbitrary fill factors (even F= 0.5) through suitable optimization. This optimization, discussed in the next section, allows us to determine a suitable input modulation and its matching filter for generating the desired high-visibility patterns.

6.4 Generalized Optimization for Light Synthesis

Efficiency is optimized when no light is projected into the designated dark regions. This requires complete destructive interference, which is possible when the interfering terms in the output have equal magnitudes. This criterion was earlier expressed using Eq. (6.12) and (6.13), which enabled us to specify optimal input modulation and filter phase shift that we illustrated for the synthesis of binary patterns. We will now outline an alternative expression of the design criterion to expand the design possibilities.

The starting point for the present approach is the output intensity expression in Eq. (6.9)

$$I(x',y') = \left| a(x',y')\exp\left[i\phi(x',y')\right] + \bar{\alpha}\left[\exp(i\theta)-1\right]g(x',y') \right|^2 . \tag{6.32}$$

As was done previously, we will illustrate the optimization procedure with a uniformly illuminated circular input aperture but rewrite the output differently

$$I(x',y') = \left| a(x',y')\exp\left[i\phi(x',y')\right] + K\bar{\alpha}\left[\exp(i\theta)-1\right] \right|^2 . \tag{6.33}$$

Compared with Eq. (6.10), this equation retains the fully complex expression for $\bar{\alpha}$ in the second term. The matching amplitude requirement for the dark background criterion is now expressed as

$$K\bar{\alpha}\left[1-\exp(\mathrm{i}\theta)\right]=-\exp(\mathrm{i}\phi_0), \tag{6.34}$$

where ϕ_0 is an input phase level that we choose to yield darkness at the output. Choosing $\phi_0 = 0$ for convenience and rearranging Eq. (6.34) to isolate $\bar{\alpha}$, we get

$$\bar{\alpha}=\bar{\alpha}_{real}+\mathrm{i}\bar{\alpha}_{imag}=\frac{1}{2K}+\frac{\mathrm{i}}{2K}\cot\left(\frac{\theta}{2}\right). \tag{6.35}$$

This equation, derived from the dark background condition, becomes the new design criterion. Examining the real and imaginary parts of this equation,

$$\begin{aligned}\bar{\alpha}_{real}&=\frac{1}{2K}\\ \bar{\alpha}_{imag}&=\frac{1}{2K}\cot\left(\frac{\theta}{2}\right)\end{aligned}, \tag{6.36}$$

we obtain two conditions that the input phase modulation and filter phase shift must satisfy to optimize performance: (1) The input phase should be modulated such that the real part of the normalized zero-order becomes $\bar{\alpha}_{real}=(2K)^{-1}$; (2) The PCF shift should match the imaginary part of the normalized zero-order according to $\bar{\alpha}_{imag}=(2K)^{-1}\cot(\theta/2)$.

When the input and SRW profiles match perfectly (K = 1) the optimum condition becomes

$$\bar{\alpha}=\bar{\alpha}_{real}+\mathrm{i}\bar{\alpha}_{imag}=\frac{1}{2}+\frac{\mathrm{i}}{2}\cot\left(\frac{\theta}{2}\right), \tag{6.37}$$

which consists of two conditions on the real and imaginary parts of the normalized zero-order: (a) $\bar{\alpha}_{real}=1/2$, and (b) $\bar{\alpha}_{imag}=1/2\cot\left(\theta/2\right)$. Compared with Eq. (6.17), we see that complete agreement is achieved for a filter phase shift of $\theta=\pi$, which was assumed in Eq. (6.17). This underscores the additional design freedom afforded by the present design criterion in terms of optimizing the input phase modulation and filter phase shift. How do we choose the filter phase shift? The new criterion shows that we can choose any phase shift, $\theta=\pi$ or $\theta\neq\pi$, for as long as we compensate for the input phase encoding to set (a) $\bar{\alpha}_{real}=1/2$, and (b) $\bar{\alpha}_{imag}=1/2\cot\left(\theta/2\right)$.

To illustrate the new approach, let us outline the procedure for finding a phase input, $\phi_{in}(x,y)$, that projects a desired intensity distribution, $I_{desired}(x',y')$. The phase input must satisfy the conditions imposed by Eq. (6.37). To satisfy the condition on the real part of Eq. (6.37), we will use Eq. (6.33) as a starting point. Applying trigonometric identities, we can rewrite Eq. (6.33) as

$$I_{desired}(x',y')=2\left[a(x',y')\right]^2\left\{1-\cos\left[\phi_{in}(x',y')\right]\right\}. \tag{6.38}$$

From this we can obtain the image of the phase input by solving

$$\cos\left[\phi_{in}\left(x',y'\right)\right]=1-\frac{I_{desired}\left(x',y'\right)}{2\left[a\left(x',y'\right)\right]^{2}}. \tag{6.39}$$

To apply the design criterion, we first evaluate $\bar{\alpha}_{real}$ using the definition of $\bar{\alpha}$ in Eq. (6.6) and, substituting the corresponding image plane quantities, obtain

$$\begin{aligned}\bar{\alpha}_{real}&=\frac{\iint a(x',y')\cos[\Delta\phi(x',y')]\mathrm{d}x'\mathrm{d}y'}{\iint a(x',y')\mathrm{d}x'\mathrm{d}y'}\\&=\frac{1}{A_0}\iint a(x',y')\left\{1-\frac{I_{desired}\left(x',y'\right)}{2\left[a\left(x',y'\right)\right]^{2}}\right\}\mathrm{d}x'\mathrm{d}y'\\&=1-\frac{1}{2A_0}\iint\frac{I_{desired}\left(x',y'\right)}{a\left(x',y'\right)}\mathrm{d}x'\mathrm{d}y'.\end{aligned} \tag{6.40}$$

where $A_0=\iint a\left(x',y'\right)\mathrm{d}x'\mathrm{d}y'$. Substituting the optimum condition, $\bar{\alpha}_{real}=1/2$, yields

$$\frac{1}{A_0}\iint\frac{I_{desired}\left(x',y'\right)}{a\left(x',y'\right)}\mathrm{d}x'\mathrm{d}y'=1. \tag{6.41}$$

Equation (6.41) imposes a normalization condition on the intensity level of the desired distribution, $I_{desired}\left(x',y'\right)$, that must be satisfied so the phase obtained from Eq. (6.39) yields optimum efficiency. To understand this normalization condition, let us consider the case of a uniformly illuminated aperture region, Ω, at the input plane. In this case, the normalization condition becomes

$$\frac{\iint_{\Omega'}I_{desired}\left(x',y'\right)\mathrm{d}x'\mathrm{d}y'}{\iint_{\Omega'}a^{2}\left(x',y'\right)\mathrm{d}x'\mathrm{d}y'}=1. \tag{6.42}$$

The denominator in Eq. (6.42) represents the total input energy while the numerator is the total output energy when the desired output intensity distribution is normalized. Thus, the normalization condition imposes that the desired output intensity level should be chosen in accordance with energy conservation principles.

When the intensity level is properly normalized, Eq. (6.39) yields the optimal input phase encoding. Retracing our analysis, we can see that the phase input image for $K\neq 1$ that satisfies the optimal condition, $\bar{\alpha}_{real}=(2K)^{-1}$, is determined from

$$\cos\left[\phi_{in}\left(x',y'\right)\right]=\frac{1}{K}-\frac{I_{desired}\left(x',y'\right)}{2K\left[a\left(x',y'\right)\right]^{2}}. \tag{6.43}$$

To fully satisfy the optimum conditions, the input phase modulation must be used in tandem with a filter phase shift, θ, that matches the imaginary component of the normalized zero-order, $\bar{\alpha}_{imag}$, according to Eq. (6.37).

One may still choose to implement a phasor-flipping procedure to avail oneself of the various advantages mentioned earlier. Phasor flipping is formally supported by the phase-to-intensity mapping in Eq. (6.43) since for every phase value that satisfies the equation at a point (x,y), the negative of such phase is also a solution. When a properly balanced flipping is implemented, the imaginary component, $\bar{\alpha}_{imag}$, vanishes and the optimal filter phase shift is $\theta=\pi$ as before. However, the present approach liberalizes the phasor flipping and no longer requires balanced flipping. If desired, one may choose to implement any phasor flipping scheme according to a desired effect at the output for as long as the filter phase shift satisfies the condition imposed by the imaginary component, $\bar{\alpha}_{imag}$.

6.5 Dealing with SRW Inhomogeneity

The GPC framework for light synthesis that we presented in Sect. 6.1 shows that the synthetic reference wave can generally exhibit a spatial variation. The fact that the SRW is reasonably homogeneous within the central region allowed us to formulate a design criterion for optimizing the synthesized output. In the previous chapter, careful consideration of the SRW spatial variations allowed us to formulate an accurate quantitative phase measurement scheme. In this section, we again consider effects from the SRW inhomogeneity and outline various approaches for coping with it in synthesis applications.

6.5.1 Filter Aperture Correction

The spatial variation of the SRW was previously described in Chapter 3, with Fig. 3.3 illustrating that such variation depends on the filter size. For the range of values of the normalized size considered, $\{0.2 \leq \eta \leq 0.63\}$, the plot showed that a large value of η produces significant curvature in the SRW, which will cause a distortion of the output interference pattern. On the other hand, a low value of η generates a flat SRW but at the cost of a reduction in the SRW amplitude. For a circular input aperture with radius Δr and circular phase-shifting filter with frequency cut-off Δf_r, the SRW spatial variation in Eq. (6.8) may be rewritten as

$$g(r')=2\pi\Delta r\int_0^{\Delta f_r} J_1(2\pi\Delta r f_r)\, J_0(2\pi r' f_r)\, \mathrm{d}f_r, \tag{6.44}$$

or, using the dimensionless filter size parameter, $\eta=(0.61)^{-1}\Delta r\Delta f_r$, we may write

$$g(r')=2\pi\Delta r\int_0^{0.61\eta/\Delta r} J_1(2\pi\Delta r f_r)\, J_0(2\pi r' f_r)\, \mathrm{d}f_r. \tag{6.45}$$

Either way, the SRW appears as a low-pass filtered image of the circular input aperture, $\mathrm{circ}(r'/\Delta r)$. Understanding that the observed SRW curvature and dampening follows directly from excluding the higher frequency components, we can expect that increasing the spatial filter bandwidth for the focused light will eventually produce an SRW profile that reasonably matches the circular input aperture. In Fig. 6.3 we plot the numerically obtained curves of the SRW for increasing values of η.

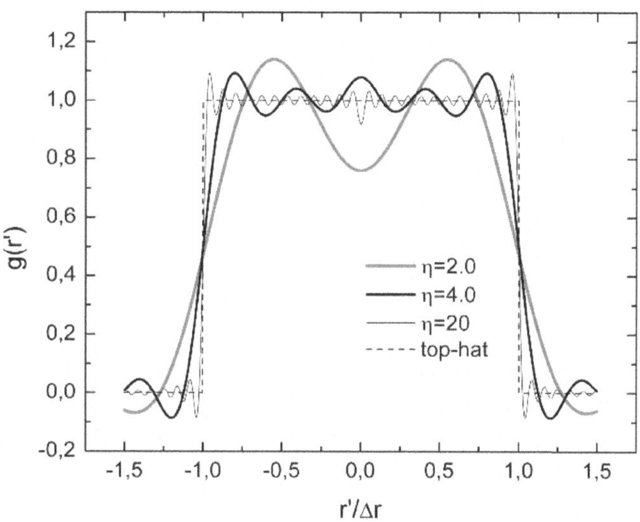

Fig. 6.3 Plot of the SRW for η= 2.0, η= 4.0, and η = 20. The SRW approaches a top-hat profile as η is increased.

Thus, one potential solution for avoiding SRW inhomogeneity is to simply use a larger filter. An important *caveat*, however, is that the SRW expression in Eq. (6.8) only approximates the more general description of Eq. (6.4), which depicts the SRW not as the low-pass filtered image of the aperture, but as the low-pass filtered image of the entire input, including the encoded phase modulation. This means that we can increase the filter bandwidth to improve homogeneity but must cautiously avoid transmitting the frequency components generated by the phase-encoded input. Thus this is a natural solution when synthesizing high-frequency outputs such as high-density spot array illumination. For patterns containing lower frequency components, we can still benefit from this approach by forcing a high-frequency input modulation through a suitable phasor flipping approach.

6.5.2 Input Phase Encoding Compensation

As a low-pass filtered image of the input illumination, the SRW spatial profile does not match the image of the input illumination, which generally introduces an intensity roll-off in the projection. For array illumination, this means that the generated spots will have non-uniform intensities. In this case, one is potentially faced with the problem that raising the lowest spot intensity to a suitable level might intensify the brightest spots beyond a tolerable operating level whereas lowering the brightest spots to acceptable levels might render the dimmer spots ineffective. One way to deal with this is to use a higher filter bandwidth, as discussed earlier. We will now describe another approach using an input phase compensation scheme to counteract potential problems from SRW curvature effects without increasing the filter bandwidth beyond the previously determined optimal values. The technique exploits the existing phase modulation capabilities of the input plane SLM to incorporate corrections directly into the input phase modulation.

First, we rewrite the design criterion described by Eq. (6.34) into

$$\bar{\alpha}\left[\exp(\mathrm{i}\theta)-1\right]=-k_0\exp(\mathrm{i}\phi_0), \tag{6.46}$$

where k_0 is a positive constant. The arbitrary constant is added to anticipate that the phase compensation might hinder us from achieving $k_0=1$ which was assumed in Eq. (6.34). This assignment maintains the π-shift between the SRW and the zero-encoded regions in the signal, which directs minimal light in the designated dark regions. Choosing $\phi_0=0$, we now write the output intensity as

$$\begin{aligned} I(x',y') &\approx \left|\exp\left[\mathrm{i}\phi(x',y')\right]-k_0 g(x',y')\right|^2 \\ &=1+k_0^2 g^2(x',y')-2k_0 g(x',y')\cos\left[\phi(x',y')\right] \end{aligned}. \tag{6.47}$$

The optimal design criterion is now

$$\bar{\alpha}=\bar{\alpha}_{real}+\mathrm{i}\bar{\alpha}_{imag}=\frac{k_0}{2}+\mathrm{i}\frac{k_0}{2}\cot(\theta/2). \tag{6.48}$$

The phase-to-intensity mapping in Eq. (6.47) shows the possibility of tuning the peak output intensity through a proper choice of the encoded phase, $\phi(x,y)$. Generating a uniform output intensity, I_{edge}, requires a phase input that maps to an image is given by

$$\phi(x',y')=\arccos\left(\frac{1+k_0^2 g^2(x',y')-I_{edge}}{2k_0^2 g(x',y')}\right), \tag{6.49}$$

where $I_{edge}=1+k_0^2 g^2(r_0,0)+2k_0 g^2(r_0,0)$ is the intensity at the edge of the phase-addressable region when the corresponding input point is π-phase-encoded.

Solving Eq. (6.49) for $\phi(x',y')$ is not as straightforward as it seems because the scaling factor, k_0, and the phase, $\phi(x',y')$, have circular dependence on each other, as evident from examining the $\bar{\alpha}$ expressions in Eq. (6.6) and Eq.(6.48). However, the uniformity error can rapidly converge to an acceptable level when iteratively solving the circularly dependent variables. This is illustrated in Fig. 6.4, which shows the non-uniformity that persists after each iteration for GPC-based array illumination. The non-uniformity, which we take to be the maximum intensity difference between the spots expressed as a percentage of the peak intensity, already drops from 40% to <1% after the first iteration. The non-uniformity is maintained at less than 1%, with minor improvements, after the second iteration.

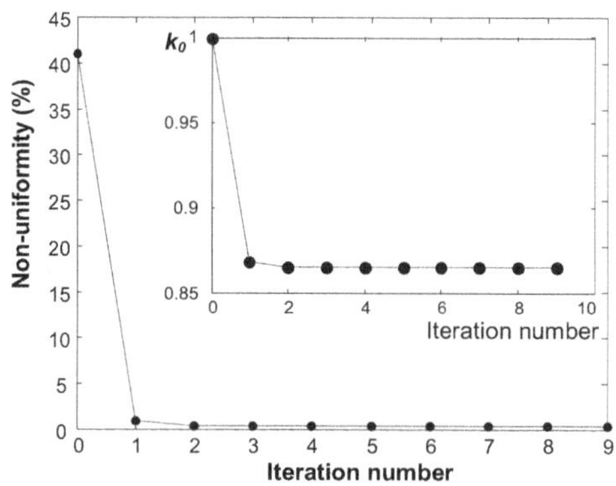

Fig. 6.4 Plot of the non-uniformity after each iteration. The non-uniformity expresses the maximum intensity difference between the spots in the generated array as a percentage of the peak intensity. Inset shows the value of the scaling factor, k_o, after each iteration.

The rapid convergence is echoed in the value of the scaling factor, k_0, as shown in the inset of Fig. 6.4, where the value of k_0 already stabilizes after the second iteration. Based on this finding, we can do away with the iteration and just use the stable value of k_0 to find the corrected input phase using Eq. (6.49) for different configurations of spot arrays. The results of this iteration-free input phase compensation scheme for periodic and pseudo-periodic configurations are shown in Fig. 6.5, where the non-uniformity is minimized for both cases (<1% for periodic, <5% for pseudo-periodic). In addition, the homogenized spot array distribution achieves the same efficiency as that obtained prior to introducing phase corrections.

6.5.3 Input Amplitude Profile Compensation

Synthesis of patterned light by phase-only modulation offers potentially high light efficiency due to minimal absorption. The GPC framework offers a simple input phase-encoding scheme where the image of the input is interfered with an internally

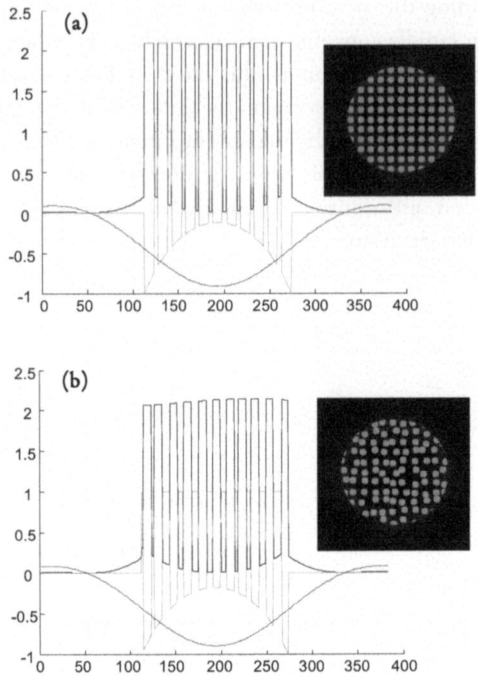

Fig. 6.5 Line scans through the GPC-based array illumination for (a) periodic and (b) pseudo-periodic configurations after input phase corrections. The insets show the generated spot array. The linescans through the intensity patterns (solid trace) show minimal uniformity error of <1% for (a) and <5% for (b) when the spatially varying SRW (dashed trace) interferes with a phase-compensated input. The dotted trace shows the real part of each linescan through the respective complex signal image fields.

synthesized reference wave to generate output light patterns that directly mimic the encoded phase. However, mismatch between the spatial amplitude profiles of the interfering components can lead the outputs with some deviation from the design. With the SRW spatial profile being a low-pass filtered image of the illumination profile, the two profiles will be inherently mismatched, especially near the regions where the input illumination exhibits sharp edges from limiting apertures. We suggested earlier expanding the filter bandwidth and, indeed as expected from Fourier theory, the match improves as the cut-off frequency is increased to transmit more Fourier components of the circular input aperture. We now take another approach to improve the match.

Owing to the high frequency components from the sharp edge of a circular input aperture, the SRW profile can achieve a very good spatial match only for a very high filter bandwidth. We can achieve a reasonable match for a smaller filter bandwidth by apodizing the input aperture. To illustrate this effect, we replace the circular input aperture with a Gaussian aperture. Physically, we can remove the aperture and replace the uniform illumination with a Gaussian beam. A Gaussian illumination with a beam waist, w_0, is equivalent to having a Gaussian aperture function given by

$$a(x, y) = a_r(r) = \exp\left[-r^2/w_0^2\right]. \tag{6.50}$$

The zero-order beam at the Fourier plane also assumes a Gaussian profile:

$$\Im\{a_r(r)\} = \pi w_0^2 \exp\left[-\pi^2 w_0^2 f_r^2\right] \tag{6.51}$$

The GPC-output still follows the same functional form as in Eq. (6.9) with the SRW profile consequently adjusted based on the new aperture function,

$$g(x', y') = g_r(r') = 4\pi^2 \int_0^{\Delta f_r} \int_0^{\infty} \exp\left(-r^2/w_0^2\right) J_0(2\pi f_r r) r\, J_0(2\pi f_r r') f_r\, dr\, df_r. \tag{6.52}$$

Figure 6.6(a) shows the radial SRW profile obtained from numerical integration of Eq. (6.52) for different PCF sizes, measured relative to the $1/e^2$ intensity beam waist of the Gaussian zero-order beam. The SRW profile approaches the amplitude of the Gaussian illumination as the PCF size is increased. For comparison, Fig. 6.6(b) shows the corresponding variations in the SRW profile with PCF size for top-hat illumination. As expected, we achieve better matching between the input and SRW amplitude profiles for Gaussian illumination since the PCF captures the relevant Gaussian Fourier components, which are close to the zero order, while a flattop illumination contains higher spatial frequencies.

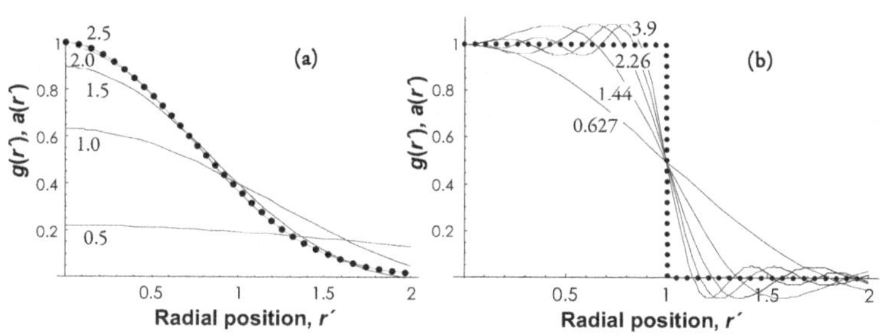

Fig. 6.6 Radial variation of the SRW profile, $g(r')$, for different PCF sizes. (a) Gaussian illumination: PCF sizes 0.5, 1.0, 1.5, 2.0, 2.5 times the Gaussian zero-order beam waist. (b) Top-hat illumination: PCF sizes 0.627, 1.44, 2.26, 3.08, 3.9 times that of the Airy central lobe. The dotted traces indicate the profile of the incident field.

For perfectly matched signal and SRW profiles, and applying the optimal conditions of Eq. (6.34), the output intensity becomes

$$I(x', y') \approx \exp\left(-2r'^2/w_0^2\right)\left|\exp\left[i\phi(x', y')\right] + \exp\left[i\left(\phi_{\bar{\alpha}} + \theta/2 + \pi/2\right)\right]\right|^2. \tag{6.53}$$

Achieving matched signal and SRW profiles, a requirement embodied in Eq. (6.34), allows us to generate a dark outer region by manipulating the signal phase to attain destructive interference. This leads to high energy efficiency as the energy channels into the central region, which is an attractive working area because of its relatively flat amplitude profile.

6.6 Generalized Phase Contrast with Rectangular Apertures

We have, for the most part, assumed a uniformly illuminated circular aperture in our derivation of design criterion for GPC-based light synthesis. However, when using a rectangular SLM as input phase modulator, inserting a circular aperture leaves a significant portion of the SLM not illuminated and not utilized. The preceding discussion on Gaussian-illuminated GPC shows that varying the input aperture can give interesting results. To correspond better with the characteristic rectangular aspect ratios of commercially available spatial light modulators, we now analyze the GPC method considering a rectangular illumination aperture together with a rectangular phase contrast filter (PCF). We derive the corresponding phase-to-intensity mapping and optimal design criterion.

Figure 6.7 illustrates the usual GPC optical system where two lenses are arranged in a standard 4*f* lens configuration. A collimated light beam illuminates a phase-only spatial light modulator (SLM) at the front focal plane of the first lens. Aside from introducing a spatial phase modulation, $\phi(x,y)$, the SLM effectively imposes a rectangular truncation window. Assuming a unit-amplitude monochromatic plane wave illumination, the spatial field distribution entering the 4*f* system takes the form

$$a(x,y)=\mathrm{rect}(x/\Delta x, y/\Delta y)\exp\left[\mathrm{i}\phi(x,y)\right], \tag{6.54}$$

where Δx and Δy are the dimensions of the SLM window.

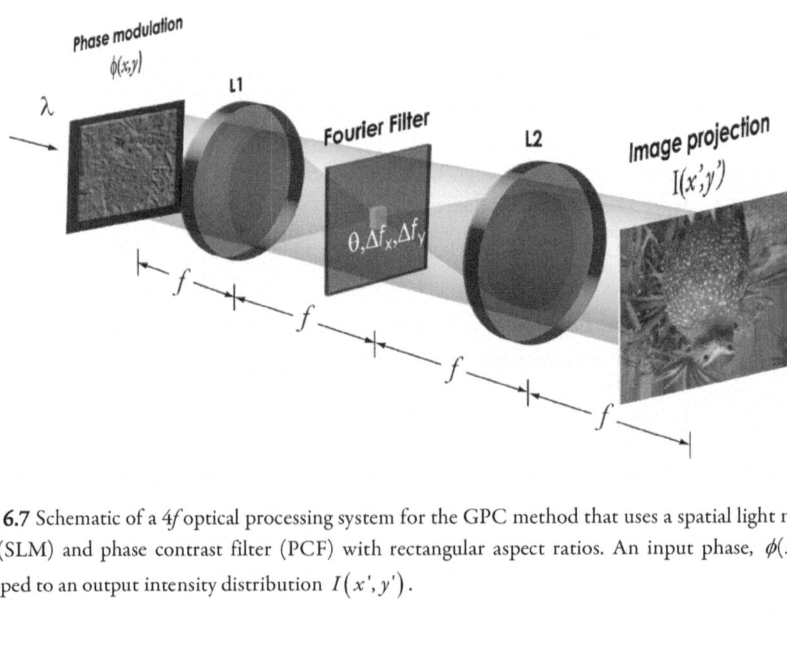

Fig. 6.7 Schematic of a 4*f* optical processing system for the GPC method that uses a spatial light modulator (SLM) and phase contrast filter (PCF) with rectangular aspect ratios. An input phase, $\phi(x,y)$, is mapped to an output intensity distribution $I(x',y')$.

To match the rectangular window of the SLM, we will consider a rectangular spatial filter at the Fourier plane. A phase-only filter, which is ideal for maximum light throughput, is described in the spatial frequency coordinates, f_x and f_y, by

$$H(f_x, f_y) = 1 + \left[\exp(i\theta) - 1\right]\mathrm{rect}\left[f_x/\Delta f_x, f_y/\Delta f_y\right], \tag{6.55}$$

where θ is the phase shift introduced by the filter onto the lowest spatial frequencies within a rectangular spectral region defined by Δf_x and Δf_y, relative to higher frequency components. We will now describe the output intensity, $\left|I(x',y')\right|^2$, that is generated by a given input phase distribution, $\phi(x,y)$, and filter $H(f_x,f_y)$.

6.6.1 Phase-to-Intensity Mapping

Within the framework of scalar diffraction theory [s1], the lenses in the 4*f* setup may be described by two Fourier transforms, $\Im\{\ \}$. For convenience, we will substitute an inverse Fourier transform, $\Im^{-1}\{\ \}$, for the second Fourier transform, subject to an inversion of the output spatial coordinates (x',y') relative to the input coordinates, (x,y). Together, the input in Eq. (6.54) and the filter in Eq. (6.55) generate a field distribution at the image plane that is expressed as (c.f. Eq. (6.3))

$$I(x',y') = a(x',y') + \left[\exp(i\theta) - 1\right]p_l(x',y'), \tag{6.56}$$

where $p_l(x',y')$ is the low-pass filtered image of the input given by

$$\begin{aligned} p_l(x',y') &= \Im^{-1}\left\{\mathrm{rect}\left(f_x/\Delta f_x, f_y/\Delta f_y\right)\Im\left\{a(x,y)\right\}\right\} \\ &= \left|p_l(x',y')\right|\exp\left[i\phi_{p_l}(x',y')\right] \end{aligned}. \tag{6.57}$$

As in earlier GPC analysis, the second term in Eq. (6.56) is conveniently interpreted as a synthetic reference wave (SRW) that interferes with the directly imaged input, $a(x',y')$, to generate an intensity distribution at the image plane. Within the bounds of the imaged SLM aperture, $\mathrm{rect}(x'/\Delta x, y'/\Delta y)$, the resulting intensity distribution is

$$\begin{aligned} I(x',y') = 1 + 4\left|p_l(x',y')\right|\sin(\theta/2)\Big\{&\left|p_l(x',y')\right|\sin(\theta/2) \\ &+ \sin\left[\phi(x',y') - \phi_{p_l}(x',y') - \theta/2\right]\Big\}. \end{aligned} \tag{6.58}$$

Achieving a high contrast output is contingent upon the capacity to produce a zero-intensity output for some input phase value imaged as $\phi_o(x',y')$. Following Eq. (6.56), this criterion can be written as

$$0 = \exp\left[i\phi_o(x',y')\right] + \left[\exp(i\theta) - 1\right]p_l(x',y'), \tag{6.59}$$

where we may choose that $\phi_o(x',y')=0$ since the output is immune to any uniform input phase offset. Equation (6.59) is satisfied if and only if $p_l(x',y')$ approaches a complex-valued constant,

$$p_l(x',y') \to p_{l,real} + \mathrm{i}p_{l,imag} = \frac{1}{2} + \frac{\mathrm{i}}{2}\cot\left(\frac{\theta}{2}\right). \tag{6.60}$$

Alternately, we may express this criterion in terms of the p_l modulus and phase as

$$\begin{aligned} |p_l(x',y')| &\to \frac{1}{2|\sin(\theta/2)|} \\ \phi_{p_l}(x',y') &\to \phi_o(x',y') - \theta/2 \pm \pi/2 \end{aligned} \tag{6.61}$$

with '$\pm$' arising due to the term $\sin(\theta/2)/|\sin(\theta/2)|$. Thus, $+\pi/2$ is taken when $\sin(\theta/2)$ is positive, and $-\pi/2$ otherwise. When this high-contrast criterion is satisfied, the output is described by a much simplified expression

$$I(x',y') = 4\sin^2\left\{\left[\phi(x',y') - \phi_o(x',y')\right]/2\right\}. \tag{6.62}$$

When designing light distributions using GPC, one can obtain the required phase inputs from its image derived using the corresponding phase-to-intensity mapping,

$$\phi(x',y') = 2\arcsin\left[\frac{\pm\sqrt{I(x',y')}}{2}\right] + \phi_o(x',y'). \tag{6.63}$$

For optimal efficiency, the intensity $I(x',y')$ to be used in Eq. (6.63) should not be entirely arbitrary, but should be constrained according energy conservation principles. In the ideal case, the total power within the output intensity distribution would be equal to the input power. Given the unit-amplitude field described by Eq. (6.54), the target pattern can be expressed as

$$I(x',y') = \Delta x \Delta y \; I_{\text{norm}}(x',y'), \tag{6.64}$$

which employs a normalized distribution, $I_{\text{norm}}(x',y')$, satisfying

$$\int_S \left|I_{\text{norm}}(x',y')\right|^2 \mathrm{d}x'\mathrm{d}y' = 1 \tag{6.65}$$

where the surface of integration, S, is defined by SLM aperture image, $\mathrm{rect}(x'/\Delta x, y'/\Delta y)$. Thus, we may rewrite the optimal phase-to-intensity mapping as

$$\phi(x',y') = 2\arcsin\left[\frac{\pm\sqrt{\Delta x \Delta y I_{\text{norm}}(x',y')}}{2}\right] + \phi_o(x',y'), \tag{6.66}$$

where $\phi_0(x',y')$ is a phase constant that defines the zero-output phase value, which may be chosen to be zero, as explained earlier. This mapping serves as a basis for designing an input phase distribution $\phi(x',y')$ to produce a desired output intensity distribution $I(x',y')$. This, of course, requires a suitable filter, whose phase shift must be determined according to the constraints expressed in Eq. (6.60) or, alternately, Eq. (6.61).

Being a low-pass filtered image of the input the term, the actual features of $p_l(x',y')$ and the subsequent synthetic reference wave are expected to be dependent on the filter bandwidth and the input phase distribution itself. The filter bandwidth physically corresponds to the aperture size of phase-shifting region on the filter. As in earlier GPC analysis [2], output optimization requires properly matching the filter characteristics to the input conditions. We will now proceed to a functional approximation of the synthetic reference wave with the aim of finding practical expressions for optimizing GPC-based synthesis of intensity distributions when using rectangular input and filter apertures.

6.6.2 Approximating the Reference Wave

As in previous analyses of the GPC method [2, 3], the following approximation for the SRW generating function is useful (c.f. Eq. (6.7)):

$$p_l(x',y') = \Im^{-1}\left\{\mathrm{rect}\left(f_x/\Delta f_x, f_y/\Delta f_y\right)\Im\left\{a(x',y')\right\}\right\} \approx \bar{\alpha}\, g(x',y'), \tag{6.67}$$

where $g(x',y')$, denoted earlier as the SRW generating function, accounts for the aperture truncation effects at the object and Fourier planes,

$$g(x',y') = \Im^{-1}\left\{\mathrm{rect}\left(f_x/\Delta f_x, f_y/\Delta f_y\right)\Im\left\{\mathrm{rect}\left(x/\Delta x, y/\Delta y\right)\right\}\right\}, \tag{6.68}$$

and the input-dependent contribution is incorporated into $\bar{\alpha}$, which is defined as the complex spatial average of the input field $a(x,y)$,

$$\bar{\alpha} = |\bar{\alpha}|\exp(\mathrm{i}\phi_{\bar{\alpha}}) = \int_S \frac{\exp\left(\mathrm{i}\phi(x,y)\right)}{\Delta x \Delta y}\mathrm{d}x\mathrm{d}y\,. \tag{6.69}$$

This approximation is most appropriate when there is a distinct separation between high and low spatial frequency components of $a(x,y)$ at the Fourier plane, as discussed in 3.2.1. It essentially assumes that only the DC component of $a(x,y)$ contributes to the phase-shifted reference wave $c(x',y')$. All other image information is then carried in higher spatial frequencies that do not pass through the PCF centre.

As shown in ref. [8], $g(x',y')$ can be evaluated and expanded into an infinite series, where the largest contribution to $g(x',y')$ is from the spatially uniform term. This leads to the approximation

$$g(x',y') \approx \frac{4}{\pi^2}\mathrm{Si}(\pi\eta_x)\mathrm{Si}(\pi\eta_y), \tag{6.70}$$

where Si( ) is the Sine Integral function [4], and η_x and η_y are dimensionless filter size parameters defined as the ratio of the respective PCF axial widths to the width of the central lobe of the emergent sinc distribution at the Fourier plane for a uniform input phase:

$$\begin{aligned}\eta_x &= \Delta x \Delta f_x / 2 \\ \eta_y &= \Delta y \Delta f_y / 2\end{aligned}\,. \tag{6.71}$$

Reconciling Eq. (6.70) with Eq. (6.67) also suggests that

$$\phi_{p_l}(x',y') \approx \phi_{\bar{\alpha}}\,. \tag{6.72}$$

In some cases it can be convenient for this phase to vanish, i.e. $\bar{\alpha}$ becomes real-valued. This can be realized, if needed, as Eq. (6.62) implies that the phase-to-intensity mapping in Eq. (6.66) is arbitrary to within the complex conjugate of $\left[\phi(x',y') - \phi_o(x',y')\right]$. Thus, there is enough flexibility to tune the input phase distribution to get $\phi_{\bar{\alpha}} = 0$. Achieving this state is ensured by requiring that the phase modulation satisfies the condition

$$\int_S \sin\left[\phi(x',y')\right] \mathrm{d}x'\mathrm{d}y' = 0\,. \tag{6.73}$$

Even if this condition were not met, the GPC method can still be optimized according to Eq. (6.60), which can now be re-expressed, using the approximation in Eq. (6.67), as

$$\bar{\alpha} \to \bar{\alpha}_{real} + \mathrm{i}\bar{\alpha}_{imag} = \frac{1}{2g(0,0)} + \frac{\mathrm{i}}{2g(0,0)}\cot\left(\frac{\theta}{2}\right). \tag{6.74}$$

These design considerations will be clarified in the following example.

6.6.3 Projection Design Illustration

Designing a GPC system for the projection of a user-defined pattern, $I(x',y')$, entails finding the appropriate phase input and the parameters for its matching phase contrast filter. The required phase input can be directly obtained from its image as specified by Eq. (6.66). Since a uniform arbitrary output phase offset does not affect the output, it is highly convenient to choose

$$\phi_0(x',y')=0. \tag{6.75}$$

A phase distribution is thus easily drawn according to

$$\phi(x',y')=2\arcsin\left[\frac{\pm\sqrt{\Delta x\Delta y\, I_{\text{norm}}(x',y')}}{2}\right]. \tag{6.76}$$

In a straightforward design, one may ignore the (–) terms in Eq. (6.76) and simply choose a phase input based on $\phi(x',y')>0$. This phase input will be used to determine the complex spatial average, $\bar{\alpha}$, using Eq. (6.69) and the matching filter parameters are then determined based its real and imaginary parts using Eq. (6.74). Specifically, the real component of $\bar{\alpha}$ specifies the correct $g(0,0)$ that, in turn, determines the filter size according to Eq. (6.70). Using this $g(0,0)$, the imaginary component of $\bar{\alpha}$ can then be used to determine the correct phase shift, θ. For example, if we get $g(0,0)=1$, then the Sine Integral functions in Eq. (6.70) specify a rectangular PCF satisfying $\eta_x=\eta_y\approx 0.62$. The matching phase shift in this case is $\theta=2\,\mathrm{arccot}(2\alpha_{imag})$.

One may also choose to use an input modulation consisting of mixed positive and negative phase values to achieve a desired effect or to match pre-existing practical constraints. An interesting case occurs when the phase input yields a real-valued $\bar{\alpha}$, which requires a filter phase shift of $\theta=\pi$. This may be achieved by taking the phase conjugate of alternate pixels. Assuming the initial phase pattern is slowly varying, as in a picture with smooth features, such a chequered pattern of phase-flipping should closely approach the desired zero-phase average. Moreover, a chequered phase pattern also provides the additional benefit of introducing a high-frequency phase modulation to the input signal. Such modulation reinforces the validity of the approximation introduced in Eq. (6.67).

6.7 Comparison of Generalized Phase Contrast and Computer-Generated Holography for Laser Image Projection

Dynamic laser image projection can serve as a generic technology for spatially controlled light-matter interaction. Aside from display applications, it has potential utility in various areas such as in materials processing [5], microscopy [6], and non-contact optical manipulation at microscopic scales [7], to name a few. Having outlined the various aspects of GPC-based light synthesis, it is of practical importance to know how it compares to other pattern projection techniques. In the following, we will attempt to benchmark its performance for dynamic greyscale planar projections with respect to computer-generated holography (CGH), a widely used technique. Generalized phase contrast is fundamentally different from CGH. In CGH, pattern construction is obtained by diffraction pattern through a modulator. Thus, design involves a global mapping where each output point derives contributions from all input points. In contrast, GPC constructs 2D images at a conjugate plane, such that input phase patterns can be directly calculated based on a simple point-to-point mapping [3, 8].

A GPC-based system uses phase-only input modulation and so we will also consider phase-only computer-generated holograms for comparison. We will consider Fourier-type CGH that generates patterns by far-field diffraction. This choice maintains generality since any Fourier-type CGH can be modified to project at an intermediate Fresnel plane by superposing a lens function into the hologram. The projection plane in a GPC-based system (bottom left of Fig. 6.8) is located at an image plane relative to the spatial light modulator (SLM) plane. This is similar to a conventional projector geometry that directly images amplitude modulation impressed upon an incident beam onto a projection plane. On the other hand, a CGH-based system (bottom right of Fig. 6.8) projects patterns at a conjugate Fourier plane relative to the SLM. This means that the required phase for encoding is coupled to a global transform (the Fourier transform) of the desired projection in contrast to the local pixel-by-pixel transform (imaging-like) phase encoding in GPC. The difference in encoding schemes in these complementary optical transforms strongly impacts their performance in projection applications.

A comprehensive assessment of laser projection techniques ultimately calls for application-specific metrics. However, some objective measures exhibit generality that can pervade across a wide spectrum of potential applications. These can serve as meaningful benchmarking criteria and can complement the other metrics in application-specific treatments. One example of such general metric is information capacity, which is considered to be invariant for a chosen optical system. The invariance of information capacity has proven useful for understanding multiple approaches for superresolution microscopy. We will adopt information capacity as the main criteria for our comparison.

6.7.1 Pattern Projection and Information Theory

The motivation for adopting an information-based criterion is illustrated in the top portion of Fig. 6.8, which depicts patterned light projection as a two-dimensional information communication system. The basic optical geometries of GPC-based and CGH-based systems are illustrated at the bottom part of Fig. 6.8. In a communication channel, the number of achievable output states serves as a meaningful assessment metric for its quality and performance. If there are N_F degrees of freedom (number of independently controllable output elements), where each degree of freedom can have M distinguishable levels, then there are altogether M^{NF} possible output states. The information capacity, N_C, is conventionally expressed in terms of the equivalent number of bits:

$$N_C = \log_2 M^{N_F} = N_F \log_2 M \tag{6.77}$$

Let us consider an optical system, whose band-limiting apertures effectively transmit spatial frequencies within $\{-0.5\Delta f_x \leq f_x \leq 0.5\Delta f_x$ and $-0.5\Delta f_y \leq f_y \leq 0.5\Delta f_y\}$, and an output projection area bounded by $\{-0.5\Delta x \leq x \leq 0.5\Delta x$ and $-0.5\Delta y \leq y \leq 0.5\Delta y\}$. Following Lukosz's approach [9], the number of degrees of freedom is given by

$$N_F = \left(1+\Delta x \Delta f_x\right)\left(1+\Delta y \Delta f_y\right) \tag{6.78}$$

This leads to the familiar space-bandwidth product, $N_F \approx \Delta x \Delta y \Delta f_x \Delta f_y$, for large N_F. The +1 term in Eq. (6.78) guarantees an available degree of freedom for $\Delta f_x \Delta f_y \rightarrow 0$ (e.g. spatially homogeneous output). Exploiting temporal modulation expands the degrees of freedom to

$$N_F = \left(1+\Delta t \Delta f_t\right)\left(1+\Delta x \Delta f_x\right)\left(1+\Delta y \Delta f_y\right) \tag{6.79}$$

where Δf_t is the temporal bandwidth and Δt is the transmission time interval.

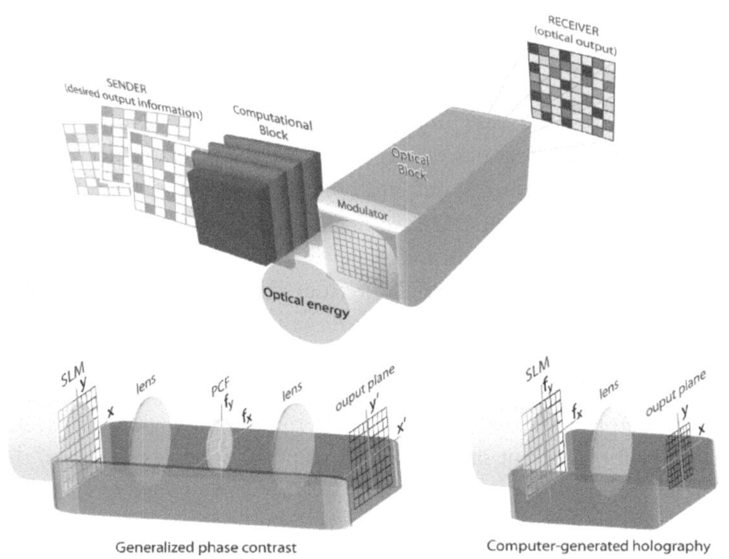

Fig. 6.8 Schematic representation of the system for rerouting of incident optical energy into multiple grey levels on a pre-defined output grid. Bottom left: optical system for a GPC-based projector; bottom right: optical system for a CGH-based laser projection system

Meanwhile, the number of distinguishable levels, M, is limited by the output signal-to-noise ratio (SNR). Assuming that the output consists of a signal, s, and an additive noise, n, (bandlimited and uncorrelated) then we can distinguish up to $M = (s+n)/n$ levels. We also note that the output signal strength depends on the energy efficiency of the projection system. Thus, we arrive at the following expression for the information capacity:

$$N_C = N_F \log_2 M = \left(1+L_t \Delta f_t\right)\left(1+L_x \Delta f_x\right)\left(1+L_y \Delta f_y\right)\log_2\left(1+\eta s_0 / n\right) \tag{6.80}$$

where η is the efficiency and s_0 is the noise-free output signal for 100% efficiency. Equation (6.55) expresses the logical result that a zero-efficiency system has no information capacity.

Following developments in superresolution microscopy, we adopt the invariance hypothesis for a system's information capacity. Lukosz originally proposed the invariance of N_F to explain that various superresolution approaches work by trading off between the different degrees of freedom [9]. Cox and Sheppard expanded the invariance to the information capacity, to which they also included degrees of freedom along the z-axis considered SNR effects [10] as done earlier by Fellgett and Linfoot for the space-bandwidth product [11]. In the following sections, we will show that projection systems are also subject to tradeoffs between the different degrees of freedom and noise.

6.7.2 Performance Benchmarks

Spatial Degrees of Freedom

As with traditional displays, laser image projection requires a good output display resolution to sufficiently render fine details in the image. The spatial degrees of freedom, $(1+L_x\Delta f_y)(1+L_x\Delta f_y)$, is a measure of output display resolution (i.e. number of independently addressable output elements). Due to the imaging geometry in GPC-based projection, the number of addressable output elements is the same as those available in the input SLM, subject to the diffraction limit from the lens numerical apertures. In a Fourier-type p-CGH, speckle problems can reduce the output display resolution.

In general, a desired far-field diffraction pattern requires a fully complex Fourier hologram, which is lossy due to the coupled amplitude modulation. Iterative hologram design uses the output phase as a free parameter to find a hologram whose amplitude modulation matches the input illumination. Except for some special cases, the design settles for an imperfect match, which means that the efficiency will be typically less than 100% [12]. The slightly mismatched amplitude amounts to applying a multiplicative spatial noise relative to the correct profile. This multiplicative noise inherently leads to intensity distortions and, possibly, even speckles at the output.

Aside from the multiplicative noise, speckles may also arise due to interference between adjacent output points. The Fourier plane output of a phase-only CGH encoded on an SLM with square pixels separated by distance d and without any interpixel dead space is given by [13]

$$T(f_x,f_y)=A(f_x,f_y)\otimes\left\{d^2\,\mathrm{sinc}(f_xd,f_yd)Q(f_x,f_y)\right\}. \tag{6.81}$$

Here $Q(f_x,f_y)$ represents the desired output that is discretized and infinitely replicated; the sinc roll-off is due to the SLM single-pixel diffraction; and $A(f_x,f_y)$ broadens each projected point due to diffraction from finite SLM apertures, which can lead to crosstalk between adjacent points. The output phase is a free parameter in iterative design resulting in random phase relationship between neighbouring points. The random interference between neighbouring output pixels generates speckled outputs. For some output patterns, the output phase may be constrained to vary smoothly during the optimization to reduce the speckles [14]. However, this requires larger arrays that increase the computational load.

A common procedure to avoid speckle noise is to designate a signal window and use the surrounding area as a so-called noise window where noise can be diverted. Choosing a signal window can be imposed by practical needs, such as avoiding a spurious zero order generated by imperfect devices. Maintaining the same square aspect ratio in the presence of a zero-order can lead to a signal window with a 4fold resolution loss. Further resolution loss is incurred when setting a minimum pixel distance to avoid crosstalk. One approach is to repetitively tile the derived phase hologram, which can require as much as 10×10 repetitions to produce completely isolated spots [15]. Also, while tiling is easily done in static micro-optics CGH implementations, it can be strongly problematic for SLMs. From the discussion above, a 512×512 image may require a 10240×10240 SLM device!

Adjacent spots can be brought closer if one can set a tolerable crosstalk level. For a uniformly illuminated circular input aperture, the separation between adjacent output spots, relative to a diffraction-limited radius, is given by [16]

$$\sigma_R = \frac{f\lambda/T}{1.22 f\lambda/D} = \frac{D}{1.22T}, \tag{6.82}$$

where f is the focal length of the Fourier lens used, λ is the illumination wavelength, T is the period of the repeated cell and D is the diameter of the illuminating aperture. A tolerable compromise may be a relative distance of $\sigma_R = 2$, where the first minima of adjacent spots coincide. This requires only 2.44 repetitions along the aperture diameter. In this case, an encoding device with at least 2498×2498 pixels suffices to project a 512×512 single quadrant image. Some noise is expected as the rings around the central spot will create a residual disturbance on the neighbouring spots.

The preceding analysis shows that the CGH-based laser projection has lower information capacity than an imaging system. Measures to reduce the speckles do not improve the information capacity but simply trade off between the different factors in Eq. (6.80). Choosing a small signal window reduces L_x and L_y and hologram tiling reduces the non-redundant Δf_x and Δf_y. The steps described above effectively reduce the information capacity by ~24 compared to an imaging system with a similar numerical aperture and band limit. The invariance of N_C implies that one may also minimize speckle effects by exploiting temporal degrees of freedom, instead of reducing the spatial degrees of freedom through the speckle reduction schemes described above. One method is to repetitively project a pattern with different speckles to an integrating detector for a time interval Δt to average out the background noise. The lower information capacity is obvious when one considers that a GPC-based projection can transmit multiple unique outputs for the same time interval.

Temporal Degrees of Freedom

Conventional optical communication aims to exploit the high temporal bandwidth afforded by optical frequencies. In a two-dimensional communication system (Fig. 6.8), bottlenecks arise from the modulator refresh rate and the computational bandwidth. As a local phase-to-intensity mapping system, GPC has virtually no computational load and can exploit the full modulator bandwidth. This enables real-time dynamic operation that can be easily scaled to higher output display resolutions when required by the application. In contrast, a phase-only CGH is burdened by optimization algorithms when finding a phase distribution that diffracts into a desired intensity pattern. The problem is geometrically compounded when scaling to higher output display resolutions. Transmitting pre-computed information can achieve modulator-limited temporal bandwidth but the lower information capacity becomes apparent in interactive operation, especially when for higher output resolutions.

Optimization algorithms such as direct binary search, genetic algorithms or simulated annealing have very low computational bandwidths that render them impractical for dynamic CGH applications. Iterative Fourier transform algorithms can yield considerably faster crude solutions but perform slower when attempting hi-fidelity outputs. For instance, including physical constraints (e.g. zero padding) in the iterations require large arrays that burden the computation. The noise from the approximate phase-only solution generally spreads into the signal window even after numerous iterations and requires a second set of iterations to divert them away from the signal window [17]. This second set of iterations may stagnate in a state with optical phase singularities or "optical vortices", where dark spots persist within the signal window [14, 18] (see Fig. 6.9). Solutions to remove these spots entail even further computations.

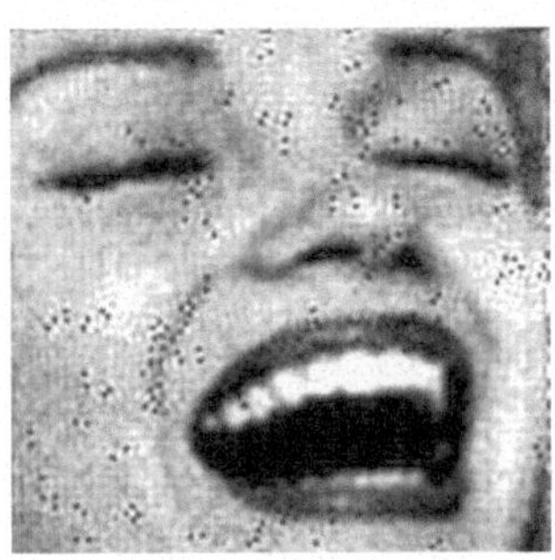

Fig. 6.9 Simulated phase-only CGH-based grey-level projection showing dark spots due to optical vortices that are impervious to further iterations. The "bumpy" nature of the reconstruction is also evident (image adapted from ref. [14])

It is easy to appreciate from the discussion above that designing a phase-only CGH to produce a grey-level image with high fidelity is burdened with computational requirements that adversely impact the effective temporal bandwidths. The computational bandwidth can restrict the temporal degrees of freedom in CGH-based projection. In contrast, GPC phase inputs are determined using straightforward and local phase-to-intensity mapping employed by the GPC-based projection method, which ensures that the temporal bandwidth afforded by the modulator is fully utilized.

Signal-to-Noise Ratio and Light Efficiency

The projection efficiency is a generally accepted performance parameter and also factors into the information capacity, as expressed by Eq. (6.80). When using a pixellated SLM to encode a phase hologram, we can expect from basic properties of Fourier transforms that pixilation generates replicated projections with an intensity roll-off arising from single pixel diffraction. An indication of image projection efficiency can be inferred from holographic single-beam scanning using a square phase-only SLM containing N×N square pixels, each with side length *d*, without any interpixel dead space. The efficiency falls for off-axis locations, down to 16% of the on-axis efficiency at the corners of the scanning region, as illustrated in the Fig. 6.10. For Fourier array illuminators a more accurate efficiency limit was determined by Arrizon and Testorf to be [15]

$$\chi_{\max} = \left[\sum_{(q,l)\in\Omega_s} \chi'_{ql} \operatorname{sinc}^{-2}(qd) \operatorname{sinc}^{-2}(ld) \right]^{-1}, \tag{6.83}$$

where χ'_{ql} describes the pointwise efficiency for achieving a grey level at a point (q,l) on a discrete output grid, separated by $f\lambda / T$, within a signal window Ω_s. Although derived for array illuminators, this also applies to general CGH-based laser projections since the projected pixels need ample separation. Equation (6.58) predicts maximum efficiency of ~52% when projecting an image to fill the entire addressable region at the projection plane. This efficiency limit holds whether the projection is a uniform grey level, a random grey image, or a typical image with a reasonable spread of grey levels like the Lena standard image. The same limit applies when the image fills only a single quadrant. Since the outer regions tend to decrease efficiency, achieving efficiency beyond 52% must exclude these regions, which consequently, comes at a price of an even lower output display resolution.

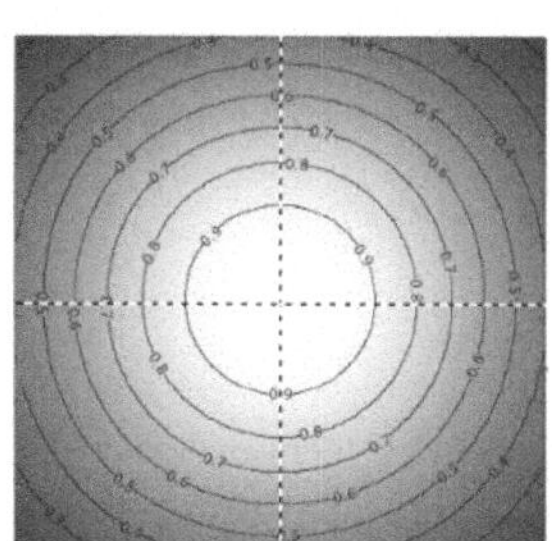

Fig. 6.10 The optimum CGH efficiency for single beam rerouting as a function of target position.

For a GPC-based projection system, the efficiency is fundamentally limited by the light that gets projected outside the addressable image region, which results in a residual intensity halo around the projected image. This halo is generated by the SRW tail and depends on the PCF size, Δf_r, as shown in Fig. 6.11 (a). Having described the SRW profile, we can account for the loss and plot the GPC efficiency limit, $\chi_{\max}$, in Fig. 6.11(b), obtained by numerically integrating

$$\chi_{\max} = 1 - \frac{2}{(\Delta r)^2} \int_{\Delta r}^{\infty} \left| s_r(r') \right|^2 r' dr' , \tag{6.84}$$

using parameters that satisfy the optimal design criterion. The efficiency limit varies with the size of the phase contrast filter as illustrated in Fig. 6.11(b).

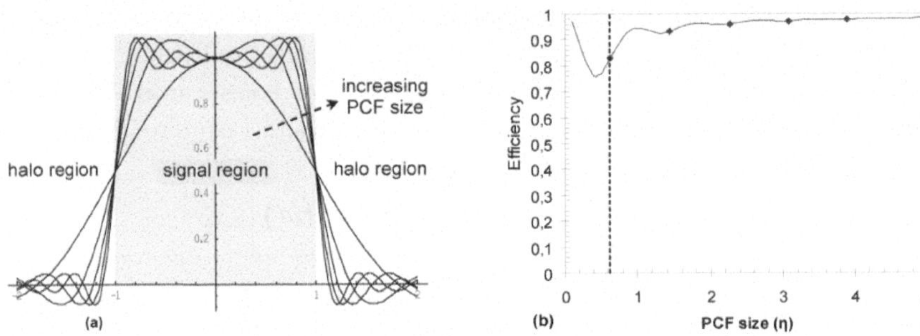

Fig. 6.11 (a) Radial SRW profiles for different PCF sizes (PCF size is expressed relative to the Airy disc generated by the circular input aperture); (b) GPC efficiency limit as a function of PCF size; indicated data points correspond to PCF sizes depicted in (a).

6.7.3 Practical SLM Devices: Performance Constraints

Fundamental properties and limitations of real-world SLMs significantly impact the encoded and projected light [19] causing realistic performance to deviate significantly from design expectations. We will now consider how practical device limitations influence the performance of GPC and phase-only CGH. We consider issues related to SLM dead space, Fourier plane zero-order beam, phase bleeding, and phase stroke requirements.

Interpixel Dead Space

Practical SLM devices can contain a non-addressable dead space between pixels that can lead to a spurious zero-order beam. Avoiding this spurious beam forces the signal window away from the optical axis, which lowers the output display resolution [20]. A large dead space constricts the pixel aperture, which reduces the energy throughput and broadens the output intensity roll-off in CGH-based projections as more energy channels to spurious higher-orders. This effect is summarized in the relation [21]

$$\chi_{\max} = F^2 \left[\sum_{(q,l)\in\Omega_s} \chi'_{ql} \operatorname{sinc}^{-2}\left(qd\sqrt{F}\right) \operatorname{sinc}^{-2}\left(ld\sqrt{F}\right) \right]^{-1} \tag{6.85}$$

where the fill factor, F, is the ratio of the addressable area to the total pixel area. Efficiency declines with decreasing F, as plotted in Fig. 6.12. The plot shows 52% efficiency for 100% fill factor and assumes a CGH projection filling an entire addressable quadrant in the output plane. Higher efficiencies may be achieved, at the expense of resolution loss, by confining images within smaller signal windows. The figure also plots the single-quadrant CGH image efficiency relative to a four-quadrant GPC-based projection. The GPC efficiency at $F = 1$ is taken to be 0.82 (the efficiency along the dotted line in Fig 6.11(b)) and linearly decreases with F. The relative efficiency plot in Fig. 6.12 indicates that GPC becomes increasingly advantageous over CGH for SLM devices with lower fill factors. (Matched performance would appear as a horizontal plot of the relative performance).

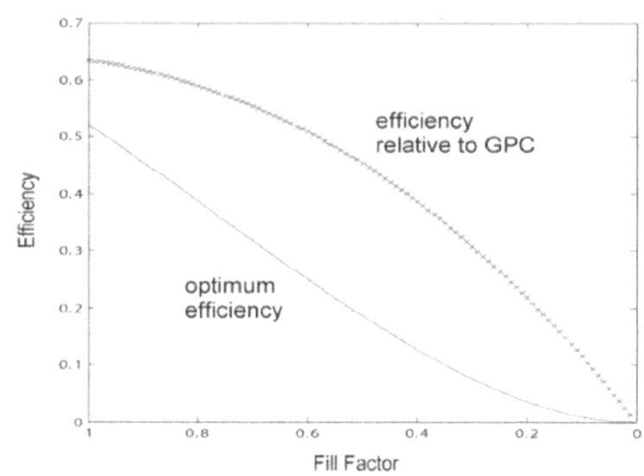

Fig. 6.12 Optimum efficiency of holographic grey-level projection for different pixel fill factors (lower trace). The upper trace shows CGH efficiency relative to GPC efficiency.

Device Modulation Transfer Function

Phase holograms encoded on SLMs are highly susceptible to the frequency-dependent diffraction efficiency, which introduces a major intensity roll-off in the projection [22]. This can arise due to factors such as electrical timing [23] and addressing crosstalk between adjacent pixels. Crosstalk between adjacent SLM pixels can occur due to fringing effects of the fields in electrical addressing [24], the nature of the photoconductive mechanism in optical addressing [25], and other mechanisms for a non-local response of the electro-optic medium. The encoded phase is blurred as the phase of each SLM pixel bleeds into neighbouring pixels. Phase bleeding yields erroneous encoding that muddles the pattern with spurious zero-order, ghost reconstructions, and noise [26]. An ideal design can easily give frustrating reconstructions on an actual SLM-device.

Phase blurring is problematic for a phase-only CGH since the phase is wrapped to within a 2π range. The 2π jumps in phase-wrapped areas demand high frequency operation, even when the pattern itself is relatively low-frequency. This is illustrated by Fig. 6.13, which shows a phase-only CGH and its reconstruction.

Fig. 6.13 Phase-only CGH and its reconstruction. The CGH shows sharp bright-to-dark transitions where the phase wraps from π to $-\pi$ (grey levels correspond to phase values).

The output resolution is further compromised when minimizing phase-wrapped regions, which demands projections close to the optical axis, while simultaneously avoiding the zero-order noise. It is unfortunate that a phase-only CGH projection avoids the optical axis where it is least susceptible to aberrations.

In contrast, GPC utilizes the zero-order beam to generate the reference wave. Phase encoding errors affecting the zero-order beam may be compensated for with a suitable choice of filter size and phase shift. The symmetric $4f$ imaging configuration offers the possibility of minimizing aberrations introduced by the first lens using a matching second lens. In a non-flipping implementation, the modulation transfer function follows directly from the device. A better MTF is required when implementing phasor flipping schemes. The next chapter illustrates experimental demonstrations of high-efficiency laser image projection implemented using practical devices.

Required Phase Stroke

A phase-only CGH generally requires 2π phase stroke, except binary phase holograms that waste half of the energy to a twin image. A wider modulation dynamic range sets a lower limit on the attainable modulator switching rates. Greyscale GPC-based image projection can operate over a flexible dynamic range and can work, for example, with a π phase stroke modulator. Thus, it can benefit from the higher refresh rates expected from lower stroke devices in general. In SLMs utilizing electro-optic materials, a lower phase stroke means that a thinner medium can be used, which comes with additional benefits such as lower susceptibility to interpixel crosstalk and aberrations introduced by the medium.

6.7.4 Final Remarks

Computer-generated holography is a highly versatile technique that is widely used in a plurality of applications. However, versatility does not guarantee best suitability for a specific application. The preceding analysis highlights the need for careful scrutiny before adopting a particular technique. From the perspective of information capacity, there are several areas where a CGH-based laser projection system can be problematic and where a GPC-based system is a more attractive option. Nevertheless, we must bear in mind that information capacity is a general measure and there are surely projection applications that give high premium on other performance measures and where a CGH-based approach will be preferable.

6.8 Wavelength Dependence of GPC-Based Pattern Projection

We now investigate the robustness of GPC-based light projection to wavelength shifts and assess its possible compatibility with multi-wavelength illumination. An incident beam of wavelength λ_0 illuminates a phase-only spatial light modulator (SLM) to generate the input plane field

$$\begin{aligned} p(x,y) &= a(x,y)\exp\left[\mathrm{i}\phi_0 + \mathrm{i}\Delta\phi(x,y)\right] \\ &= a(x,y)\exp\left[\mathrm{i}\frac{2\pi d_{opt:s}}{\lambda_0} + \mathrm{i}\frac{2\pi\Delta d_{opt:s}(x,y)}{\lambda_0}\right], \end{aligned} \tag{6.86}$$

where $d_{opt:s}$ is the bias optical path length through an unmodulated SLM, $a(x,y)$ describes the incident (read-out) field amplitude profile and $d_{opt:s}(x,y)$ is the user-definable SLM optical path length modulation. This is mathematically equivalent to a system with a uniform incident illumination where the spatial variation of $a(x,y)$ represents an aperture function at the SLM plane.

The phase contrast filter (PCF) at the common focus between the Fourier lenses is typically an optical flat that introduces a different optical path length within an aperture region centred on the zero-order beam through a different thickness or refractive index. We can mathematically describe the extent of the phase-shifting region by an aperture function, $S(f_x,f_y)$. For example, the aperture function for a PCF that contains a circular phase shifting region is mathematically described as $S(f_x,f_y)=\mathrm{circ}(f_x,f_y)$. The spatial frequency response of the PCF is mathematically represented as

$$\begin{aligned} H(f_x,f_y) &= \left\{1+\left[\exp(\mathrm{i}\theta)-1\right]S(f_x,f_y)\right\}\exp(\mathrm{i}\phi_f) \\ &= \left\{1+\left[\exp\left(\mathrm{i}\frac{2\pi\Delta d_{opt:f}}{\lambda_0}\right)-1\right]S(f_x,f_y)\right\}\exp\left(\mathrm{i}\frac{2\pi d_{opt:f}}{\lambda_0}\right), \end{aligned} \tag{6.87}$$

where $d_{opt:f}$ is the optical path length through the substrate used for fabricating the PCF, and remains the optical path length outside the phase-shifting aperture region, $S(f_x,f_y)$. The diffraction-limited zero-order beam transmitted through the aperture $S(f_x,f_y)$ is θ shifted with respect to the other Fourier components. The filter in Eq. (6.87) transmits a uniformly shifted signal and synthesizes a reference wave, $r(x',y')$, in the output plane:

$$r(x',y') \approx \left[\exp\left(i\frac{2\pi\Delta d_{opt:f}}{\lambda_0}\right)-1\right]\exp\left(i2\pi\frac{d_{opt:f}+d_{opt:s}}{\lambda_0}\right)\Im^{-1}\left\{\bar{\alpha}S(f_x,f_y)\Im\{a(x,y)\}\right\}, \tag{6.88}$$

where $\Im\{\cdot\cdot\}$ and $\Im^{-1}\{\cdot\cdot\}$ denote scaled forward and inverse Fourier transforms, respectively. The output is the interference between the image of the phase-modulated input and the synthesized reference wave (SRW):

$$\begin{aligned} I(x',y') &\approx \left|p(x',y')+r(x',y')\right|^2 \\ &= \left|a(x',y')\exp\left[i\frac{2\pi\Delta d_{opt:s}(x',y')}{\lambda_0}\right]+\bar{\alpha}\left[\exp\left(i\frac{2\pi\Delta d_{opt:f}}{\lambda_0}\right)-1\right]g(x',y')\right|^2 \\ &= \left|a(x',y')\exp\left[i\Delta\phi(x',y')\right]+\bar{\alpha}\left[\exp(i\theta)-1\right]g(x',y')\right|^2, \end{aligned} \tag{6.89}$$

where the expanded expression omits the uniform SLM and PCF phase offsets, ϕ_0 and ϕ_f, which is the same for $p(x,y)$ and $r(x,y)$, and do not contribute to the output intensity modulation. The spatially varying component of the SRW is

$$g(x',y') = \Im^{-1}\left\{S(f_x,f_y)\Im\{a(x,y)\}\right\}. \tag{6.90}$$

This describes the projected field when an input field having limited spatial extent, described by $a(x,y)$, generates a diffraction-broadened spot, $\Im\{a(x,y)\}$, that subsequently diffracts through the phase shifting aperture $S(f_x,f_y)$ and propagates to the far-field region, and simply omits a common phase and the amplitude factor as normalization:

$$g(x',y') \propto \frac{1}{i\lambda_0 f}\iint S(\xi,\eta)A(f_x,f_y)\exp\left[-\frac{i2\pi}{f\lambda_0}(x'\xi+y'\eta)\right]d\xi d\eta, \tag{6.91}$$

where f is the focal length of the Fourier lens and the spatial coordinates (ξ,η) on the PCF plane correspond to spatial frequencies $f_x = \xi/(\lambda_0 f)$ and $f_y = \eta/(\lambda_0 f)$, respectively and where

$$A(f_x,f_y) = \frac{1}{i\lambda_0 f}\iint a(x,y)\exp\left[-\frac{i2\pi}{f\lambda_0}(x\xi+y\eta)\right]dxdy. \tag{6.92}$$

Equation (6.89) enables us to conveniently generate patterns by tailoring the input phase to exploit constructive and destructive interference at the output. For example, when the input and SRW profiles match up to a multiplicative real-valued scaling factor K, we can rewrite Eq. (6.89), using trigonometric identities, in the form

$$I(x',y')=\left[a(x',y')\right]^2\left|\exp\left[\mathrm{i}\Delta\phi(x',y')\right]+K|\bar{\alpha}|\sin\left(\frac{\theta}{2}\right)\exp\left[i\left(\frac{\theta+\pi}{2}+\phi_{\bar{\alpha}}\right)\right]\right|^2 . \quad (6.93)$$

To consider the effects of changing the illumination wavelength, let us start with binary intensity patterns. We calibrate the input modulator to encode binary phase (i.e. either 0 or π) within a subregion b(x,y) inside an aperture defined by a(x,y) that is uniformly illuminated at wavelength λ_0. This phase modulation generates a purely real zero-order and, from Eq. (6.36), the PCF parameters must be chosen to shift the zero-order phase by $\theta=\pi$ since $\bar{\alpha}_{imag}=0$. Furthermore, let us consider that out of the total input aperture area that is available for phase modulation a fraction, F, is π-encoded while the remaining region, representing 1−F of the total area, is 0-encoded. Under these conditions, evaluating the normalized zero order from Eq. (6.6) is straightforward and yields a real-valued result, $\bar{\alpha}=1-2F$. Furthermore, a fully optimized system will have matching illumination and SRW profiles, $a(x',y')=g(x',y')$, corresponding to K = 1. Thus, we can obtain optimal contrast, $\bar{\alpha}_{real}=1/2$, when F = 0.25. When this system is illuminated at a different wavelength, λ, the output is given by

$$I(x',y')\approx\left|a(x',y')\exp\left[\mathrm{i}\pi\, b(x',y')\frac{\lambda_0}{\lambda}\right]+\bar{\alpha}\left[\exp\left(\mathrm{i}\pi\frac{\lambda_0}{\lambda}\right)-1\right]g(x',y')\right|^2 . \quad (6.94)$$

The normalized zero-order, $\bar{\alpha}$, and the SRW profile, $g(x,y)$, are both wavelength dependent. The wavelength dependence of the zero-order follows from the input phase scaling

$$\bar{\alpha}=|\bar{\alpha}|\exp(\mathrm{i}\phi_{\bar{\alpha}})=\frac{1}{A_0}\iint a(x,y)\exp\left[\mathrm{i}\Delta\phi(x,y)\frac{\lambda_0}{\lambda}\right]\mathrm{d}x\mathrm{d}y . \quad (6.95)$$

To determine the wavelength dependence of $g(x',y')$, we first propagate the aperture field, $a(x,y)$, to the Fourier plane:

$$\begin{aligned}A(f_x,f_y;\lambda)&=\frac{1}{\mathrm{i}\lambda_0 f}\frac{\lambda_0}{\lambda}\iint a(x,y)\exp\left[-\frac{\mathrm{i}2\pi}{f\lambda_0}\left(x\xi\frac{\lambda_0}{\lambda}+y\eta\frac{\lambda_0}{\lambda}\right)\right]\mathrm{d}x\mathrm{d}y\\&=\frac{\lambda_0}{\lambda}A\left(f_x\frac{\lambda_0}{\lambda},f_y\frac{\lambda_0}{\lambda},\lambda_0\right).\end{aligned} \quad (6.96)$$

The field that diffracts through the PCF aperture is Fourier-propagated by the second lens and generates the output field which forms the basis for $g(x',y')$

$$g(x',y';\lambda)\propto\frac{1}{\mathrm{i}\lambda_0 f}\frac{\lambda_0^2}{\lambda^2}\iint S(\xi,\eta)A\left(f_x\tfrac{\lambda_0}{\lambda},f_y\tfrac{\lambda_0}{\lambda};\lambda_0\right)\exp\left[-\frac{\mathrm{i}2\pi}{f\lambda_0}\left(x'\xi\tfrac{\lambda_0}{\lambda}+y'\eta\tfrac{\lambda_0}{\lambda}\right)\right]\mathrm{d}\xi\mathrm{d}\eta$$
$$=\frac{1}{\mathrm{i}\lambda_0 f}\iint S\left(\xi_\lambda\tfrac{\lambda}{\lambda_0},\eta_\lambda\tfrac{\lambda}{\lambda_0}\right)A\left(f_{x,\lambda},f_{y,\lambda};\lambda_0\right)\exp\left[-\frac{i2\pi}{f\lambda_0}\left(x'\xi_\lambda+y'\eta_\lambda\right)\right]\mathrm{d}\xi_\lambda\mathrm{d}\eta_\lambda, \tag{6.97}$$

where the new variables, with subscript λ, are scaled by λ_0/λ relative to the old variables (e.g. $\xi_\lambda = \xi\,\lambda_0/\lambda$, $f_{x\lambda} = f_x\,\lambda_0/\lambda$, etc.). Comparing with Eq. (6.91) we can see that, as a result of the change in variables in Eq. (6.97), the wavelength dependence of $g(x',y')$ is obtained by using the original wavelength and applying a reciprocal wavelength-scaling on the PCF aperture size instead. Physically, the use of a different wavelength broadens or shrinks the diffraction-limited zero-order beam by λ_0/λ. With an inverse scaling from the second Fourier lens, the net effect on the SRW spatial profile, $g(x',y')$, is equivalent to using the original wavelength and assigning a reciprocal λ/λ_0 shrinking or broadening to the filter size. This equivalent picture of a wavelength-dependent filter size is particularly convenient as the effect of filter size variations on the SRW is a well-described aspect of GPC [2, 27].

When the feature sizes of a projected pattern are small relative to the input aperture size, such as those employed in optical lattices, a relatively large filter size can be used [27]. Moreover, the SRW profile, $g(x',y')$, remains relatively unaffected as the wavelength is scanned over a reasonably wide range (e.g. the entire visible spectrum). We can investigate the effect of wavelength change by examining how the interfering terms in Eq. (6.94) are affected. The signal predictably suffers from phase scaling. The effect on the SRW, on the other hand, arises from two factors: (1) the normalized zero-order, $\bar{\alpha}(\lambda)$; and (2) the filter parameter, $\exp[\mathrm{i}\theta(\lambda)]-1$. The wavelength dependence of these factors and the resulting SRW are shown in Fig. 6.14. The figure shows that the two contributions have opposing tendencies that tend to reduce the wavelength dependence of the overall SRW. For instance, the SRW phase is maintained close to π even after a doubling of the wavelength. This is desirable since destructive interference in the zero-phase regions of the signal can be optimized over a range of illumination wavelengths. However, the mismatched amplitudes mean that these regions cannot be completely dark, thereby reducing the contrast and the efficiency.

The magnitude and phase of the SRW allow us to determine the contrast ratios that result upon interference with the signal, defined in Eq. (6.86). Furthermore, information on the contrast ratio and the fraction, F, of the bright regions within the projection allows us to predict how the efficiency will vary with wavelength. The wavelength dependence of the contrast ratio and the efficiency are shown in Fig. 6.15. The results show that contrast ratios of ~20 is obtained at the extremes of a wavelength doubling range ($0.75\lambda_0$ to $1.5\,\lambda_0$). Thus, efficiencies exceeding 85% of the peak efficiency can be maintained over this wide wavelength range.

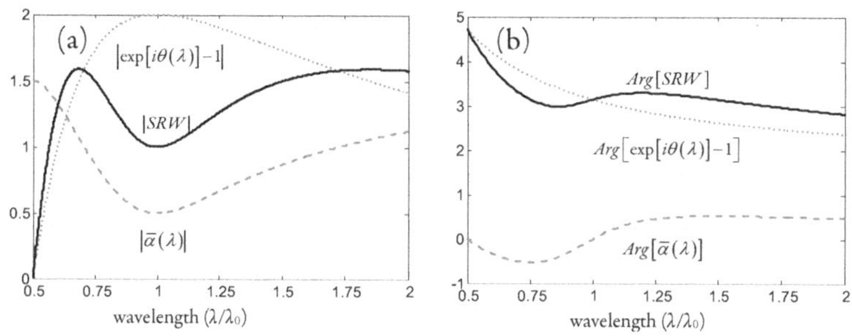

Fig. 6.14 (a) Magnitude and (b) phase of the normalized zero-order, $\bar{\alpha}$ (dashed trace), filter parameter, $[\exp(i\theta)-1]$ (dotted trace), and the SRW (solid trace) at different illumination wavelengths.

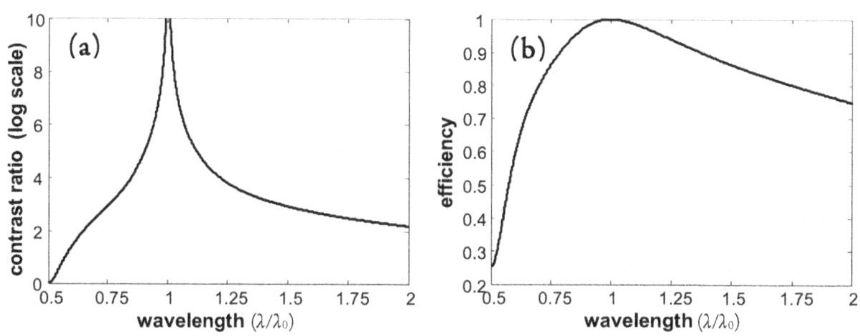

Fig. 6.15 Performance metrics at different wavelengths: (a) contrast ratio (logarithmic scale); (b) efficiency.

So far, we have considered cases where the SRW spatial profile, $g(x',y')$, perfectly matches the illumination amplitude profile, $a(x',y')$, and where $g(x',y')$ remains unaffected over the wavelengths considered. The estimates of performance metrics that we obtained based on these assumptions are expected to be fairly correct for cases that allow the use of large filters, as in the generation of optical lattices [27]. In the next section we present and discuss the results of numerical experiments that include these effects to investigate the wavelength dependence of more general GPC-based pattern generation using a numerical implementation of the GPC setup.

Earlier, our theoretical analysis assumed that the dominant diffractive effects arise due to the limiting apertures of the SLM and the phase shifting region of the PCF. This implicitly assumes negligible losses through the lens apertures, which we exploited to conveniently revert between the input and image plane variables. In laboratory experiments employing practical SLM devices, the discrete nature of the device introduces artefacts such as spurious tiled replicas at the PCF plane. This can be particularly problematic for techniques that project patterns to the Fourier plane [13]. For GPC, the inverse effect of the second Fourier lens serves to rectify these artefacts, hence improving the projection. However, it is important that the lenses must collect most of the energy from the propagating light. Here the GPC benefits from the transverse intensity rolloff due to the finite size SLM pixels, which minimizes the energy in the higher-order replicas and thus allows reasonably sized lenses to capture most of the energy. In the following numerical experiments we examine the efficiency and related performance metrics of various GPC-based pattern projection applications such as array illumination, Gaussian beam shaping, and image projection. The numerical model assumes negligible losses through the Fourier lenses.

6.9 Summary and Links

In this chapter we harnessed our generalized understanding of common-path interferometry to veer away from traditional phase contrast applications for imaging unknown phase perturbations. We turned our attention towards GPC implementations where the input phase modulation is actively controlled to engineer desired outputs. We built upon the analytical framework developed earlier and adapted it to derive equations for optimising light efficiency and visibility of the generated outputs. Starting with a model for arbitrary analogue phase-encoded wavefronts, we adopted simple and practical encoding schemes such as ternary-phase or even binary phase for synthesizing light efficient binary intensity patterns. Knowing the input means that we can now choose matching filter parameters to optimize the output. We can also specify appropriate input parameters to match given filter constraints. We also exploited design freedoms in the filter and the input to discuss various approaches for dealing with the synthetic reference wave (SRW) inhomogeneity to improve the output. We will apply these results in Chapter 7 when we describe specific applications such as optical lattice generation and Gaussian laser beam shaping. We also described optimization criteria for rectangular aspect ratios, which is of practical relevance considering the common geometry of practical modulators. Moreover, we expanded the basic formulation to include effects of the illumination wavelength in preparation for possible broadband applications. It is easy to appreciate that the encoding simplicity in GPC can be advantageous over global mapping systems and benchmarking with computer generated holography showed that GPC can benefit better from available device information capacity. This is a vital advantage in dynamic projections and becomes a highly desirable feature when synthesizing light patterns for optical trapping and manipulation, as discussed in Chapter 8.

References

1. J. W. Goodman, "*Introduction to Fourier Optics* 2nd ed." (San Francisco: McGraw-Hill, 1996).
2. J. Glückstad and P. C. Mogensen, "Optimal Phase Contrast in Common-Path Interferometry," Appl. Opt. **40**, 268-282 (2001).
3. J. Glückstad, "Phase contrast image synthesis," Opt. Commun. **130**, 225–230 (1996).
4. Gautsch W and Cahill W F 1964 *Handbook of Mathematical Functions with Formula, Graphs, and Mathematical Tables* ed Abramowitz M and Stegun I A (New York: Dover) p 231
5. C. David, J. Wei, T. Lippert, and A. Wokaun, "Diffractive grey-tone phase masks for laser ablation lithography," Microelectron. Eng. **57-8**, 453–460 (2001).
6. M. G. L. Gustafsson, "Nonlinear structured-illumination microscopy: Wide-field fluorescence imaging with theoretically unlimited resolution," P. Natl. Acad. Sci. USA **102**, 13081–13086 (2005).
7. M.P MacDonald, G.C. Spalding, K. Dholakia, "Microfluidic sorting in an optical lattice," Nature **426**, 421–4, (2003).
8. C.A. Alonzo, P.J. Rodrigo, & J. Glückstad, "Photon-efficient grey-level image projection by the generalized phase contrast method," New J. Phys. **9**, 132 (2007).
9. W. Lukosz, "Optical systems with resolving powers exceeding classical limit," J. Opt. Soc. Am. **56**, 1463–1472 (1966).
10. I. J. Cox and C. J. R. Sheppard, "Information capacity and resolution in an optical system," J. Opt. Soc. Am. A **3**, 1152–1158 (1986).
11. P. B. Fellgett and E. H. Linfoot, "On the assessment of optical images," Phil. Trans. R. Soc. London Ser. A **247**, 369–407 (1955).
12. F.Wyrowski, "Upper bound of the diffraction efficiency of diffractive phase elements.", Opt. Lett. 16, 1915–1917, (1991).
13. D. Palima and V. R. Daria, "Effect of spurious diffraction orders in arbitrary multi-foci patterns produced via phase-only holograms," Appl. Opt. **45**, 6689–6693 (2006).
14. H. Aagedal, M. Schmid, T. Beth, S. Teiwes, and F. Wyrowski, "Theory of speckles in diffractive optics and its application to beam shaping," J. Mod. Opt. **43**, 1409–1421 (1996).
15. V. Arrizon and M. Testorf, "Efficiency limit of spatially quantized Fourier array illuminators," Opt. Lett. **22**, 197–199 (1997).
16. A. J. Waddie and M. R. Taghizadeh, "Interference Effects in Far-Field Diffractive Optical Elements," Appl. Opt. **38**, 5915–5919 (1999).
17. F. Wyrowski, "Diffractive optical elements: iterative calculation of quantized, blazed phase structures," J. Opt. Soc. Am. A **7**, 961–969 (1990).

18. P. Senthilkumaran, F. Wyrowski, H. Schimmel. "Vortex Stagnation problem in iterative Fourier transform algorithms," Opt. Laser Eng. **43**, 43–56 (2005).
19. RW Cohn, "Fundamental properties of spatial light modulators for the approximate optical computation of Fourier transforms: a review," Opt Eng **40**, 2452–2463 (2001).
20. D. Palima and V. R. Daria, "Holographic projection of arbitrary light patterns with a suppressed zero-order beam," Appl. Opt. **46**, 4197–4201 (2007).
21. V. Arrizón, E. Carreón, M. Testorf, "Implementation of Fourier array illuminators using pixelated SLM: efficiency limitations," Opt. Commun. **160**, 207–213 (1999).
22. A. Márquez, C. Iemmi, I. Moreno, J. Campos, and M. Yzuel, "Anamorphic and spatial frequency dependent phase modulation on liquid crystal displays. Optimization of the modulation diffraction efficiency," Opt. Express **13**, 2111–2119 (2005).
23. ML Hsieh, KY Hsu, EG Paek and CL Wilson, "Modulation transfer function of a liquid crystal spatial light modulator," Opt. Commun. **170**, 221–227, (1999).
24. E. Hällstig, T. Martin, L. Sjöqvist, and M. Lindgren, "Polarization properties of a nematic liquid-crystal spatial light modulator for phase modulation," J. Opt. Soc. Am. A **22**, 177–184 (2005).
25. G. Moddel and L. Wang, "Resolution limits from charge transport in optically addressed spatial light modulators," J. Appl. Phys. **78**, 6923-6935 (1995).
26. M. Duelli, L. Ge, and R. W. Cohn, "Nonlinear effects of phase blurring on Fourier transform holograms," J. Opt. Soc. Am. A **17**, 1594–1605 (2000).
27. P. Rodrigo, V. Daria, and J. Glückstad, "Dynamically reconfigurable optical lattices," Opt. Express **13**, 1384–1394, 2005.

Chapter 7
Shaping Light by Generalized Phase Contrast

Phase contrast methods are most often associated with optical microscopy of weak phase specimens. Generalized phase contrast breaks away from the small phase restrictions of traditional formulations, which enables not only accurate quantitative phase measurements, but also an optimal design criterion for synthesizing specified light distributions. Light possessing user-defined intensity distributions has wide utility ranging from macroscopic profilometry [1] to wide-field optical sectioning [2] and nanoscopic imaging [3], from passive array illumination [4] to interactive optical micromanipulation [5, 6], and a wide variety of materials processing applications [8, 9, 10], among many others. GPC-based light shaping makes an interesting addition in the toolbox of techniques for light shaping. Amplitude masks are straightforward but lossy and can be dynamically reconfigured in spatial light modulator (SLM) implementations [8, 9]. The promise of lossless light redistribution makes phase-only modulation techniques particularly attractive when situations call for efficient conversion. Energy remapping using refractive and reflective approaches [11] and related lenslet array implementations [12] are efficient and robust to shift in wavelength but can achieve a limited set of patterns due to fabrication constraints and are mostly suited for static applications. Techniques using multiple beam interference [10] and special diffraction effects such as Talbot array generators [13, 14, 15] are limited to periodic patterns and have limited reconfigurability. Computer-generated diffractive approaches [16] are reconfigurable but can suffer from noise and computational load.

The convenience of phase-encoding enabled by the tandem of straightforward design and availability of programmable phase-only spatial light modulators (SLM) makes GPC a highly favourable method for efficient projection of arbitrary and reconfigurable images. In this chapter, we apply the various design principles outlined in the previous chapter to show practical demonstrations of GPC-based synthesis of a wide variety of patterns including binary images, optical lattices, and arbitrary greyscale images. We also present numerical experiments to demonstrate other exciting GPC-based possibilities such as Gaussian beam shaping and achromatic light shaping and image projection.

GPC exhibits robustness to wavelength in various light patterning applications, which makes for an interesting ingredient in innovative integration of multi-wavelength approaches into patterned illumination that can enable new broad-band optical applications. One area where GPC has been applied with great success is in the projection of light patterns through arbitrary-NA microscope objectives for real-time three-dimensional (3D) manipulation of microscopic particles. The substantial contribution of GPC to the promising and actively growing field of optical micromanipulation merits greater attention and will be taken up in the next chapter.

Throughout this chapter, we will constantly need to refer back to the generic optical setup for GPC-based light patterning. Thus, we reproduce the schematic illustration of this setup to show its basic elements (see Fig. 7.1).

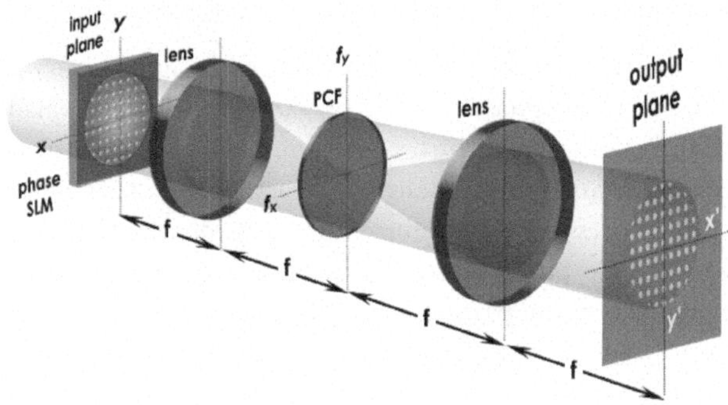

Fig. 7.1 Schematic illustration of a typical optical setup for GPC-based synthesis of patterned light.

7.1 Binary Phase Modulation for Efficient Binary Projection

Devising a new phase-only imaging method that provides for the most efficient, simple and robust use of available photons radiated from a given light source is fundamentally challenging and practically appealing. Avoiding photon dissipation prevents heat generation and potential damage in the optical hardware. This leads to efficient photon transfer and utilization at a desired target. It maintains its relevance in the face of commercially available compact and affordable powerful fibre lasers since many spectral gaps remain where users must contend with lower peak powers from traditional sources. Using a weaker light source in an application that normally requires a more powerful laser has many practical advantages. An energy-efficient system is also attractive for high-power laser applications as it allows the realization of even higher power densities and minimizes concerns about deteriorating effects due to absorption.

7.1.1 Experimental Demonstration

After laying down its theoretical foundations [17], a GPC-based pattern projection system based on Fig. 7.1 was implemented using a programmable spatial light modulator and a custom-fabricated phase contrast with support from Hamamatsu Photonics and the Danish Technical Research Councils [18]. Binary phase patterns were addressed onto a reflection type, optically addressed PAL-SLM from Hamamatsu Photonics. A simple beam-splitter configuration was used to read-out the phase patterns by a spatially filtered and expanded 5 mW He-Ne laser. This beam-splitting configuration is not energy efficient but was adopted in the early demonstrations to provide a practical "proof-of-principle" verification of the encoding technique, assuring accurate phase encoding and minimizing possible parasitic effects from off-axis illumination.

The phase contrast filter was fabricated by removing a 60 μm diameter circular region of Indium Tin Oxide (ITO) from a coated glass plate by use of a Hamamatsu C4540 excimer-laser micro processing unit. A reflection microscope image of the phase contrast filter is shown in Fig. 7.2(a). Typically 8–10 excimer-laser exposures were needed to completely remove the ITO coating. The ITO-coating was fabricated to provide a π-phase-shifting for the 633 nm wavelength used in the experimental system. The phase shift accuracy of the ITO coating was verified with a Mach-Zender

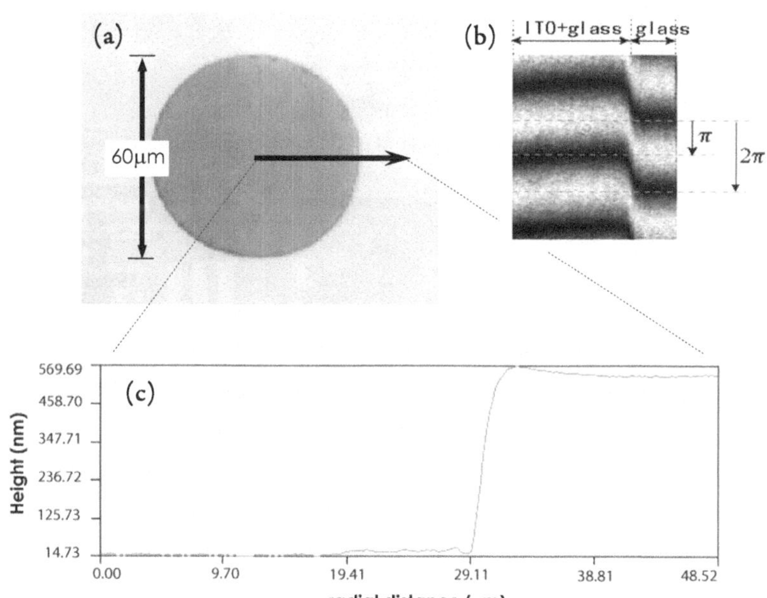

Fig. 7.2 Phase contrast filter: (a) Reflection microscope image; (b) Mach-Zender interferometric measurement of the ITO coating-glass transition at the filter edge; (c) radial topographic profile obtained by atomic force microscopy (filter fabricated by T. Ooii).

interferometer as shown by the interferogram in Fig. 7.2(b). A line profile through the edge of the fabricated filter was obtained by atomic force microscopy (AFM) as shown in Fig. 7.2(c). This provided for an alternative phase shift verification by use of the relation: $\theta = 2\pi\Delta n d / \lambda$ with $\Delta n = 0.7$ and $d \approx 500\,\text{nm}$ where d is the ITO layer thickness and Δn is the difference in refractive index between glass and ITO.

Without a phase contrast filter, the 4*f* system simply transmits the illuminating laser beam, although a low-contrast image of the encoded pattern on the PAL-SLM may appear due to finite aperture effects, as shown in Fig. 7.3(a). The input π-phase pattern used in Fig. 7.3(a) modulates approximately 25% of the input frame area (13 by 13 mm) giving the correct spatial average value, $\overline{\alpha} = 1/2$, for optimal visibility and contrast

(a)

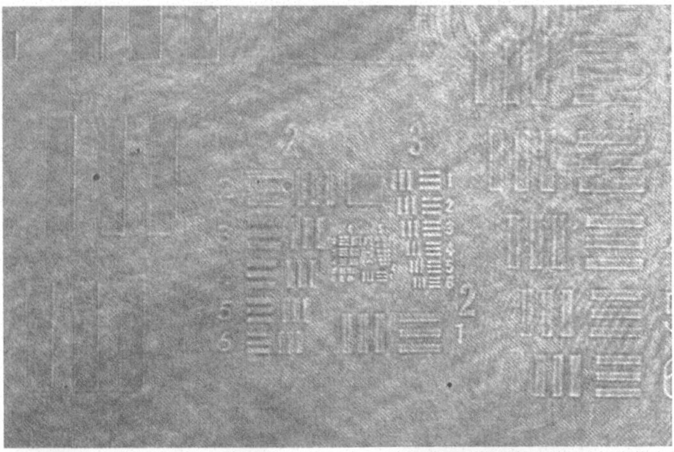

(b)

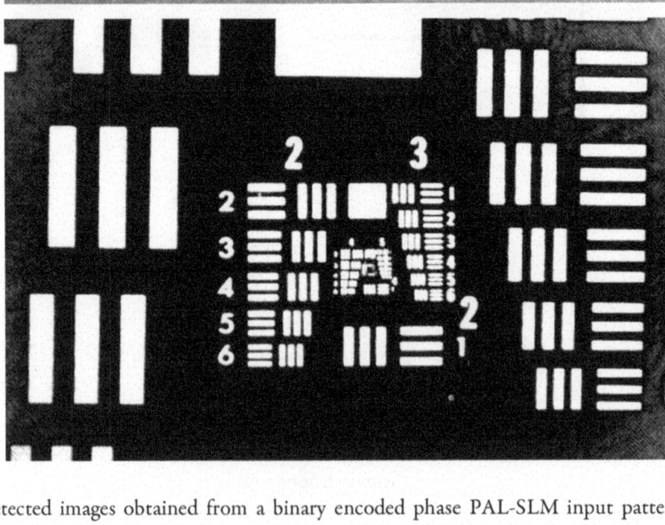

Fig. 7.3 Detected images obtained from a binary encoded phase PAL-SLM input pattern: (a) simple imaging without applying phase contrast filter; (b) imaging obtained when situating a π-phase-shifting phase contrast filter at the Fourier plane. The bright regions in (b) are over 3.5 times brighter than corresponding regions in (a).

when using a π-phase-shifting phase contrast filter. This is illustrated by the high contrast output image shown in Fig. 7.3(b), obtained after the phase contrast filter is aligned into place. Output intensity levels exceeded 3.5 times the input beam in the brightest regions of the projected images, close to the theoretical fourfold increase for 25% area-modulated binary patterns. Close to 90% energy efficiency was measured. The loss was partly due to the SRW tail that creates a weak halo around the projected image. The demonstrated efficiency is notable, considering that practical aspects such as the Gaussian distribution of the laser beam, SLM non-uniformity and encoding deviations, together with PCF fabrication errors and a small absorption in the ITO coating also contributed energy losses. Furthermore, this was achieved using binary phase elements at both the input and at the filter, which provides for a rather robust phase modulation regime with a relatively large tolerance against small phase deviations. The visualized phase also reveals the encoding characteristics of the SLM and equally lends itself to device characterization.

7.2 Ternary-Phase Modulation for Binary Array Illumination

An array illuminator provides a way of generating multiple bright spots or a periodic structure from a single laser beam [19]. A cross-grating can create a regular array of multiple spots at positions predetermined by the grating spacing. A simple but inefficient arbitrary array illuminator could be implemented using an opaque screen with apertures in the regions where light is required. We are therefore interested in efficient ways of dividing a light beam into an arbitrary array of secondary light beams without energy loss. Periodic arrays can be realized using the Talbot effect where the Fresnel diffraction pattern through a phase-only grating repeats at certain propagation distances, thus producing array illumination with relatively high efficiency [13, 14, 15]. Other alternatives use phase-only diffractive structures with the desired array illumination formed in the far-field or at a Fourier plane [20, 21, 22]. This technique offers a wide range of array patterns, the efficiency of which depends on the fine structure of the diffractive element. One draw-back or limitation is that the rising complexity of the diffractive element with the number of spots, spot shapes and configuration, leading to a reduced efficiency arising from the finite space bandwidth product of the system. This is of particular relevance when a phase-only spatial light modulator is used to generate the diffractive phase structure [23].

The generalized phase contrast method can be used for array illumination with flexibility on spot shape and array configuration with the usual convenience of direct correspondence between encoded phase and output intensity, and optimizes the phase contrast approach to array illumination [24, 25, 26]. The result presented in Fig. 7.3 may be considered as a GPC-based array illumination based on different spot shapes using binary phase encoding. In the succeeding sections, we present experimental demonstrations of GPC-based array illumination using ternary-phase encoding. A ternary phase encoding scheme increases the range of possible compression factors for the array illuminator, compared to binary phase encoding, while maintaining high visibility and efficiency [27]. This implementation of an array illuminator could offer considerable advantages where the generation of a dynamic intensity profile is required, in for example reconfigurable illumination of arrays in photonic devices, the generation of dynamic laser tweezer arrays and also the coding of structured light in machine vision applications [28, 29, 4]. We show experimental results for binary intensity arrays using ternary-phase modulation with different modulation patterns and filters. The phase pattern generating the output intensity pattern is displayed on a computer-controlled phase-only spatial light modulator.

7.2.1 Ternary-Phase Encoding

Array illumination entails high efficiency generation of binary output intensity distributions. To achieve a wide range of compression factors we consider the ternary phase encoding approach where the input phase pattern consists of three phase levels as discussed in Section 6.2.1. A design procedure derived from the theoretical framework can be as follows:

1. Choose the desired compression factor σ and phase filter parameter θ.
2. Calculate the phase $\Delta\phi$ from Eq. (6.27).
3. Calculate the fractional areas F_1 and F_2 from Eq. (6.26).
4. Address the spatial layout of the array illuminator with the calculated fractional areas F_1 and F_2 with the phase difference $\Delta\phi$.

Let us illustrate the procedure using particular array generators. The first example is for an array with compression factor σ =2 using a PCF with $\theta = \pi$. Performing the required calculations as detailed above gives:

$$\begin{aligned} \Delta\phi &= \pi/2 \\ F_1 &= 0.25 \\ F_2 &= 0.25 \end{aligned} \tag{7.1}$$

Ternary phase encoding also makes it possible to generate a binary pattern with the same compression factor σ =2, but using a different PCF, $\theta = 3\pi/4$, where we get

$$\begin{aligned} \Delta\phi &= \pi/2 \\ F_1 &= 8^{-1/2} \cong 0.35 \\ F_2 &= 1/2 - 8^{-1/2} \cong 0.15 \end{aligned} \quad . \tag{7.2}$$

These two examples are theoretically described and demonstrated graphically using graphical phasor charts in ref. [27].

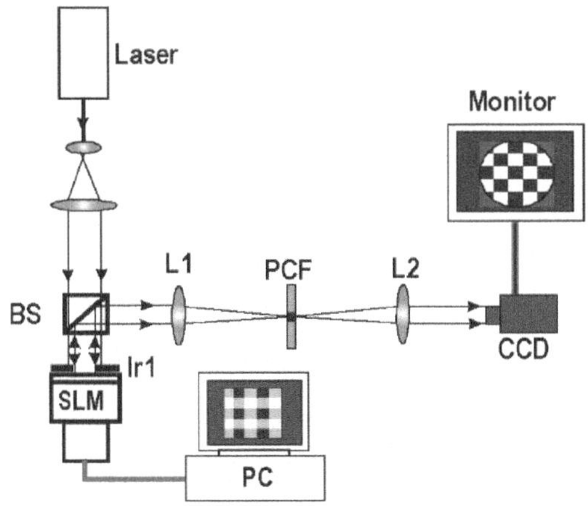

Fig. 7.4 The experimental set-up. A filtered, expanded and collimated diode laser beam is incident on the PO-SLM with an aperture defined by the iris (Ir1). A beam splitter (BS) directs the modulated light into the $4f$ system (formed by L1 and L2) containing the phase contrast filter (PCF). A high-contrast intensity distribution, which is directly related to the pattern addressed on the PO-SLM, is generated in the image plane containing the CCD camera.

7.2.2 Experimental Results

The experimental set-up is shown in Fig. 7.4. Light from an 830 nm diode laser is circularized, spatially filtered, expanded and collimated to generate a plane wavefront at the spatial light modulator. The Phase-only SLM (PO-SLM) is a parallel-aligned nematic liquid crystal Hamamatsu X8279 SLM optimized for operation at 830 nm. It is optically addressed with an XGA-resolution (1024×768 pixel) liquid crystal projector element and is capable of a phase modulation of at least 2π at 830nm [30]. The SLM generates a two dimensional three-level phase pattern to modulate the light that goes into the phase contrast system. The reflection-type SLM is again read out through a beam-splitter to achieve accurate phase encoding. This liquid crystal based SLM exhibit some tolerance to oblique incidence and this can be exploited in practical applications to avoid a significant power loss. The $4f$ system formed by the lenses L1 (f=200 mm) and L2 (f=200 mm) images the aperture defined by iris Ir1. Images were acquired using a standard black and white CCD camera (Pulnix TM-765) as a spatial intensity detector in the image plane of the $4f$ set-up and a frame-grabber (data translation DT2867). The PCFs we use are made of photoresist deposited on a glass optical flat. Experiments were undertaken using two filters with phase shifts of $\theta = \pi$ and $\theta = 3\pi/4$ at a wavelength of λ=830 nm, both with the same size, 60 μm in diameter.

In order to obtain a value, $K=1$, it is necessary to match the size of the iris in front of the PO-SLM to the PCF size. A very large η is desirable when the signal frequency components are well outside the main Airy lobe such as for a periodic phase structure. Otherwise, it is advisable to work with $\eta=0.627$, the smallest filter size giving $K=1$, to minimize intensity variations over the image aperture. Given the PCF diameter, one finds the matching input iris size using the following relation, based on the Airy disc size [31].

$$D_{Iris} = 1.53\frac{\lambda f}{D_{PCF}} \tag{7.3}$$

In this equation D_{Iris} and D_{PCF} are the diameters of the iris limiting the modulation area of the PO-SLM and the filter, respectively, f is the focal length of the Fourier transforming lens in the $4f$ set-up and λ is the wavelength of the laser source. The optimum iris diameter of, $D_{Iris}=4.2\ mm$, was used in the experiments. A detailed discussion of how the output intensity in a GPC system depends on the choice of the value η has previously been undertaken [32] and was outlined in Chapter 6.

The experiments maintain a compression factor, σ =2, corresponding to a 50% fill factor for the binary intensity distribution. A periodic test pattern is encoded into the SLM where each period consists of four rectangles having different sizes and phase values. The dynamic range of the PO-SLM, $2\Delta\phi$, is set to π and calibration yields grey levels 1, 180 and 255 as corresponding to phase values 0, $\pi/2$ and π respectively.

A schematic representation of an input test pattern and the desired output intensity distribution is shown in Fig. 7.5. Only a small section of the pattern consisting of four unit cells are shown, where each unit cell compromises four quadrants in which the

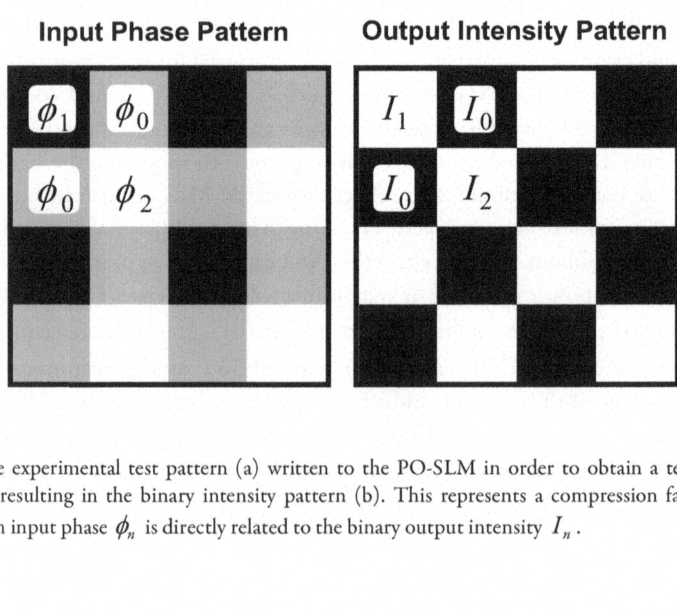

Fig. 7.5 The experimental test pattern (a) written to the PO-SLM in order to obtain a ternary-phase modulation resulting in the binary intensity pattern (b). This represents a compression factor of two where a given input phase ϕ_n is directly related to the binary output intensity I_n.

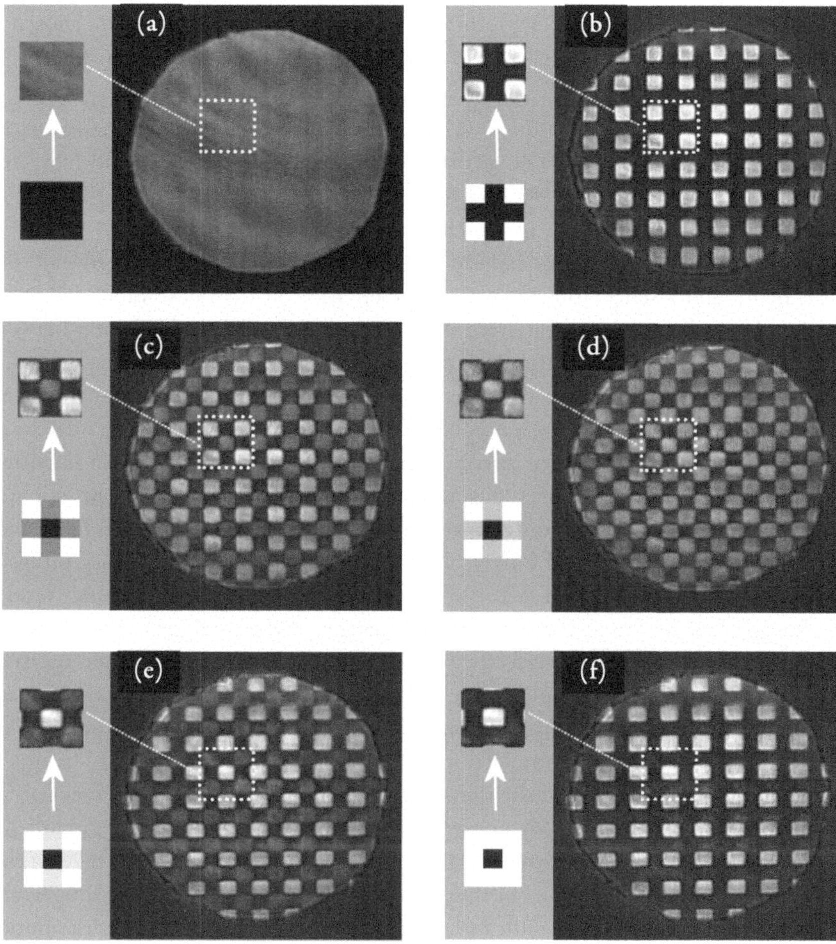

Fig. 7.6 A set of images showing the experimental demonstration of binary and ternary-phase array illumination with a π -phase shifting PCF. The image (a) shows the input intensity distribution imaged to the detector camera with zero phase modulation (illustrated with the "phase" rectangle in the lower left corner). Figures (b) to (f) show the images obtained with varying phase modulation of ϕ_0. Figure (d) is the case of ternary phase modulation resulting in binary intensity pattern. Figures (b) and (f) are the special cases where the input is a binary-phase modulation and figures (c) and (e) show the intensity patterns for unbalanced phase modulation of ϕ_0 ($\phi_0 = \pi/4$ and $\phi_0 = 3\pi/4$ respectively).

phase values ϕ_0, ϕ_1 and ϕ_2 are addressed. The aim of the ternary-phase encoding approach is to convert the phase distribution represented by Fig. 7.5(a) into the intensity distribution shown in Fig. 7.5(b). In this case the values of ϕ_1 and ϕ_2 are chosen so that they generate equal intensity levels I_1 and I_2 in the binary output intensity pattern as described by Eq. (6.20).

In Fig. 7.6 we show the experimentally measured intensity distributions for the test pattern shown in Fig. 7.5(a) for a range of different phase values in the region characterized by ϕ_0 (the phase values $\phi_1 = 0$ and $\phi_2 = \pi$ are fixed). For each figure we show an image of the complete aperture in which a region of interest, enclosed by the white dotted line, is selected. An enlargement of this region is shown on the top left-hand side of each figure and a schematic grey-scale representation of the phase values being addressed is shown below this, facilitating the interpretation of the relationship between the input phase level and the output intensity distribution.

Figure 7.6(a) shows an image of the available intensity distribution through the iris without the PCF in place and with no phase modulation at the SLM. Some inhomogeneity is apparent in this intensity distribution, which is primarily due to an imperfect spatial filtering of the laser diode. This available intensity distribution is redistributed through the PCF system depending on the phase modulation in the SLM. Fig. 7.6(b) and Fig. 7.6(f) show the case of binary phase modulation resulting in binary intensity modulation with a compression factor of $\sigma = 4$. Here the theoretical intensity level is four times that of the input distribution in a quarter of the total area. The measured intensity distributions are not perfectly homogeneous over the entire output aperture; this is due to limitations in the phase level addressing of the SLM module as well as the non-uniform input intensity distribution. There is also a reduction in the visibility towards the edges of the aperture, arising from the fact that the factor K is only a constant in the central region of the output plane and therefore not invariant over the aperture [32]. An experimental demonstration of successful ternary-phase modulation resulting in a binary intensity pattern is shown in Fig. 7.6(d). Firstly, it can be seen that the intensity of the bright regions in this figure are equal and secondly the intensity of the light in these regions is significantly higher than that in the input distribution shown in Fig. 7.6(a) (in theory this should be twice as intense over half the area). For comparison, Fig. 7.6(c) and (e) show image distributions with two non-zero intensity levels. The two unbalanced intensity levels are obtained with phase values of ϕ_0 equal to approximately $\pi/4$ and $3\pi/4$ respectively.

Figure 7.7 shows the intensity levels for the input and the binary and ternary input phase modulation outputs shown in Figs. 7.6(b) and (d) respectively. The intensity scale is normalized to the average of the intensity input level. The input intensity profile fluctuates about its average value by approximately 5%. The three curves in Fig. 7.7 were obtained by averaging over 10 pixels in the vertical direction in a 150 pixel wide horizontal strip in the central region of the appropriate images from Fig. 7.6. The CCD camera used for these measurements has a pixel size of 11x11 μm so the measurement field width of Fig. 7.7 is approximately 1.65 mm.

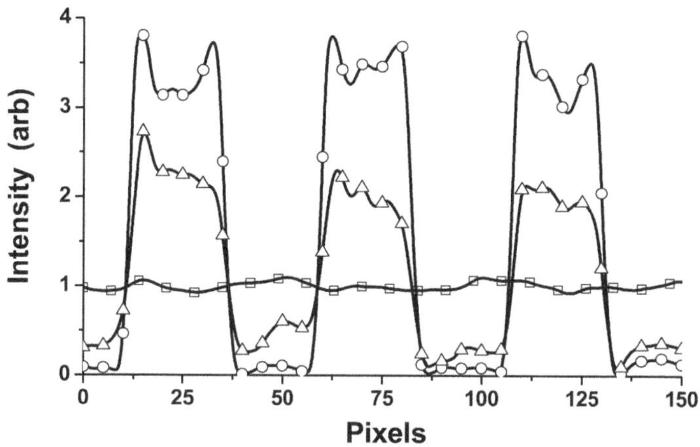

Fig. 7.7 A plot of the output intensity across a portion of the central region of the image aperture for: no input phase modulation (□), binary phase modulation (○) and ternary phase modulation (△). The intensity scale is normalized to the average input intensity and the line profiles are the average of 10 adjacent line scans with a length of 150 pixels.

The experimental results show reasonable correspondence with the theoretically expected optimal values. The data shown in Fig. 7.7 for the binary case shows an average maxima of 3.4 ± 0.2 with an average minima of 0.1 ± 0.03, giving an average visibility of approximately 0.95. The optimum theoretical visibility is 1 with a peak intensity of 4 within a 0 dark background. Similar calculation for ternary modulation reveals an average maxima of 2.11 ± 0.2 and an average minima of 0.34 ± 0.1, leading to a visibility of approximately 0.72. The optimal theoretical result is a dark background of zero and an average maximum value of 2.0. There are a number of reasons for the discrepancies between the real and ideal values. The noisy ringing effect in the linescans, especially at the edges, can come from a small defocusing in the imaging system. The precision and uniformity with which the phase modulation of the PO-SLM can be controlled is a performance-limiting factor to which the output intensity pattern is susceptible. The overall visibility of the intensity pattern is in fact directly dependent on the phase addressing precision, as the spatial average of the input wavefront is influenced by this factor. From the qualitative and quantitative data presented in Fig. 7.6 and Fig. 7.7, it is clear that the dark background condition for the ternary case is not completely fulfilled and that the resulting intensity is non-uniform. This is primarily due to noise from the PO-SLM addressing module when working with a phase modulation of $\pi/2$ around which the visibility of any fluctuations is likely to be enhanced by the GPC system. Perfect matching of the iris size to the PCF size is also critical in obtaining high output visibility for the generated arrays and a small experimental deviation from the ideal matching conditions is not unlikely. Finally, the non-uniformity of the input intensity distribution would also influence the output intensity distribution from the PFC system.

To assess the flexibility of the ternary-phase approach we attempted to generate the same compression factor, $\sigma=2$, using a PCF with a different phase shift, in this case $\theta=3\pi/4$. The test pattern programmed on the PO-SLM is shown in Fig. 7.8(a) and the corresponding theoretical binary intensity pattern is shown in Fig. 7.8(b). The fractional areas and phase level of each region in the test pattern are the same as those calculated at the end of section 2. In Fig. 7.8(c) the input intensity distribution is shown and Fig. 7.8(d) shows the results obtained with ternary-phase array illumination used to generate a binary intensity distribution. As the measured intensity pattern clearly shows, the brightness of both regions is equal although the inhomogeneities previously discussed are still present. Once again it can be seen that although the contrast of the central region in Fig. 7.8(d) is very high it decreases somewhat towards the edges if the aperture. This radial fall off in the intensity is more pronounced than was the case with a PCF with phase shift $\theta=\pi$, see Fig. 7.6(d). We ascribe this to a non-perfect matching the size of the iris to the PCF size as specified by Eq. (7.3).

These results demonstrate that a ternary-phase array illuminator based on the generalized phase contrast method can be used to generate binary intensity patterns.

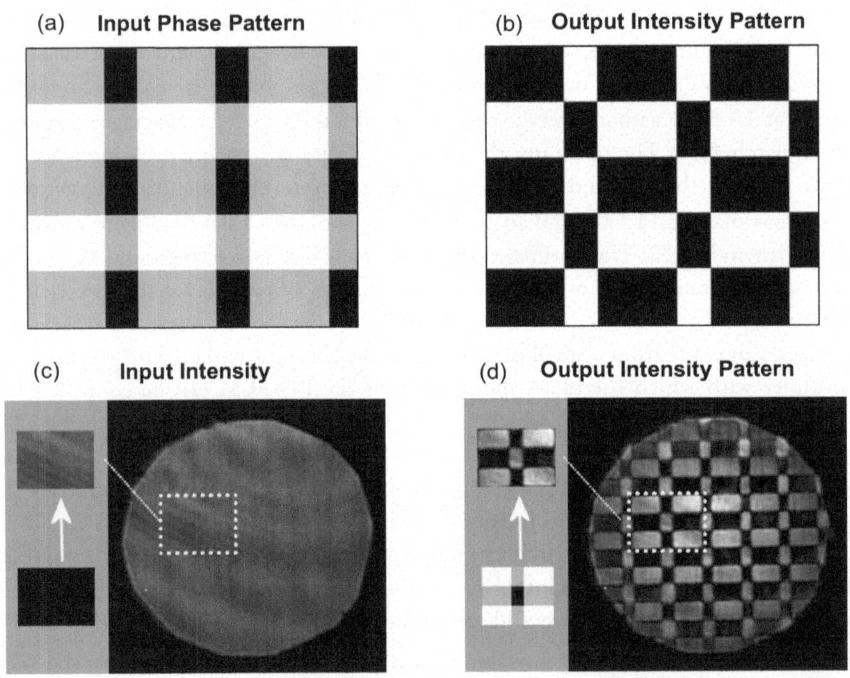

Fig. 7.8 Experimental results for a ternary phase-array illuminator with a PCF that phase shifts by $\theta=3\pi/4$: (a) ternary phase input test pattern programmed on the PO-SLM; (b) binary intensity pattern expected for the phase input in (a); (c) detected output for zero input phase modulation (image of the input intensity); (d) the output intensity pattern for the case of ternary phase modulation.

7.3 Dynamically Reconfigurable Optical Lattices

Optical lattices – light beams arranged in periodic arrays – have wide applications that extend from optical rerouting of microflow-driven materials to periodic pinning of viscously damped particles. For microfluidics or lab-on-a-chip systems, optical lattices can supplement the functionalities of dynamically reconfigurable optical traps to achieve a greater degree of control over the activities within miniaturized fluid channels and chambers [33, 34, 35, 36, 37, 38, 39, 40]. Aside from aiding in colloidal crystallization of dielectric microparticles [41], arrays of light spots may also be used as dipole traps for grouping cold atoms at multiple locations [42, 43, 44, 45].

Multi-beam interference can sculpt lattice-like optical landscapes that can be used for separating fluid borne particles with differing optical properties. This particle separation scheme, called "optical fractionation" [34], has recently been demonstrated experimentally by MacDonald and co-workers [35]. In optical fractionation, the trajectories of co-flowing particles with distinct optical properties are influenced by the topology of an applied optical lattice. In their work, an optical lattice was generated by a five-beam interference pattern. When mixed particles flowed across the optical interference pattern, a particular lattice topology induced deflection of selected particles from their original trajectories while others passed almost straight through. This demonstrated the ability to separate microscopic biological matter by size, specifically by separating out protein microcapsules. Exponential size selectivity was pointed out as inherently possible with a dense particle solution driven through this optical lattice. The authors also demonstrated sorting of equally sized particles of differing refractive indices, namely, silica and polymer microspheres.

Light patterning by generalized phase contrast can easily generate reconfigurable optical lattices using two-dimensional phase patterns encoded onto a programmable phase-only spatial light modulator (SLM). The lattices can have adjustable periodicity and can be sculpted with high output resolution. It is even possible to generate so-called optical obstacle arrays that are very hard to produce by other means. The periodic nature of the required input phase pattern means we can optimize the Fourier filter to achieve nearly 100% light efficiency.

7.3.1 Dynamic Optical Lattice Generation

Using the generic GPC optical system illustrated in Fig. 7.1, periodic phase patterns can be dynamically encoded onto a phase-only SLM to produce desired optical intensity lattices. The earlier sections demonstrate binary and ternary input phase encoding using the smallest filter size that yields K=1. As we showed in Chapter 6, the SRW can vary spatial variations that can be minimized by using bigger filters, which is possible when the zero-order is sufficiently separated from the signal frequency components. Patterns like optical lattices exhibit this desirable frequency separation and can utilize bigger filters to achieve almost perfect conversion efficiency. Figure 7.9(a) illustrates this nearly 100% light-efficient result using a PCF with η=4 for a binary phase input consisting of a square array, having identical horizontal and vertical periodicity, where 75% of the total input area is encoded with $\phi = 0$ while the remaining 25% is encoded with $\phi = \pi$. This results in a real-valued complex average, $\overline{\alpha} = 0.75\exp(i0) + 0.25\exp(i\pi) = 0.5$. The output area is filled by intensity peaks, each of which is approximately four times the incident intensity (Fig. 7.9(c)), sum up to 25% of the total aperture area. Moreover, the intensity peaks of this optical lattice are spatially discrete in all lattice symmetries.

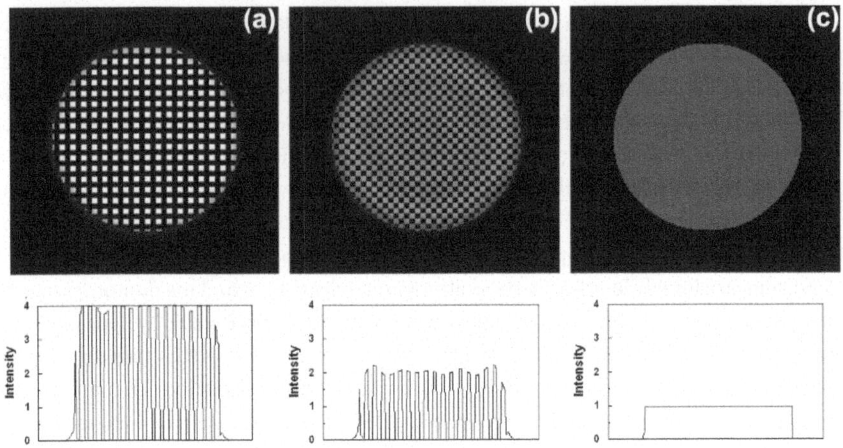

Fig. 7.9 Optical lattices generated using a phase-only filter $\{A=1, B=1, \theta=\pi, \eta=4\}$ with (a) binary (75% zero and 25%. levels) and (b) ternary (50% $\phi_0 = \pi/2$, 25% $\phi_1 = 0$, and 25% $\phi_2 = \pi$) phase input, respectively. (c) Intensity plot of incident beam to the phase input. Line scan along the horizontal direction in (a) and in (b) shows maximum intensity four times and twice, respectively, the maximum incident intensity (c) indicating light efficiency of close to 100%.

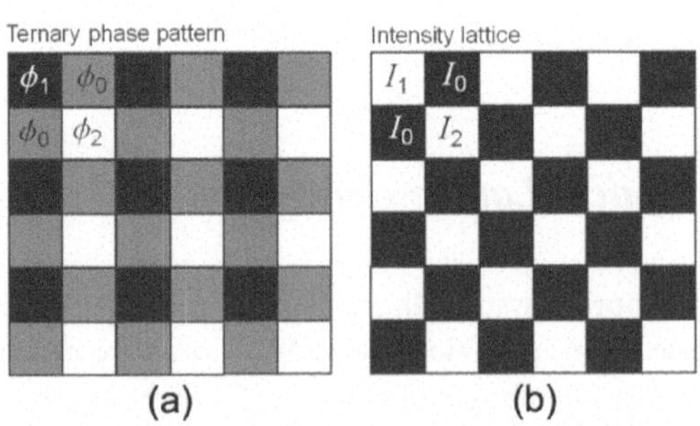

Fig. 7.10 (a) Input ternary phase pattern and (b) corresponding high contrast intensity lattice at the observation plane. Setting $\phi_0 = \pi/2$, $\phi_1 = 0$, and $\phi_2 = \pi$, results in $I_0 = 0$ and $I_1 \approx I_2 > 0$.

Another optical lattice with quasi-binary intensity is achieved by using the same filter parameters but with a periodic ternary-phase-encoded input (see Fig. 7.10). Using the ternary encoding we have developed, the phase corresponding to zero intensity is found to be $\phi_0 = \pi/2$ (over 50% of input aperture area), and bright regions with $\phi_1 = 0$ (25% of input aperture), and $\phi_2 = \pi$ (25% of input aperture).

We find that the intensity maxima for the case of this ternary phase are interconnected along the ±45° symmetry directions and are approximately twice that of the incident beam, which again indicates a virtually 100% efficiency (Fig. 7.9(b)). A filter with $\eta = 4$ ensures that high-order spatial frequency components of the binary or the ternary phase patterns used are well beyond a Δf_r radius. This condition must be kept in mind if a new grating period is chosen.

7.3.2 Dynamic Optical Obstacle Arrays

For micro-fluidics or lab-on-a-chip systems, optical lattices can supplement the functionalities of dynamically reconfigurable optical traps to achieve a greater degree of control over the activities within miniaturized fluid channels and chambers. Aside from aiding in colloidal crystallization of dielectric microparticles, arrays of light spots may also be used as dipole traps for grouping cold atoms at multiple locations.

Optical obstacle arrays provide an alternative approach to passive microfluidic sorting through laser-projected optical patterns. Optical obstacle arrays are composed of dark sites set periodically within a bright background. In such intensity landscapes, the optical forces will pull high-index microparticles towards regions of greater light intensity. Hence, the dark sites effectively behave as obstacles that deflect microparticles. In contrast to trap arrays, the motion of microparticles through obstacle arrays is primarily determined by the geometry of the potential landscape, independent of thermal effects [46].

A pertinent issue for optical obstacle arrays is finding a light-efficient method for synthesizing the appropriate intensity distributions. The methods commonly used for constructing bright spots in optical lattice patterns, such as multi-beam interference [47], computer-generated holography [48], and Talbot diffraction [49], are generally not well suited for constructing a continuous bright background field. Fortunately, the generalized phase contrast (GPC) method provides a convenient means of achieving the desired optical landscape. Earlier reports on GPC have focused on the construction of bright optical traps, similar to those composing optical lattice patterns [40, 50]. However, more recent work [51, 52] has shown GPC to be also useful for projecting broadly illuminated areas.

Figure 7.11 shows experimentally acquired samples of laser-projected intensity patterns generated using GPC [53]. An expanded and collimated beam from a continuous wave Ti:Sapphire laser (Spectra Physics 3900, 1 W, λ= 830 nm) was used to illuminate a reflection-type, optically-addressed phase-only SLM (Hamamatsu PPM X7550, 41.6 μm/pixel) through a circular aperture (diameter = 11.5 mm). The SLM lay at the object plane of a standard 4*f* imaging setup ($f1$ = 300 mm, $f2$ = 150 mm) and imprinted user-defined phase modulation upon the reflected beam while a CCD camera (Pulnix TM-765, 11 μm/pixel) at the image plane captured the resulting intensity patterns. An optical flat with a circular pit (diameter = 33 μm) at the common focal plane between the two lenses was used as a phase contrast filter (PCF) to provide π phase shift between the zero-order and the higher order Fourier components.

The images in Fig. 7.11 cover a 5.5 mm×5.5 mm area and are shown with the same scale. The apparent size variation between the patterns is a consequence of maintaining the same 25% fill factor for all cases. The bright background field in Fig. 7.11(a) covers a larger fractional area relative to the periodic dark sites, thus the pattern was cropped with a dark frame to achieve the specified fill-rate. Implementing the same approach in Figs. 7.11(b) and (c) resulted in bigger patterns as the relative areas covered by the periodic dark sites increased. The patterns may be expanded to cover the entire projection aperture for sufficiently large dark sites. There is sufficient flexibility since additional optical components can be easily added to the imaging system to rescale the projection to an appropriate size. Furthermore, although a common fill factor was maintained in these examples, laser projections from an optimized GPC scheme can have close to 100% optical throughput for optical lattices over a wide fill factor operating range. It is reasonable to believe that the reconfigurability achieved using a spatial light modulator can improve the level of control in optical fractionation procedures.

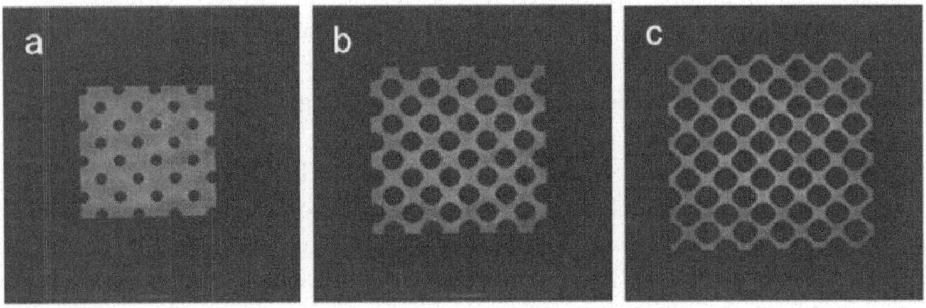

Fig. 7.11 Examples experimentally acquired, dynamically laser-projected intensity patterns constructed with GPC. When coupled through a microscope objective into a microfluidic system, these patterns can act as optical obstacle arrays for fluid-borne microparticles. The illuminated areas vary from (a) to (c) as the specific GPC constraints implemented require an aperture fill-rate of 25%.

7.4. Photon-Efficient Grey-Level Image Projection

Despite many advances in diffractive optics and computer-generated holography (CGH) [54], determining the appropriate phase pattern that will yield a specific output intensity distribution remains a non-trivial task. Generalized phase contrast (GPC) provides an alternative approach to the phase-to-intensity problem by combining phase-only modulation with a non-absorbing Fourier phase filter [32]. Combining principles of phase contrast microscopy with phase-shifting interferometry can enable quantitative phase imaging [55]. Conversely, the quantitative phase-to-intensity mapping in a system can be used to design an appropriate phase pattern for a desired output intensity distribution. Subject to a minimal set of conditions and approximations, GPC theory provides such a quantitative mapping [17, 56]. A new analysis of GPC is used to design phase patterns that are applicable within the practical constraints of a dynamically addressable spatial light modulator (SLM).

The key advantages of GPC for real-time projection of two-dimensional (2D) grey-level images are efficient utilization of the incident light through phase-only modulation, and rapid update of the generated light patterns due to the low computational overhead [57]. Phase-only modulation can create intensity patterns without blocking or absorbing light. It is therefore attractive as an inherently more efficient alternative to amplitude modulation based light-valve systems for image and video projection. Computer generated holography (CGH) is perhaps the most popular approach to image projection by phase-only modulation. However, the CGH method is notorious

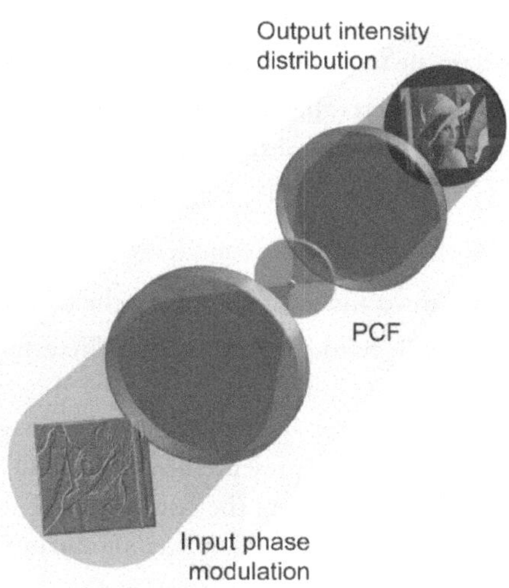

Fig. 7.12 Generic GPC system for image construction.

for being computationally expensive [58] and demanding in terms of the required high space-bandwidth product. Some researchers have even gone to the extent of designing specialized hardware to deal with the cumbersome iterative phase-retrieval algorithms [59]. Others have taken the route of exchanging spatial bandwidth in the SLM, for temporal bandwidth during hologram computation [60]. With the GPC method however, it is possible to address the appropriate phase patterns directly, without need for such compromises.

An optical system for GPC-based image construction is schematized in Fig. 7.12. A phase modulated light beam is decomposed into Fourier components using a lens. A small, non-absorbing wave-retarder or phase contrast filter (PCF) at the centre of the Fourier plane shifts the phase of the lowest spatial frequencies relative to the higher frequency components. Interference between frequency components upon recombination by a second lens creates the desired intensity distribution.

7.4.1 Matching the Phase-to-Intensity Mapping Scheme to Device Constraints

The optical field, $o(x,y)$, at the output plane of a GPC system can be described by a simple interference between two terms:

$$o(x,y)=\exp\left[\mathrm{i}\phi(x,y)\right]+c(x,y)\left[\exp(\mathrm{i}\theta)-1\right], \tag{7.4}$$

where $\phi(x,y)$ is the phase distribution at the modulator and $c(x,y)$ describes a synthetic reference wave profile determined from Fourier components that have propagated through the phase-shifting region of the PCF with phase shift θ. If all AC components of $\exp\left[\mathrm{i}\phi(x,y)\right]$ fall outside the PCF at the Fourier plane, the approximation $c(x,y)\approx\bar{\alpha}\, g(x,y)$ is appropriate. $\bar{\alpha}$ is the complex-valued spatial average of the input field, and $g(x,y)$ is a real-valued function describing the composite diffractive effects due to the modulator and PCF hard-edge apertures. These functions have been described previously in greater detail [17, 32, 56].

Phasor diagrams are an intuitive way of analyzing light propagation in the GPC system. The three terms in Eq. (7.4) are represented by phasors **O**, **P**, and **R**, respectively (see Fig. 7.13). **P** is a unit phasor depicting the complex field at a single point on the modulator, and $\bar{\alpha}$ is the average of all **P** at the modulator plane. The identity: $\exp(\mathrm{i}\theta)-1=2\left|\sin(\theta/2)\right|\exp\left[\mathrm{i}(\theta+\pi)/2\right]$, shows that the reference wave, **R** is always angularly displaced from $\bar{\alpha}$ by $(\theta+\pi)/2$. The output field, **O**, is then constructed simply by taking the sum of each **P** with **R**. For any particular **P**, its complex conjugate, **P'**, with respect to **R** will yield the same output intensity.

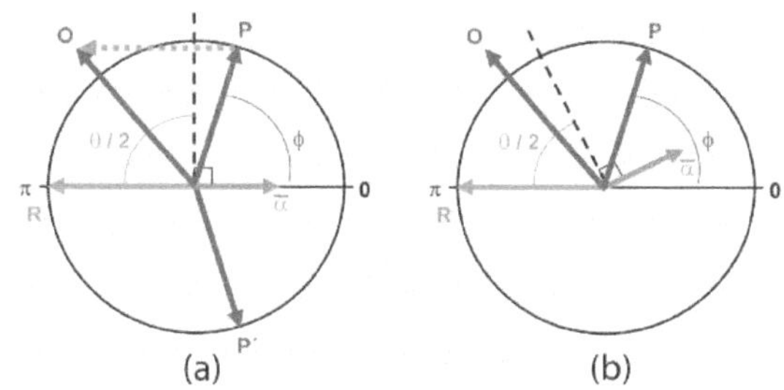

Fig. 7.13 Phasor diagrams of interfering light components in a GPC system. (a) Alternate-pixel conjugation scheme with $\theta = \pi$. (b) Arbitrary scheme requiring only 0 to π encoding.

In the next section, we will show how to utilize this degeneracy in a phase conjugation scheme for constructing arbitrary grey-level images, which can be understood from Fig. 7.13(a). The aim is to position the phase of $\overline{\alpha}$ and **R** to 0 and π, respectively, by alternately conjugating pixels in a slowly varying phase pattern and choosing a PCF with $\theta = \pi$. If the PCF size is appropriately chosen such that $g(0,0)=1$, the output image follows the simple mapping:

$$\left|o(x,y)\right|^2 = 4\left|\sin\left[\phi(x,y)/2\right]\right|^2. \tag{7.5}$$

Although Eq. (7.5) neglects a slight modulation envelope in $g(x,y)$ that causes a decrease in contrast towards the outer edges of the image, very good image quality with up to four-fold gain in peak intensity is still achievable. Implementing a phase-conjugation scheme requires an SLM device with very good modulation transfer function (MTF) to properly render rapidly varying phase between adjacent pixels. It also requires consistent phase encoding performance is over a complete 2π phase stroke.

The phasor diagram in Fig. 7.13(b) depicts an alternative encoding scheme that works for most available devices, which are characterized by a moderate MTF. Rather than forcing the phase of $\overline{\alpha}$ to 0 through phase-conjugation, **R** can be independently brought to by matching θ to the existing $\overline{\alpha}$. The output intensity will still map according to Eq. (7.5), but without the need for encoding conjugate phases. Thus, both MTF and phase stroke requirements on the implementing device are greatly relaxed. The reduced phase stroke requirement is particularly significant as it would allow the adoption of faster SLM devices. We note, however, that $\overline{\alpha}$ is image dependent, i.e. it varies with the histogram of the encoded phase pattern. An ideal system should then utilize a dynamic PCF whose phase shift can be adjusted as arbitrary images are introduced to the projection system.

A similar projection scheme may also be applied with a fixed PCF if a dynamic filter is not available. Instead of adjusting the phase shift, the histogram of the encoded pattern may be modified such that the correct phase for $\overline{\alpha}$ and consequently for **R**, is achieved. In particular, histogram equalization ensures that the average phase is consistent and independent of any specific image information. The principal image can then be circumscribed with a frame of zero-phase pixels to shift the effective $\bar{\alpha}$ to a smaller phase value. To find the optimal frame size, we must apply the design criterion that relates $\bar{\alpha}$ and PCF shift, θ. The optimum fill-factor, *F*, of the principal image with respect to the entire illuminated area is then derived from the complex-valued expression

$$\bar{\alpha}' = \frac{1-2(1-F)g(0,0)+\mathrm{i}\cot(\theta/2)}{2Fg(0,0)}, \tag{7.6}$$

where $\bar{\alpha}'$ indicates the complex spatial average of the principal image before the zero-phase frame is applied.

This framing technique allows the $\bar{\alpha}$ of different framed grey-level images to be consistently matched to a fixed PCF with arbitrary θ. Equation (7.6) ensures that we satisfy the condition depicted in Fig. 7.13b, where the zero-padding frame appears dark in the output plane and energy is diverted almost entirely into the central region containing the principal image.

7.4.2 *Efficient Experimental Image Projection Using Practical Device Constraints*

In comparison to previously known phase-only imaging techniques this dramatically simplifies the synthesis of arbitrary intensity patterns and the requirements of the space bandwidth product are also significantly reduced compared to that of e.g. phase-only holography. This makes it more feasible to utilize a dynamic and relatively coarse-grained spatial light modulator as input phase modulating device without seriously compromising on the image reconstruction quality. The experimental demonstration of laser image projection was implemented using a typical GPC imaging system. A telescopic lens pair (f = 35 mm, f = 1000 mm) expanded the 830-nm output from a Ti:Sapphire laser (Spectra Physics 3900) to illuminate a reflection-type optically-addressed phase-only SLM (Hamamatsu PPM X7550, 41.6 m/pixel) through a circular aperture (diameter = 11.5 mm). The SLM is at the front focal plane of a standard 4*f* imaging setup (f1 = 300 mm, f2 = 150 mm) with a CCD camera (Pulnix TM-765, 11 μm/pixel) located at the back focal plane of the second lens. A phase contrast filter (PCF) sits at the confocal plane between the two lenses. The PCF consists of a circular pit (diameter = 33 μm) etched on an optical flat. The pit is aligned at the optical axis and provides θ~2 radians phase shift between the zero-order and the higher order Fourier components.

Figure 7.14(a) presents the intensity distribution captured by a CCD at the output image plane of the GPC system. The projected image is an excerpt from the well-known "Lena" image [61]. Figure 7.14(b) shows the output of the system when the PCF is removed from the optical train. In an ideal system, phase-only modulation implies that this image should be of uniform intensity. It is obvious, however, that some amplitude modulation has been introduced by the SLM and other optical components.

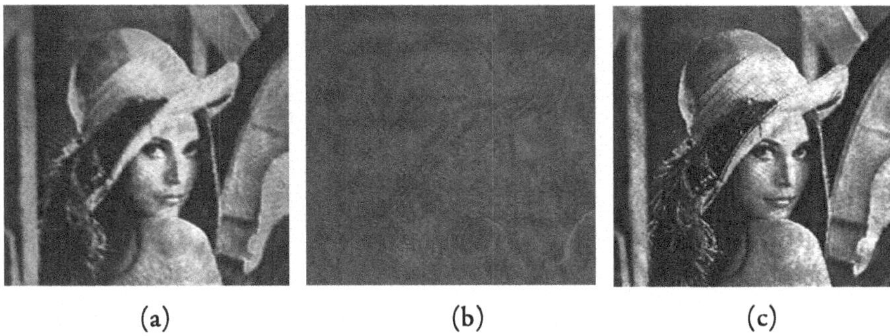

Fig. 7.14 Image projection of a GPC system. (a) Image captured with a CCD at the output plane. (b) Intensity distribution at the output plane when the PCF is removed. (c) Numerically simulated image projection incorporating amplitude modulation crosstalk.

A simulated projection taking into consideration such crosstalk was numerically calculated and is shown in Fig. 7.14(c). Most of the observable intensity artefacts in the captured image are reproduced here. These errors are thus more indicative of device limitations and read-out beam quality, rather than fundamental limitations of the projection scheme. The principal difference between the simulated and actual projections is a softening of edge information. Such loss of high spatial frequency content is attributed to the limited MTF of the SLM, as well as the finite lens apertures in the Fourier lens system.

Perhaps the strongest advantage of GPC over other image-construction methods is in light efficiency. Following a previous work by Arrizón and Testorf [62], it can be shown that, considering only pixilation in an otherwise ideal SLM device, the maximum conversion efficiency achievable using CGH is limited to 52%. Other constraints such as phase-stroke, bit-depth, and phase level errors will degrade this optimal performance. During our experimental demonstration of GPC grey-level image projection, 74% of the light transmitted through the system, Fig. 7.14(b), was efficiently redistributed into the desired output image, Fig. 7.14(a). Numerical simulations predict that efficiencies for GPC image projection can reach as high as 87% with an ideal device and a uniform input beam.

This novel demonstration of GPC laser projection shows GPC to be a genuinely applicable method for the light-efficient construction of arbitrary images with existing dynamic SLM devices. This shows GPC has the potential to play a major role as a light-efficient laser projection technique in a diverse range of photonics applications – from laser imaging for parallel patterning of material surfaces [63, 64], phase-only encryption and data storage in optical information systems [65, 66], to optical addressing of surface plasmons [67] and other optical manipulation schemes in all-optical lab-on-a-chip devices [68, 69, 33]. GPC image construction holds several key advantages over other phase modulation techniques, particularly in terms of light efficiency, image quality, and device performance requirements. Further, GPC is more readily able to leverage future advances in SLM device technology. For example, direct, i.e. non-iterative, addressing of GPC input phase patterns implies greater scalability to high-resolution modulators without compromise on speed. The GPC method represents an exciting enabling technology that can drive further development in a multitude of photonics applications that require high throughput illumination with arbitrary and dynamic intensity distributions.

7.4.3 Photon-Efficient Grey-Level Image Projection with Next-Generation Devices

GPC perhaps represents the most feasible solution to the problem of real-time two-dimensional grey-level dynamic image projection by phase-only modulation. The continuous development of spatial light modulators yields devices with improved modulation transfer function. These are more suitable for implementing other GPC phase-encoding solutions such as the phasor flipping approach. In the following, we present numerical simulations to demonstrate excellent image reconstruction by GPC. We implement phase flipping with the GPC formulation for rectangular input and PCF apertures in Sect. 6.6, which aims to utilize all the available pixels in a rectangular SLM device. The images produced using different PCF sizes are compared in terms of image fidelity and light efficiency. Optical throughput is predicted to be as high as 87% for optimum image quality, and may reach 98% with some trade-off in image fidelity.

Eight-bit grey-level images, each 256×256 pixels in size, were used to find appropriate phase patterns using the phase-to-intensity mapping derived in Sect. 6.6. Chequered phase-flipping was applied to satisfy (20). The phase distributions were quantized to 256 phase-levels between $[0,2\pi]$ to emulate the finite bit-depth of a practical SLM device. The GPC optical system was simulated using an efficient Fast Fourier Transform (FFT) algorithm [70] using a PCF with $\theta=\pi$. Finally, the mean-square error (MSE = [*original pixel value* – *rescaled output pixel value*]^2 / *total number of pixels*) of each output image was measured with respect to the corresponding original image. The brightness of output images was rescaled before MSE-comparison such that the integrated brightness matched that of the corresponding original images.

Figures 7.15(a) and 7.15(b) show some representative output images. The corresponding original images are shown in Figs. 7.15(c) and 7.15(d), respectively. Presented images employed 9×9-pixel PCFs. Given the 256-pixel image aperture and 2048×2048 pixels FFT array employed, this PCF size corresponded to $\eta=0.5625$. The excellent visual similarity is consistent with the low MSE scores. Light projection was also very efficient, with 87% of the input light contributing to the final image in both examples. These two images were particularly well-suited to the GPC image projection method. The many sharply contrasting details in the images translated to dominantly high spatial frequency content, thus making the approximation in (14) highly applicable.

Other GPC image projections under the similar conditions are shown in Figs. 7.16(a) and 7.16(b), with corresponding original images shown in Figs. 7.16(c) and 7.16(d). These images featured larger flat areas and a lack of sharply contrasting edges. Consequently, reconstruction quality was expected to be poorer as significant image information was not well separated from the DC spatial frequency component. MSE scores were indeed nominally higher, but subjective image quality still remained very good. Errors manifested mainly as degraded contrast, particularly close to edges of the images. Diffraction efficiency was also slightly reduced to 85% (Fig. 7.16(a), 'flower') and 84% (Fig. 7.160(b), 'mallet').

Figures 7.17(a)–(d) demonstrate the effect of PCF size on image quality. Larger PCF sizes were used in Figs. 7.17(a) ($\eta = 2.5625$, 41×41 pixels), and 7.17(b) ($\eta = 0.9375$, 15×15 pixels), in comparison to the best-quality image in Fig. 7.17(c) ($\eta = 0.5625$, 9×9 pixels). Figure 7.17(d) utilized the smallest PCF ($\eta = 0.1875$, 3×3 pixels). Visual comparison revealed some image quality degradation, primarily contrast errors, introduced by using PCF sizes much larger or much smaller than the optimal size predicted by (21), i.e. $\eta \approx 0.62$.

Variations were further emphasized by the corresponding absolute error (|*original pixel value – rescaled output pixel value*|) maps in Fig 7.18. Perhaps the most interesting details presented in these figures were the dark outlines in the background of the image, indicating robust reconstruction of edge features. Edge information is carried by high spatial frequency components, and fits well under the approximations of the previous derivations. On the other hand, Figs. 7.18(a) and 7.18(b) revealed the limitations of the planar reference wave assumption for larger PCF apertures.

Figure 7.19(a) plots the MSE behaviour as a function of η for the four images presented. In all cases, the lowest MSE score was achieved by the PCF size closest to $\eta = 0.62$. This was 9×9 pixels with $\eta = 0.5625$. Only odd-sized filters were used to preserve the symmetry of the system, leading to coarse steps between η-values. Finer steps are possible by either reducing the image size, or increasing the FFT array, e.g. to 4096×4096 pixels. However, the configuration used sufficiently demonstrated high-quality reconstruction of the images.

An interesting system trade-off between image quality and power throughput is suggested by the previous graph and the plot of image gain against PCF size in Fig. 7.19(b). Here, gain is defined as the ratio of integrated brightness of an image projected by GPC phase modulation, to the expected brightness of the same image projected by pure amplitude modulation, assuming the same incident light power. This value represented the light efficiency of the GPC method compared to a traditional light-valve type projection system.

Fig. 7.15 Simulated projections of images (a) 'bird' and (b) 'lettuce' by the GPC method using 9×9-pixel PCF sizes, i.e. $\eta = 0.5625$. Mean-square errors were calculated with respect to the corresponding original images, (c) and (d). Images are 256×256 pixels, with 8-bit grey-level depth.

Fig. 7.16 Simulated projections of images (a) 'flower' and (b) 'mallet' by the GPC method using 9×9-pixel PCF sizes, i.e. $\eta = 0.5625$. Mean-square errors were calculated with respect to the corresponding original images, (c) and (d). Images are 256×256 pixels, with 8-bit grey-level depth.

Fig. 7.17 Simulated image projections by the GPC method using various PCF sizes to illustrate trade-off between image quality (MSE) and light efficiency (gain). (a) $\eta = 2.5625$ (41×41 pixels); MSE = 2528, gain = 2.17; (b) $\eta = 0.9375$ (15×15 pixels), MSE = 2365, gain = 2.10; (c) $\eta = 0.5625$ (9×9 pixels), MSE = 495, gain = 1.75; and (d) $\eta = 0.1875$ (3×3 pixels), MSE = 3825, gain = 2.24.

(a) $\eta = 2.5625$ (b) $\eta = 0.9375$

(c) $\eta = 0.5625$ (d) $\eta = 2.5625$

Fig. 7.18 Absolute error maps of image projections by the GPC method at various PCF sizes. *absolute error* = |*original pixel value – rescaled output pixel value*|.

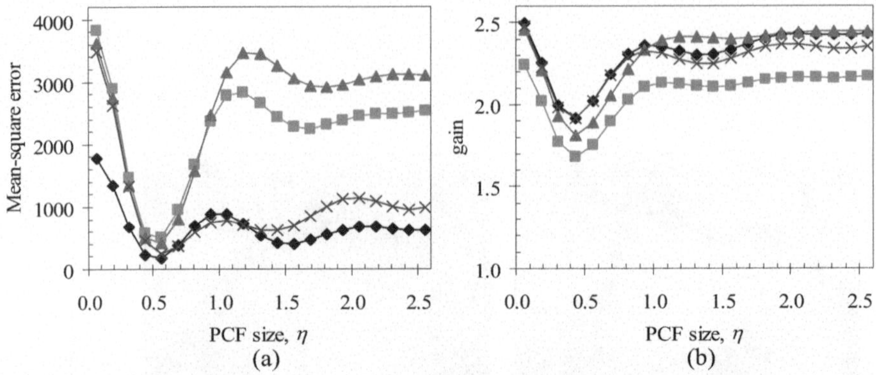

Fig. 7.19 (a) Mean-square error and (b) light efficiency of simulated GPC image projections for different PCF sizes (▲ 'flower', ■ 'mallet', × 'lettuce', ◆ 'bird'). Efficiency is measured as the ratio of the integrated brightness at the output to that expected from purely amplitude modulation. Curves are drawn as visual guides.

As seen in the plot, gain increased as PCF size was increased over the best image quality condition. Adjusting the PCF size was analogous to tuning the beam ratio in an interferometer to increase fringe visibility. Gain reached values as high as 2.44, corresponding to a diffraction efficiency of 98%. Although the MSE-score rapidly worsened for the larger PCF sizes, the image was still clearly recognizable, as in Figs. 7.17a and 7.17b. This may prove useful in some applications where light efficiency and power throughput might be considered primary over image fidelity, for example in fully parallel laser marking. Notably, very small PCF sizes also indicated relatively large gain values, but only because these represented minimal perturbation of the input field. Figures 7.17d and 7.18d show the poor image reconstruction with a PCF width of 3 pixels ($\eta = 0.1875$).

7.5 Reshaping Gaussian Laser Beams

Many laser applications require beams with uniform intensities within specified transverse distributions, thus fuelling research interest on techniques to homogenize and shape the Gaussian profile emitted by most lasers [71, 72]. One of the oldest tricks is to expand and truncate a Gaussian beam, which is still popular choice when the application favours simplicity over energy efficiency. There are other straightforward lossy approaches such as inhomogeneous absorptive filters that attenuate the central beam parts more than the peripheral portions to get a homogenized beam [73, 74].

A straightforward but energy-efficient approach would be useful over a wider spectrum of laser applications. Geometric optics can help design refractive or reflective optics systems that redirect portions of an incident Gaussian beam into a homogenized distribution [11, 75, 76]. Similar energy rerouting may be implemented using lenslet arrays that initially split an incident Gaussian beam into discrete beams that are later recombined into homogeneous distributions [77]. Geometric solutions map each point on an incident beam to target locations on the output plane to homogenize the energy distribution. These designed phase profiles may be phase-wrapped and implemented as diffractive elements when fabrication technologies allow for continuous surface relief. Diffractive optical design [78, 79, 80, 81] using physical optics can work with very limited phase levels using a global mapping designed by iterative optimization [54].

Refractive beam-shaping solutions [71, 11, 75] promise lossless conversion over a wide range of illumination wavelengths. However, commercially available refractive elements have a very limited set of intensity profiles. Microlens arrays, which divide an incident beam into beamlets that are later recombined in beam integrators, can produce beams with very poor homogeneity, especially under coherent illumination [77]. These integrators are likewise limited in terms of achievable intensity patterns. Diffractive optical approaches [71, 78, 79, 80, 81] offer capacity for producing a variety of beam shapes, with some compromise on phase homogeneity, and are rich in design algorithms that promise theoretical conversion efficiencies in the upper 90% range. However, fabrication errors can degrade the efficiency and uniformity of the generated patterns [82]. Since phase errors can easily give rise to a spurious zero-order beam, diffractive designs commonly avoid the optical axis, which is the more attractive reconstruction region in terms of optimizing efficiency and minimizing aberrations.

The GPC method can achieve high efficiency using a very straightforward design of the needed optical element. The input simply requires an easy-to-fabricate binary phase mask (0 and π) that is patterned after the desired intensity distribution. Thus, static applications of GPC-based beam-shaping are less susceptible to fabrication errors and can as well provide excellent output phase homogeneity unlike that of diffractive approaches. Compared to diffractive optical elements, the GPC phase mask generally contains fewer locations with phase jumps and, hence, suffers less from scattering losses. Additionally, the GPC intensity projections can be centred on the optical axis to minimize aberration effects.

In the following, we will illustrate how GPC can be utilized to efficiently convert an incident Gaussian beam into different beam patterns. This applies the GPC formulation for Gaussian illumination described in the previous chapter as a correction scheme for coping with SRW inhomogeneity. In this case, GPC operates like a phase-only aperture that channels energy from intended dark regions in a Gaussian beam into designated transverse intensity distributions. Artefacts akin to the incident Gaussian rolloff can remain in the illuminated regions of the output. We will thus follow up with an input phase compensation scheme to homogenize the output. This involves a pointwise phase correction to dim the central brighter region, which redirects energy towards the dimmer edges to homogenize the output while avoiding significant energy loss.

The GPC capacity for generating exotic shapes at rapid reconfiguration rates is particularly attractive, since the current literature is focused on static and simple patterns, owing to practical constraints in the other methods. The simplicity of designing binary phase inputs in the GPC-based approach lends itself to dynamic pattern reconfiguration that is limited only by the frame-rate of the encoding device (e.g. this can reach up to kilohertz in ferroelectric liquid crystals). This high refresh rate is achieved without compromising issues associated with speckle, spurious higher orders and zero-order effects that are expected in computer generated phase holograms [83, 84], especially when the number of iterations is compromised for faster computation.

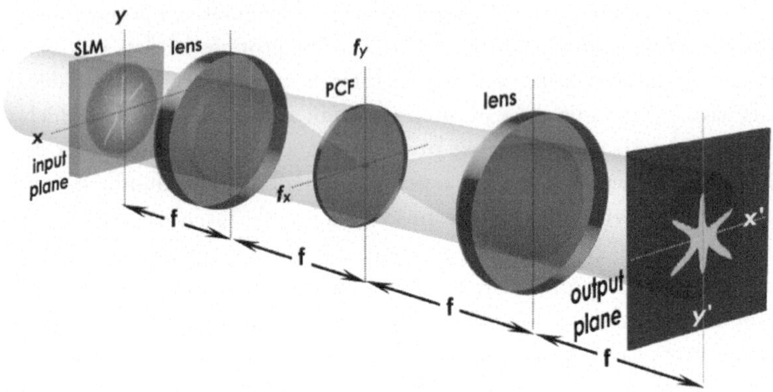

Fig. 7.20 Optical setup for GPC-based patterning of Gaussian beams.

7.5.1 *Patterning Gaussian Beams with GPC as Phase-Only Aperture*

In the previous chapter, we saw that under Gaussian illumination with a beam waist, w_0,

$$a(x,y) = a_r(r) = \exp\left[-r^2/w_0^2\right], \tag{7.7}$$

the PCF size may be tweaked to such that the SRW profile approaches that of the Gaussian illumination. For optimized parameters that yield perfectly matched signal and SRW profiles, the output intensity is

$$I(x',y') \approx \exp\left(-2r'^2/w_0^2\right)\left|\exp\left[i\phi(x',y')\right] + \bar{\alpha}\left[\exp(i\theta)-1\right]\right|^2. \tag{7.8}$$

Equation (7.8) prescribes a method for spatially modulating the output intensity by modulating the input phase to exploit interference effects. Matching signal and SRW profiles can generate a dark outer region by manipulating the signal phase to get destructive interference. This channels energy into the central region, which is an attractive working area because of its relatively flat amplitude profile.

Using a π-phase shifting PCF, a GPC-based phase-only aperture can be implemented by encoding the signal beam with a π-phase at the intended bright regions and encoding a zero-phase where darkness is desired. This phase-only aperture promises a higher efficiency since the energy from the truncated portions of the Gaussian beam can be diverted into the transmitted region and will not be lost. This is confirmed in Fig. 7.21(a), which shows the efficiency of a GPC-based phase-only aperture relative to a simple truncation for a Gaussian input beam. The 86% efficiency obtained for an aperture radius of $0.36w_0$ is 2.33 times better than the throughput of an equally-sized hard aperture. While the efficiencies of GPC apertures decrease with size, this actually represents an increasing gain when referenced to the energy throughput of truncating apertures of corresponding sizes. The energy throughput is only $1-\exp(-2A^2/w_0^2)$ when using a circular aperture of radius A to truncate a Gaussian beam having a $1/e^2$ radius of w_0. This is equal to the relative centre-to-edge intensity difference of the transmitted beam.

Residual light that is not diverted into the main spot can be easily blocked by an exit aperture in applications that cannot tolerate stray light. Additionally, some improvement in relative flatness is gained using a GPC aperture. For example, the centre-to-edge intensity difference is 23% for aperture truncation but only 17% for GPC-based truncation for the output illustrated in Fig. 7.21(b). Also, a flat phase profile is maintained throughout the illuminated region.

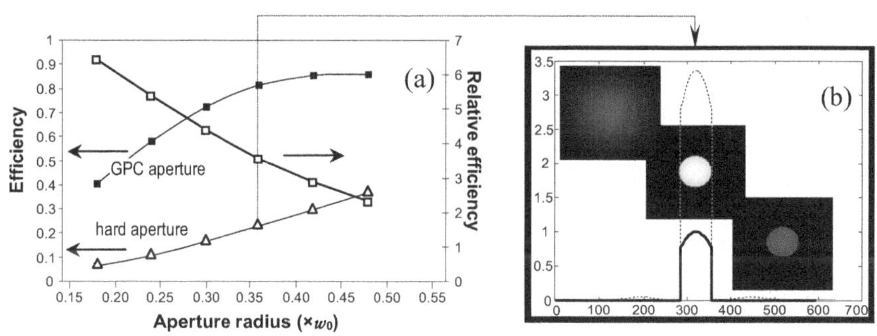

Fig. 7.21 (a) Effect of aperture size on efficiency for Gaussian beam truncation. GPC aperture (–■–); hard aperture (–△–); GPC gain (–□–). (b) Outputs for an aperture with radius = $0.36w_0$. dashed: line-scan across the GPC output; solid: line-scan across the hard aperture output. Insets: Gaussian input (left), GPC output (center), and aperture output (right).

More sophisticated aperture functions can be implemented using programmable phase-only spatial light modulators. The outputs of several GPC-based phase-only apertures are illustrated in Fig. 7.22 with their corresponding efficiencies. The efficiencies when using amplitude masks to accomplish the task are also presented for comparison. As in the case of circular apertures, GPC-based phase-only apertures show superior energy efficiency over their amplitude-based counterparts. In addition, a flat phase profile is maintained throughout the illuminated region.

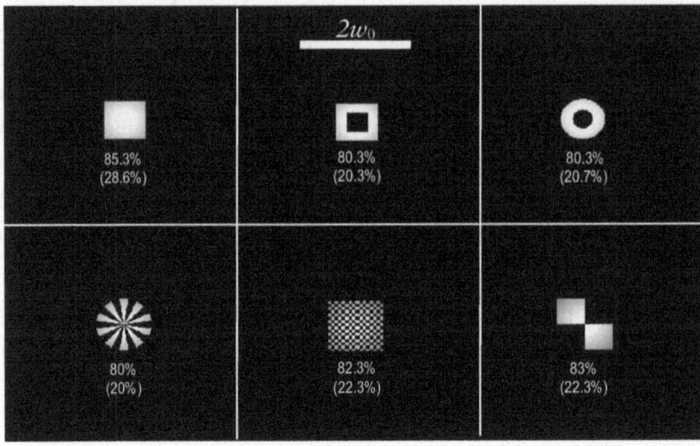

Fig. 7.22 Output of numerical experiments implementing various GPC-based phase-only apertures. The efficiencies of the patterns are shown below each pattern, followed by the efficiencies of corresponding amplitude masks (in parenthesis). The scale bar indicates the $1/e^2$ width of the Gaussian beam relative to the patterns.

These results illustrate the convenience of using GPC to generate arbitrary lateral beam patterns from an incident Gaussian beam. It is straightforward to design phase masks, which are likewise simpler to fabricate. The close match between the signal and SRW profiles enables efficient diversion of energy from designated dark regions into desired intensity distributions. Various aperture functions can be implemented with high energy efficiency using a simple binary phase mask that is patterned to mimic the desired intensity distribution. The sharp bright-to-dark transition in patterns obtained using GPC-designed apertures can be highly suitable in applications where a soft intensity gradient can produce undesirable effects. In the next section, we will discuss how to improve the output homogeneity.

7.5.2 Homogenizing the Output Intensity

Equation (7.8) shows the possibility to eliminate the Gaussian roll-off at the output by producing a reciprocal profile from the interference term:

$$\left|\exp\left[i\phi(x',y')\right]+\bar{\alpha}\left[\exp(i\theta)-1\right]\right|^2 = I_0\exp\left(2r'^2/w_0^2\right)A(x',y') \qquad (7.9)$$

where $A(x',y')$ is the desired intensity profile with uniform intensity I_0 that is determined according to energy conservation constraints. The correction principle is graphically illustrated in Fig. 7.23(a), which shows the SRW phasors, $-a_1$ and $-a_2$, and signal phasors, $a_1e^{i\phi_1}$ and $a_2\,e^{i\phi_2}$, at two points in the output plane with mismatched amplitudes, $a(x_1,y_1) = a_1$ and $a(x_2,y_2) = a_2$. To achieve interference with homogenized intensity, we adopt a corrective phase encoding scheme where points having smaller amplitudes are encoded closer to π and points with larger amplitudes are conversely encoded.

The dark background criterion,

$$\bar{\alpha} = \bar{\alpha}_{real} + i\bar{\alpha}_{imag} = \frac{1}{2} + \frac{i}{2}\cot(\theta/2), \tag{7.10}$$

specifies a practical range for $\bar{\alpha}$ as illustrated in the phasor diagram of Fig. 7.23(b) for the upper semi-circle (the lower semi-circle offers another set of symmetric solutions). The dotted lines indicate the requirement set by Eq. (6.37) that the real part of $\bar{\alpha}$ be ½ and the maximal magnitude of the imaginary part is $\sqrt{3}/2$. Furthermore, Eq. (7.10) allows us to determine the required phase shift, θ, from the imaginary part of $\bar{\alpha}$:

$$\theta = 2\cot^{-1}\left(2\bar{\alpha}_{imag}\right) = 2\cot^{-1}\left[2\int a(x,y)\sin[\phi(x,y)]\mathrm{d}x\mathrm{d}y \Big/ \int a(x,y)\mathrm{d}x\mathrm{d}y\right]. \tag{7.11}$$

Choosing the filter phase shift, θ, ensures that the SRW is π-shifted with respect to the projected image of the zero-phase encoded input, allowing us to designate dark regions and ensure that minimal energy is lost to these regions.

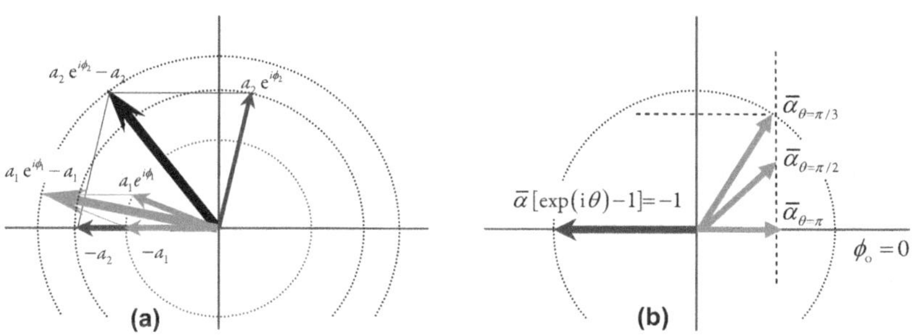

Fig. 7.23 (a) Phasor illustration of how superpositions having matching amplitudes is achieved by encoding corrected phases, ϕ_1 and ϕ_2, to compensate for the amplitude mismatch between a_1 and a_2. (b) Phasors of the normalized zero-order, $\bar{\alpha}$, for PCF phase shifts, θ= π/3, π/2, and π. Vertical dashed line indicates $\bar{\alpha}_{real}$ = ½ criterion; horizontal dashed line shows the corresponding maximum value of $\bar{\alpha}_{imaginary} = \sqrt{3}/2$. Using matched $\bar{\alpha}$ and θ achieves the dark background condition: $\bar{\alpha}[\exp(i\theta)-1]=-1$.

Applying the condition in Eq. (7.10) into Eq. (7.8) leads to the simplified relation

$$\left|\exp\left[i\phi(x',y')\right]-1\right|^2 = I_0\exp\left(2r'^2/w_0^2\right)A(x',y'). \tag{7.12}$$

Substituting the trigonometric identity $\left|\exp(i\phi)-1\right|^2 = 2-2\cos(\phi)$ into Eq. (7.12) yields

$$\cos\left[\phi(r')\right] = 1-\frac{I_0}{2}\exp\left(2r'^2/w_0^2\right)A(x',y'). \tag{7.13}$$

Solving this equation yields the phase input that yields a homogenized output.

Encoding spatial phase information on an incident Gaussian beam perturbs the spatial frequency components close to the zero-order beam at the Fourier plane, except with phase information confined to high frequencies as in optical lattices. This sets an upper limit on practical PCF sizes. The Fourier relation between the PCF and output planes means that a smaller PCF broadens the SRW. When constrained to use a smaller PCF, we rewrite the GPC output as

$$I(x',y') \approx \left|\exp\left(-r'^2/w_0^2\right)\exp\left[i\phi(x',y')\right]+g(x',y')\bar{\alpha}\left[\exp(i\theta)-1\right]\right|^2 \quad (7.14)$$

to account for the mismatched profiles.

Mismatched amplitude profiles can no longer guarantee complete darkness will always be achieved at any arbitrarily chosen output point. However, interference can create high contrast even with amplitude mismatch. For example, the interference of two beams with 10% amplitude mismatch yields a minimum intensity that is less than 0.28% of the maximum intensity. The conditions are even more favourable in beam shaping tasks such as Gaussian-to-circular flattop conversion that require darkness only in certain peripheral regions – these can be achieved with minor losses since these peripheral regions have minimal intensities to begin with.

Choosing, for convenience, $\phi=0$ as the phase input for minimum output intensity sets the SRW phase to π, which requires the condition

$$\bar{\alpha}\left[\exp(i\theta)-1\right]=-k, \quad (7.15)$$

where the constant k is not necessarily equal to unity. We can again rewrite the relationship between the normalized zero-order and the PCF phase shift as

$$\bar{\alpha}=\bar{\alpha}_R+i\bar{\alpha}_I=\frac{k}{2}+i\frac{k}{2}\cot(\theta/2). \quad (7.16)$$

For this we can see that $k = 2\alpha_R$ and the matching phase shift is $\theta = 2 \operatorname{arccot} (\alpha_I / \alpha_R)$.

Homogenizing the output for mismatched illumination and SRW profiles involves similar corrections to the encoded phase to compensate for the spatially varying amplitude. Substituting Eq. (7.15) into Eq. (7.14) and expanding the result yields

$$I(x',y') \approx a_0^2(x',y') + k^2 g^2(x',y') - 2ka_0(x',y')\, g(x',y')\cos(\phi), \quad (7.17)$$

where $a_0(x',y')$ describes the Gaussian input profile. The phase input is then obtained using the image-plane phase

$$\phi(x',y') = \arccos\left[\frac{-I(x',y')+a_0^2(x',y')+k^2 g^2(x',y')}{2ka_0(x',y')g(x',y')}\right] \quad (7.18)$$

Solving for the input phase in Eq. (7.18) is nontrivial since it requires knowledge of the profile $g(x',y')$, and constant k, which both depend on the input phase. Moreover, the target homogeneous level for $I(x',y')$ is likewise determined based on the constraints from the input and SRW profiles. However, we may treat the phase correction as perturbations to the binary phase inputs that generate inhomogeneous outputs. This allows us to use the k and $g(x',y')$ obtained using binary inputs as suitable approximations. We can then use the phase inputs obtained to iteratively correct the SRW parameters and improve the phase input. We illustrate these design principles in the next section where we consider GPC-based projection of a circular flattop from a Gaussian input.

7.5.3 Gaussian-to-Flattop Conversion

To examine the expected performance of beam-shaping systems based on the design principles outlined above, we performed numerical experiments using a Fourier optics-based model of the GPC optical system. The results for Gaussian-to-circular flattop beam conversion are illustrated in Fig. 7.24. For comparison, we present in Fig. 7.24(d) the generated output when the binary (0 and π) phase input, shown in Fig. 7.24(a), is used with a π-phase shifting PCF. The PCF size, chosen to optimize the output efficiency while minimizing output distortions, is 1.1 times the conjugate beam waist parameter of the zero-order beam in the Fourier plane.

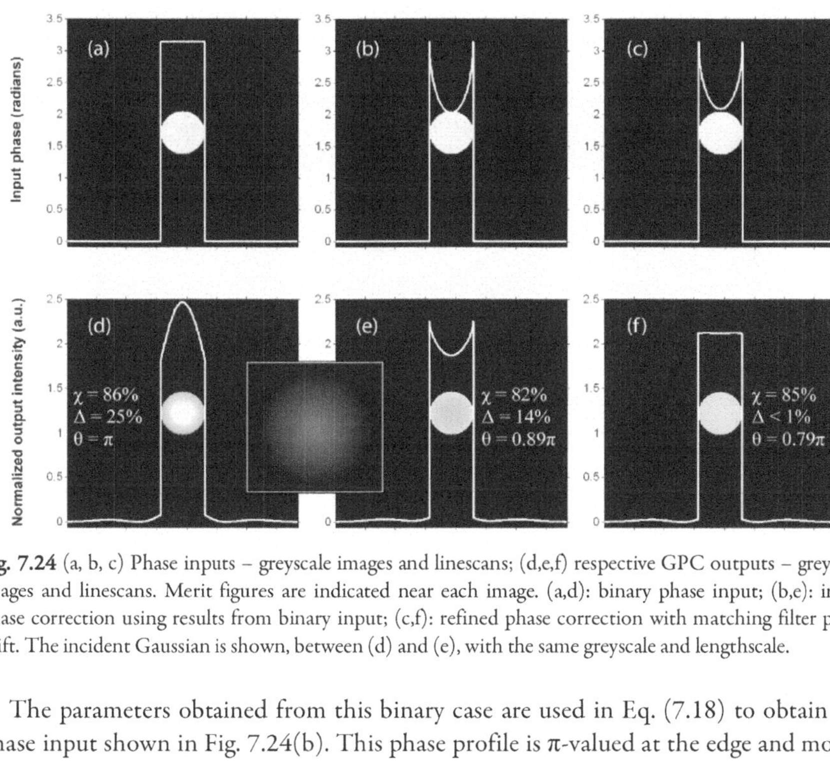

Fig. 7.24 (a, b, c) Phase inputs – greyscale images and linescans; (d,e,f) respective GPC outputs – greyscale images and linescans. Merit figures are indicated near each image. (a,d): binary phase input; (b,e): initial phase correction using results from binary input; (c,f): refined phase correction with matching filter phase shift. The incident Gaussian is shown, between (d) and (e), with the same greyscale and lengthscale.

The parameters obtained from this binary case are used in Eq. (7.18) to obtain the phase input shown in Fig. 7.24(b). This phase profile is π-valued at the edge and monotonically decreases towards the centre. As a result, the image of the phase-encoded input is in phase with the SRW at the edges and gradually becomes out of phase towards the centre. The output pattern that results from the interference is shown in Fig. 7.24(e). Comparing with the initial output, the new output illustrates that we can introduce corrections to the input phase to suppress the intensity in the central region while enhancing the intensity in the outer region to improve the homogeneity. In effect, it works similar to the inhomogeneous absorptive filters [73, 74] that attenuate the central intensities but with the major difference that it redistributes the energy towards the outer regions instead of wasting it.

While it improves the homogeneity, using the parameters obtained from the binary case in Eq. (7.18) yields a phase input that overcompensates for the inhomogeneity. This reduces the efficiency due to the over-attenuated central region. By refining the input profile (see Fig. 7.24(c)) and using a matched PCF phase shift to set the correct SRW phase, we are able to obtain the homogenized output with improved efficiency as illustrated in Fig. 7.24(f). Prior to correction the output intensity profile monotonically rolls off away from the centre and attenuates by as much as 25% at the edge. The corrected output exhibits a flattop profile whose maximum peak-to-peak fluctuation, Δ, is only 0.4% of the peak intensity within the target region and with minimal intensity loss compared to the initial inhomogeneous output. Applying similar compensation schemes on the input phase makes it possible to generate other profiles with homogenized intensity from an incident Gaussian beam. Some examples obtained from numerical experiments are illustrated in Fig. 7.25.

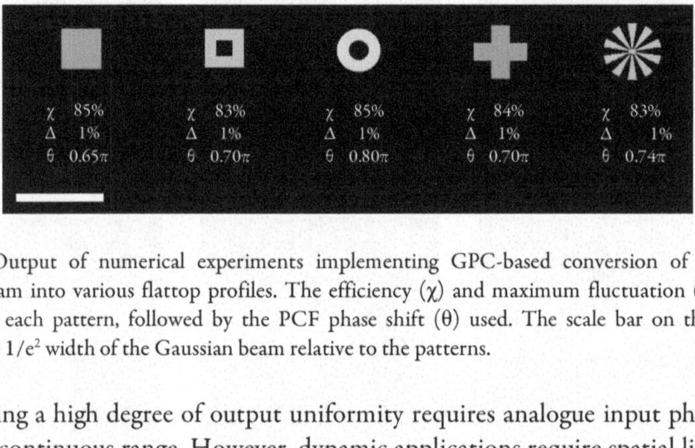

Fig. 7.25 Output of numerical experiments implementing GPC-based conversion of an incident Gaussian beam into various flattop profiles. The efficiency (χ) and maximum fluctuation (Δ) are indicated below each pattern, followed by the PCF phase shift (θ) used. The scale bar on the lower left indicates the $1/e^2$ width of the Gaussian beam relative to the patterns.

Achieving a high degree of output uniformity requires analogue input phase encoding over a continuous range. However, dynamic applications require spatial light modulation (SLM) devices that usually quantize the encoded phase into discrete levels. To assess the impact of phase quantization on the output uniformity, we implemented various phase quantization levels in the numerical experiments. Figure 7.26 shows the maximum fluctuation, Δ, for various phase quantization levels when generating a circular pattern from a Gaussian incident beam. The inset shows the projected pattern when encoding the phase with 16 quantization levels, where the biggest fluctuation is 10% of the peak intensity. The fluctuations decrease monotonically with an increasing number of phase quantization levels and is reduced to less than 3% when encoded with only 64 phase quantization levels. Thus, the required phase quantization requirements for generating outputs with acceptable intensity fluctuations can be sufficiently met by commercially available devices. However one must exercise caution with regard to the dead spaces between SLM pixels as these can map to the output pattern and degrade uniformity. This can be avoided by using so-called "non-pixellated SLMs" such as optically addressed devices. If using pixellated devices, one can block the tiled higher-order replicas at the PCF plane to improve the output uniformity by avoiding pixilation at the output with some efficiency tradeoff.

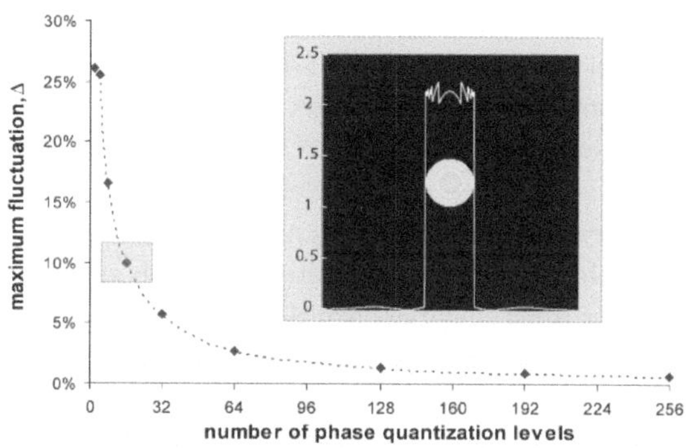

Fig. 7.26 Plot of the maximum fluctuation, Δ, for different phase quantization levels. The inset shows the projected pattern and intensity linescan for 16 quantization levels.

The $4f$ optical processing setup illustrated in Fig. 7.20 can be modified to suit the requirements of particular applications. For example, GPC has been implemented using planar integrated micro-optics for a compact optical decryption device [85]. It is also possible to reduce the number of optical components if required by the application. One can incorporate the lens phase into the phase input to eliminate the first Fourier lens and, similarly, combine the phase-only PCF and the second Fourier lens into a single phase element. This reduces the GPC system into just two optical elements. Other combinations with diffractive implementations can also be considered. However, one has to cautiously examine whether the advantages of having fewer components outweigh the side effects. When merging the Fourier lens with the input phase, one must keep in mind that SLM-based diffractive lenses can generate spurious secondary lenses [86]. Incorporating the lens phase into the PCF must be weighed against the simplicity of fabricating the standard and simple binary-phase PCF. A binary diffractive lens has a much lower efficiency while a multilevel implementation must contend with fabrication issues [82]. Furthermore, GPC achieves high reconfiguration rates by exploiting the degrees of freedom afforded by working with two conjugate planes. Consequently, the input phase simply mimics the spatial features of the desired output patterns and do not require the computationally expensive algorithms used in single plane designs such as computer generated holograms.

These results show that the generalized phase contrast method (GPC) serves as a useful tool for shaping a Gaussian beam into flattop beams having arbitrary lateral profiles. GPC efficiently diverts energy from designated dark regions into desired intensity distributions where they can be homogenized by intuitive corrections on the input phase. This helps expanding the repertoire of techniques for getting more out of the fundamental Gaussian laser output through beam homogenization.

7.6 Achromatic Spatial Light Shaping and Image Projection

The analysis presented in Sect. 6.8 shows that GPC-based patterned projection exhibits robustness to shift in wavelength due to the reciprocal effects from input and Fourier planes that tend to compensate for each other. This opens the possibility for achromatic light shaping using GPC. Patterned multi-wavelength illumination has obvious utility in display technology. Beyond display applications, the creative integration of patterned illumination with the solutions offered by multi-wavelength approaches can yield tools with potentially exciting functionalities. Multi-wavelength techniques, on the other hand, are invaluable in spectroscopy [87] and provide beneficial enhancements in interferometry [88]. The significance of the well-established fields of spectroscopy and interferometry cannot be overemphasized. Exploiting the wavelength-dependent material response also allows for controlled photo-excitation, targeted monitoring [89], and even simultaneous excitation and monitoring using multi-wavelength techniques in pump-probe geometries [90].

Let us first consider GPC-based pattern generation for array illumination. For illustration, we study a GPC system that projects an 11×11 periodic array of circular spots when uniformly-illuminated at wavelength λ_0. Binary phase inputs that encode π-phase at the desired locations of the bright spots are designed for this wavelength. The PCF is likewise chosen to π-shift the phase for the same wavelength. It was previously shown in ref. [50] that one can use PCFs with bigger phase shifting areas when projecting lattice-like patterns using GPC. For the simulations we use a square PCF with side length that is 4 times larger than the diameter of the central zero-order spot. We perform numerical experiments to investigate how the efficiency is affected when the illumination wavelength, λ, is scanned over the range $\lambda_0/2\leq\lambda\leq2\lambda_0$. We also execute numerical experiments for quasi-periodic pattern projection where the spots are randomly displaced from their regular lattice sites and the wavelength is scanned over the same range.

The efficiencies obtained for array illumination are shown in Fig. 7.27. The efficiency variation with wavelength obtained from numerical simulations follows a similar trend to that of the predicted efficiency when $a(x',y')$ and $g(x',y')$ are perfectly matched. However, the numerically obtained efficiencies saturate at a slightly lower value of ~92%. This occurs due to ripples in $g(x',y')$ that mismatches it slightly with $a(x',y')$ (see Fig. 7.27 inset showing $a(x',y')$ and $g(x',y')$ for the periodic array at λ_0). The mismatch elevates the efficiency at other wavelengths. As a positive side effect, a fairly uniform high efficiency is maintained over a wide wavelength range. The image insets show the array illumination generated at different wavelengths (assuming that $\lambda_0 = 550$ nm, these would correspond to the primary colours red, green and blue). These images illustrate that the characteristic length scales of the projections are maintained at the different illumination wavelengths. This is in contrast with diffractive projection where the length scales vary depending on the illumination wavelength and a prominent zero-order can appear.

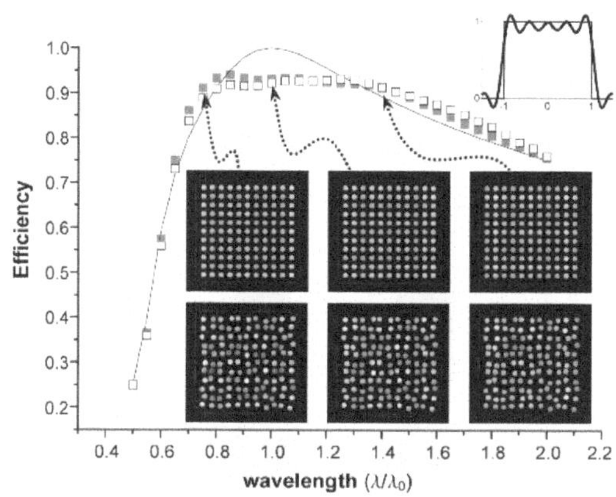

Fig. 7.27 Efficiency of GPC array illuminator as the wavelength λ is tuned away from the design wavelength λ_0 for periodic array (■) and aperiodic array (□). Solid line shows the efficiency for matched illumination and SRW profiles. Inset pictures: outputs at different $0.8\lambda_0$, $1.0\lambda_0$, and $1.2\lambda_0$ (where λ_0 is the original design wavelength). Inset plot: illumination profile (thin) and SRW profile (thick) at λ_0.

We next consider the effect of the illumination wavelength in GPC-based generation of transverse shapes from a Gaussian beam. A GPC-based design requires binary phase elements that are highly compatible with dynamic applications and are simpler to fabricate for static applications. A beam shaper with a wide operating wavelength range is desirable not only for recycling the same phase element at different wavelengths, but for possible concurrent multi-wavelength operation. Phase inputs corresponding to different output transverse profiles are designed for $\lambda_0 = 550$ nm. When no phase modulation is introduced at the input, an incident Gaussian input generates a Gaussian beam at the PCF plane with a characteristic $1/e^2$ beam waist, w_f. In the simulations, we choose the PCF parameters to provide π-phase shift within an axially centred circular region whose radius that is $1.1w_f$. Conversion efficiencies are calculated for different illumination wavelengths from 400 nm to 700 nm.

The results of the numerical experiments for shaping a Gaussian beam are presented in Fig. 7.28. The wavelength variation of the efficiency is fairly similar for the different target shapes. The GPC-based projections exhibit consistently high efficiency and projection quality for the different patterns over a wavelength range that can span the entire visible spectrum. For instance, GPC-based beam shaping can maintain efficiencies to within 10% of the peak efficiency from 400 nm to 675 nm. While we use λ_0 = 550 nm to cover the familiar visible spectrum, the results generally apply to other choices of λ_0 and over wavelengths ranging from 0.73 λ_0 to 1.27 λ_0. The inset images of the projected shapes again illustrate that the characteristic length scales are maintained at different illumination wavelengths.

Finally, we consider the effects of wavelength on GPC-based grey image projection [52]. We histogram equalize two standard images to get a fairly uniform spread of grey levels. These grey images then serve as basis for phase inputs that have equalized phase distributions from 0 to π at λ_0= 550 nm. The normalized zero-order for histogram-

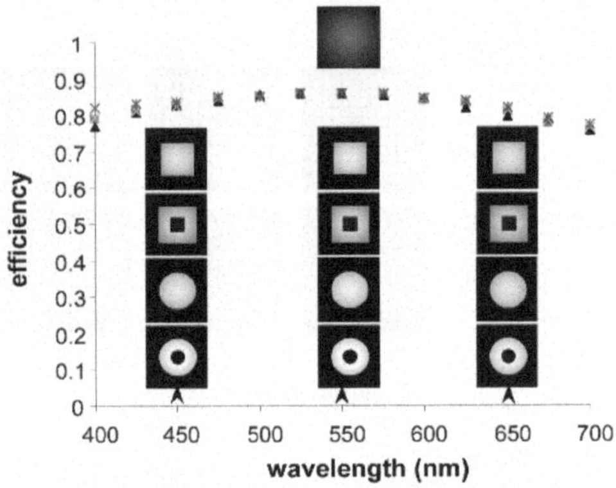

Fig. 7.28 Wavelength dependence of the efficiency when generating shapes using GPC illuminated with a Gaussian beam. Insets show the incident Gaussian illumination and the generated patterns at λ = 450 nm, 550 nm, and 650 nm drawn with the same length- and greyscale.

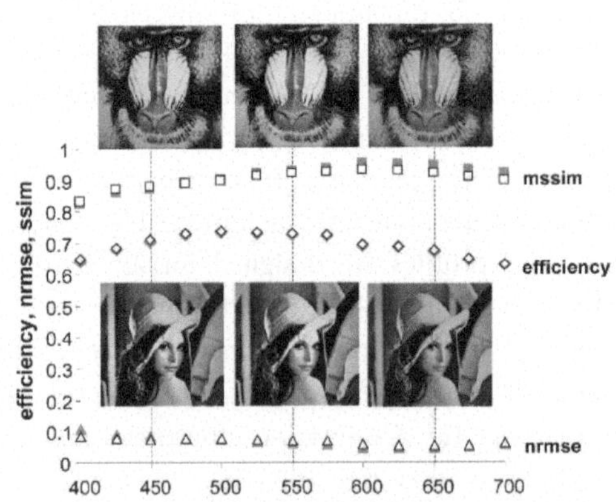

Fig. 7.29 Wavelength dependence of the efficiency (◇), normalized root mean square error (*nrmse,* △) and structural similarity (*mssim,* □) of greyscale images projected using GPC (■ ◆▲ – Lena; □ ◇△ – Mandrill). Insets show the generated images at λ = 450 nm, 550 nm, and 650 nm.

equalized phase input is purely imaginary, $\overline{\alpha} = 2i/\pi$, prompting the use of a PCF with θ = 2 rad based on the optimum condition in Eq. (6.37). To mimic the experimental demonstration in ref. [52], the numerical simulations use a circular input aperture that includes a phase deadspace frame with zero-phase around the active SLM square region where the image is phase encoded. With its zero phase, the surrounding deadspace frame ideally projects as a dark frame in the image plane. The size of the zero-phase frame is chosen to set the real part of $\overline{\alpha}$ to its optimal value, according to Eq. (6.37). The PCF size used is 0.63 times that of the Airy disc produced by an unmodulated circular aperture illuminated at λ_0.

The image projection efficiencies for wavelengths ranging from 400 nm to 700 nm are shown in Fig. 7.29. The insets in Fig. 7.29 show the projected images at λ = 450 nm, 550 nm, and 650 nm (i.e., wavelengths that produce the colours blue, green and red, respectively). Some residual light is projected in the area surrounding the image (not shown). As a measure of efficiency, we determined what fraction of the total energy in the output plane is projected to greyscale image region. To gauge the projection quality, we compared the images generated at different wavelengths with the expected outputs in terms of mean structural similarity, *mssim* [91], and normalized root mean square error, *nrmse*. These additional figures of merit are also plotted in Fig. 7.29. Overall, the results show that the grey image projections maintain decent efficiencies, structural similarity, and minimal errors over a range of wavelengths.

These results illustrate the robustness of GPC-based pattern projection to wavelength change. The GPC-based projections maintain high quality and efficiency over a doubling of wavelength in Gaussian beam shaping and greyscale image projections and for an even wider wavelength range for array illumination. Viewed against the existing GPC applications, the capacity to encompass multiple wavelengths opens new avenues for even more exciting applications. For example, using multi-spectral light sources such as supercontinuum lasers in dynamic optical trapping will efficiently incorporate spectroscopic capabilities into these platforms. This can enhance single-particle trapping-and-spectroscopy systems [92] and extend into massively parallel and dynamic systems that resolve single-particle effects, rather than yielding ensemble averages. Using a GPC system, one can generate strong trap beams at a non-invasive wavelength and introduce other wavelengths at the trap sites with tuned intensities for initiating light-induced processes, such as photoexcitation, that activate relevant mechanisms including fluorescence or other photochemical or photobiological effects. Furthermore, other relevant wavelengths can be simultaneously introduced for monitoring. Patterned illumination allows the response to be selectively triggered only at the intended regions, thereby boosting photon efficiency while reducing noise and other spurious side effects. When properly exploited, this functionality may yield new insights in various areas of physical, chemical, biological, and medical research. Integrating multifunctional wavelengths into other areas of patterned illumination, such as microscopy, can also provide promising enhancements. The consistent GPC performance for grey-scale images at red, green and blue wavelengths also makes it an attractive candidate for light-efficient colour image projection.

7.7 Summary and Links

In this chapter we have continued the application of GPC beyond the imaging of unknown phase distributions, which we started in Chapter 6. We discussed various light projection experiments and applications that utilized and validated the formulation developed in the previous chapter. We showed efficient experimental projections of binary intensity distributions that were conveniently implemented using ternary and binary phase-encoded inputs. We also presented dynamically reconfigurable optical lattices as well as optical obstacle arrays that can be utilized for microscopic particle manipulation. Laser projections are typically associated with image display, an application that we also covered by presenting experimental GPC-based greyscale image projection, inspired by previous results promising high reconfiguration rates. Moreover, we elaborated on the wavelength dependence considered in the previous chapter and demonstrated the impressive achromaticity of the GPC performance in various light shaping tasks. Beyond laser light shows and displays, programmable light distributions are promising tools in spatially modulated light-matter interactions. Particular applications of GPC-based miniaturized dynamic laser projections in optical trapping and manipulation of microscopic objects are discussed in Chapter 8.

References

1. F. Chen, G. M. Brown, and M. Song, "Overview of three-dimensional shape measurement using optical methods," Opt. Eng. 39, 10–22 (2000).
2. M. A. A. Neil, R. Juskaitis, and T. Wilson, "Method of obtaining optical sectioning by using structured light in a conventional microscope," Opt. Lett. 22, 1905–1907 (1997).
3. M. G. L. Gustafsson, "Nonlinear structured-illumination microscopy: Wide-field fluorescence imaging with theoretically unlimited resolution," P. Natl. Acad. Sci. USA 102, 13081–13086 (2005).
4. J. Glückstad, "Adaptive array illumination and structured light generated by spatial zero-order self-phase modulation in a Kerr medium," Opt. Commun. 120, 194–203 (1995).
5. P. J. Rodrigo, V.R. Daria, and J. Glückstad, "Real-time three-dimensional optical micromanipulation of multiple particles and living cells," Opt. Lett. 29 2270–2272 (2004).
6. V.R. Daria, R.L. Eriksen and J. Glückstad, "Dynamic optical manipulation of colloidal structures using a spatial light modulator," J. Mod. Opt. **50**, 1601-1614 (2003).

7. Rodrigo P J, Gammelgaard L, Bøggild P, Perch-Nielsen I R and Glückstad J 2005 Actuation of microfabricated tools using multiple GPC-based counterpropagating-beam traps Opt. Express 13 6899-904
8. S. Singh-Gasson, R. D. Green, Y. J. Yue, C. Nelson, F. Blattner, M. R. Sussman, and F. Cerrina, "Maskless fabrication of light-directed oligonucleotide microarrays using a digital micromirror array," Nat. Biotechnol. 17, 974–978 (1999).
9. S. E. Chung, W. Park, H. Park, K. Yu, N. Park, and S. Kwon, "Optofluidic maskless lithography system for real-time synthesis of photopolymerized microstructures in microfluidic channels," Appl. Phys. Lett. 91, 041106 (2007).
10. S. Shoji, H. B. Sun, and S. Kawata, "Photofabrication of wood-pile three-dimensional photonic crystals using four-beam laser interference," Appl. Phys. Lett. 83, 608–610 (2003).
11. B. R. Frieden, "Lossless conversion of a plane laser wave to a plane wave of uniform irradiance," Appl. Opt. 4, 1400–1403 (1965).
12. X. M. Deng, X. C. Liang, Z. Z. Chen, W. Y. Yu, and R. Y. Ma, "Uniform illumination of large targets using a lens array," Appl. Opt. 25, 377–381 (1986).
13. A. W. Lohmann and J. A. Thomas, "Making an array illuminator based on the Talbot effect", Appl. Opt. **29** (29), (1990) 4337-4340.
14. J. R. Leger and G. J. Swanson, "Efficient array illuminator using binary-optics phase plates at fractional-Talbot planes," Opt. Lett. **15**, 288-230 (1990).
15. W. Klaus, Y. Arimoto, K. Kodate, "High-performance Talbot array generators", Appl. Opt. **37** (20), (1998) 43574365.
16. F. Wyrowski, "Diffractive optical elements – iterative calculation of quantized, blazed phase structures," J. Opt. Soc. Am. A 7, 961–969 (1990).
17. J. Glückstad, "Phase contrast image synthesis," Opt. Commun. **130**, 225-230 (1996).
18. J. Glückstad, L. Lading, H. Toyoda and T. Hara, "Lossless light projection", Optics Letters 22 (1997) 1373-1375.
19. S. Sinzinger and J. Jahns, *Microoptics*, (Wiley-VCH Verlag, 1999)
20. D. Mendlovic, Z. Zalevsky, G. Shabtay and E. Marom, "High-efficiency arbitrary array generator", Appl. Opt. **35** (35), (1996) 6875-6880.
21. S. J. Walker, and J. Jahns, "Array generation with multilevel phase gratings", J. Opt. Soc. Am. A. **7** (8), (1990) 1509-1513.
22. J-N. Gillet and Y. Sheng, "Irregular spot array generator with trapezoidal apertures of varying heights", Opt. Commun. **166** (1999) 1-7.
23. V. Arrizon, E. Carreon and M. Testorf, "Implementation of Fourier array illuminators using pixelated SLM: efficiency limitations", Opt. Commun. **160** (1999) 207-213.
24. C. Zhou and L.Liu, "Zernike array illuminator", Optik **102**, 2, (1996) 75-78.
25. P. Xi, C. Zhou, S. Zhao and L. Liu, "Phase-contrast hexagonal array illuminator", Opt. Commun. **192** (2001) 193-197.
26. A. W. Lohmann, J. Schwider, N. Streibl and J. Thomas, "Array illuminator based on phase contrast", Appl. Opt. **27**, 14 (1988) 2915-2921.

27. J. Glückstad and P. C. Mogensen "Reconfigurable ternary-phase array illuminator based on the generalised phase contrast method", Opt. Commun. **173** (2000) 169-175.
28. J. Liesener, M. Reicherter, T. Haist and H. J. Tiziani, "Multi-functional optical tweezers using computer-generated holograms", Opt. Commun. **185** (2000) 77-82.
29. P. C. Mogensen and J. Glückstad, "Dynamic array generation and pattern formation for optical Tweezers", Opt. Commun. **175** (2000) 75-81.
30. Y. Kobayashi, Y. Igasaki, N. Yoshida, N. Fukuchi, H Toyoda, T. Harab and M. H. Wu "Compact High-efficiency Electrically-addressable Phase-only Spatial Light Modulator", Proceedings of SPIE vol. **3951**, 158-165 (2000).
31. J. W. Goodman, Introduction *to Fourier Optics* (McGraw-Hill, San Francisco, 2nd edition, 1996).
32. J. Glückstad and P. C. Mogensen, "Optimal phase contrast in common-path interferometry," Appl. Opt. 40, 268-282 (2001).
33. J. Glückstad, "Microfluidics: sorting particles with light," Nature Materials 3, 9-10 (2004).
34. M. Pelton, K. Ladavac and D. G. Grier, "Transport and fractionation in periodic potential-energy landscapes," Phys. Rev. *E* **70**, 031108 (2004).
35. M. MacDonald, G. Spalding and K. Dholakia, "Microfluidic sorting in an optical lattice," Nature **426,** 421–424 (2003).
36. D. Grier, "A revolution in optical manipulation," Nature **424,** 810–815 (2003).
37. R. L. Eriksen, V. R. Daria and J. Glückstad, "Fully dynamic multiple-beam optical tweezers," Opt. Express **10**, 597-602 (2002).
38. P. J. Rodrigo, R. L Eriksen, V. R. Daria and J. Glückstad, "Interactive light-driven and parallel manipulation of inhomogeneous particles," Opt. Express **10,** 1550–1556 (2002).
39. V. Daria, P. J. Rodrigo and J. Glückstad, "Dynamic array of dark optical traps," Appl. Phys. Lett. **84**, 323-325 (2004).
40. P. J. Rodrigo, V. R. Daria and J. Glückstad, "Four-dimensional optical manipulation of colloidal particles," Appl. Phys. Lett. **86,** 074103.1-074103.3 (2005).
41. M. M. Burns, J. M. Fournier and J. A. Golovchenko, "Optical matter: crystallization and binding in intense optical fields," Science **249**, 749 (1990).
42. G. Grynberg, B. Lounis, P. Verkerk, J.-Y. Courtois and C. Saloman, "Quantized motion of cold cesium atoms in two- and three-dimensional optical potentials," Phys. Rev. Lett. **70**, 2249-2252 (1993).
43. B. P. Anderson, T. L. Gustavson and M. A. Kasevich, "Atom trapping in nondissipative optical lattices," Phys. Rev. A **53**, R3727-R3730 (1996).
44. M. Greiner, I. Bloch, O. Mandel, T. W. Hänsch and T. Esslinger, "Exploring phase coherence in a 2D lattice of Bose-Einstein condensates," Phys. Rev. Lett. **87**, 160405 (2001).
45. S. Bergamini, B. Darquié, M. Jones, L. Jacubowiez, A. Browaeys and P. Grangier, "Holographic generation of microtrap arrays for single atoms by use of a programmable phase modulator," J. Opt. Soc. Am. B **21,** 1889-1894 (2004).

46. AM Lacasta, et al., "Sorting on periodic surfaces," Phys. Rev. Lett. 94, 160601, 2005.
47. MP MacDonald, GC Spalding, K Dholakia, "Microfluidic sorting in an optical lattice," Nature 426, pp 421-424, 2003.
48. K Ladavac, K Kasza, and DG Grier, "Sorting mesoscopic objects with periodic potential landscapes: optical fractionation," Phys. Rev. E 70, 010901, 2004.
49. YY Sun, et al, "Large-scale optical traps on a chip for optical sorting," App. Phys. Lett. 90, 031107, 2007.
50. P. J. Rodrigo, V. R. Daria, and J. Gluckstad, "Dynamically reconfigurable optical lattices," Opt. Express 13, 1384–1394 (2005).
51. C.A. Alonzo, P.J. Rodrigo, J.Glückstad, "Photon-efficient greylevel image projection by the generalized phase contrast method," New J. Phys. 9, 132, 2007.
52. J. Glückstad, D. Palima, P. J. Rodrigo, and C. A. Alonzo, "Laser projection using generalized phase contrast," Opt. Lett. 32, 3281–3283 (2007).
53. C.A. Alonzo, J. Glückstad, "Microparticle sorting using optical obstacle arrays," in Proceedings of 26th Samahang Pisika ng Pilipinas Physics congress (2008).
54. V. A. Soifer, Methods for Computer Design of Diffractive Optical Elements (John Wiley & Sons, New York, 2002).
55. N. Lue, W. Choi, G. Popescu, T. Ikeda, R. Dasari, K. Badizadegan, and M. S. Feld, "Quantitative phase imaging of live cells using fast Fourier phase microscopy," Appl. Opt. 46, 1836-1842 (2007).
56. C. A. Alonzo, P. J. Rodrigo, and J. Glückstad, "Photon-efficient grey-level image projection by the generalized phase contrast method," New J. Phys. 9, 132 (2007).
57. Perch-Nielsen I R, Rodrigo P J, Alonzo C A and Glückstad J 2006 Autonomous and 3D real-time multi-beam manipulation in a microfluidic environment Opt. Express 14 12199-205
58. Munjuluri B, Huebscham M L and Garner H R 2005 Rapid hologram updates for real-time volumetric information displays Appl. Opt. 44 5076-85
59. Ito T, Masuda N, Yoshimura K, Shiraki A, Shimobaba T and Sugie T 2005 Special-purpose computer HORN-5 for a real-time electroholography Opt. Express 13 1923-32
60. Buckley E, Cable A, Lawrence N and Wilkinson T 2006 Viewing angle enhancement for two- and three-dimensional holographic displays with random suppereso-lution phase masks Appl. Opt. 45 7334-41
61. D. C. Munson, Jr., "A note on Lena," IEEE Trans. on Image Processing 5 (1996).
62. V. Arrizon and M. Testorf, "Efficiency limit of spatially quantized Fourier array illuminators," Opt. Lett. 22, 197-199 (1997).
63. G. Kerner and M. Asscher, "Buffer layer assisted laser patterning of metals on surfaces," Nano Lett. 4, 1433-1437 (2004).
64. Y. Liu, S. Sun, S. Singha, M. R. Cho, and R. J. Gordon, "3D femtosecond laser patterning of collagen for directed cell attachment," Biomaterials 26, 4597-4605 (2005).
65. P. C. Mogensen, and J. Glückstad, "Phase-only optical encryption," Opt. Lett. 25, 566-568 (2000).

66. D. Psaltis, "Coherent optical information systems," Science 298, 1359-1363 (2002).
67. M. Righini, A. S. Zelenia, C. Girard, and R. Quidant, "Parallel and selective trapping in a patterned plasmonic landscape," Nature Phys. 3, 477-480 (2007).
68. N. Arneborg, H. Siegumfeldt, G. H. Andersen, P. Nissen, V. R. Daria, P. J. Rodrigo, and J. Glückstad, "Interactive optical trapping shows that confinement is a determinant of growth in a mixed yeast culture," FEMS Microbiol. Lett. 245, 155-159 (2005).
69. G. Milne, D. Rhodes, M. MacDonald, and K. Dholakia, "Fractionation of polydisperse colloid with acousto-optically generated potential energy landscapes," Opt. Lett. 32, 1144-1146 (2007).
70. Frigo M and Johnson S G 2005 The design and implementation of FFTW3 Proc. of the IEEE 93 216-31
71. F. M. Dickey and S. C. Holswade, Laser Beam Shaping: Theory and Techniques (Marcel Dekker, New York, 2000).
72. F. M. Dickey, S. C. Holswade, & D. L. Shealy, eds., Laser. Beam Shaping Applications (CRC Press, 2005).
73. M. A. Karim, A. M. Hanafi, F. Hussain, S. Mustafa, Z. Samberid, and N. M. Zain, "Realization of a uniform circular source using a two-dimensional binary filter," Opt. Lett. 10, 470–471 (1985).
74. S. P. Chang, J. M. Kuo, Y. P. Lee, C. M. Lu, and K. J. Ling, "Transformation of Gaussian to Coherent Uniform Beams by Inverse-Gaussian Transmittive Filters," Appl. Opt. 37, 747–752 (1998).
75. J. A. Hoffnagle and C. M. Jefferson, "Design and performance of a refractive optical system that converts a gaussian to a flattop beam," Appl. Opt. 39, 5488–5499 (2000).
76. P. H. Malyak, "Two-mirror unobscured optical system for reshaping the irradiance distribution of a laser beam," Appl. Opt. 31, 4377–4383 (1992).
77. F. Wippermann, U. D. Zeitner, P. Dannberg, A. Bräuer, and S. Sinzinger, "Beam homogenizers based on chirped microlens arrays," Opt. Express 15, 6218–6231 (2007).
78. M. T. Eismann, A. M. Tai, and J. N. Cederquist, "Iterative design of a holographic beamformer," Appl. Opt. 28, 2641–2650 (1989).
79. C. Y. Han, Y. Ishii, and K. Murata, "Reshaping collimated laser beams with Gaussian profile to uniform profiles," Appl. Opt. 22, 3644–3647 (1983).
80. T. Dresel, M. Beyerlein, and J. Schwider, "Design and fabrication of computer-generated beam-shaping holograms," Appl. Opt. 35, 4615–4621 (1996).
81. J. S. Liu and M. R. Taghizadeh, "Iterative algorithm for the design of diffractive phase elements for laser beam shaping," Opt. Lett. 27, 1463–1465 (2002).
82. A. J. Caley, M. Braun, A. J. Waddie, and M. R. Taghizadeh, "Analysis of multimask fabrication errors for diffractive optical elements," Appl. Opt. 46, 2180–2188 (2007)

83. D. Palima and V. R. Daria, "Effect of spurious diffraction orders in arbitrary multi-foci patterns produced via phase-only holograms," Appl. Opt. 45, 6689–6693 (2006).
84. D. Palima and V. R. Daria, "Holographic projection of arbitrary light patterns with a suppressed zero-order beam," Appl. Opt. 46, 4197–4201 (2007)
85. V. R. Daria, P. J. Rodrigo, S. Sinzinger, and J. Glückstad, "Phase-only optical decryption in a planar-integrated micro-optics system," Opt. Eng. 43 2223–2227 (2004).
86. E. Carcole, J. Campos, and S. Bosch, "Diffraction theory of Fresnel lenses encoded in low-resolution devices," Appl. Opt. 33, 162–174 (1994).
87. A. E. Cerussi, D. Jakubowski, N. Shah, F. Bevilacqua, R. Lanning, A. J. Berger, D. Hsiang, J. Butler, R. F. Holcombe, and B. J. Tromberg, "Spectroscopy enhances the information content of optical mammography," J. Biomed. Opt. 7, 60–71 (2002).
88. Y. Y. Cheng and J. C. Wyant, "Multiple-wavelength phase-shifting interferometry," Appl. Opt. 24, 804–807 (1985).
89. E. L. Heffer, and S. Fantini, "Quantitative oximetry of breast tumors: a near-infrared method that identifies two optimal wavelengths for each tumor," Appl. Opt. 41, 3827–3839 (2002).
90. Y. C. Chen, N. R. Raravikar, L. S. Schadler, P. M. Ajayan, Y. P. Zhao, T. M. Lu, G. C. Wang, and X. C. Zhang, "Ultrafast optical switching properties of single-wall carbon nanotube polymer composites at 1.55 μm," Appl. Phys. Lett. 81, 975–977 (2002).
91. Z. Wang, A. C. Bovik, H. R. Sheikh, and E. P. Simoncelli, "Image quality assessment: From error visibility to structural similarity," IEEE Trans. Image Process. 13, 600–612 (2004).
92. N. Kitamura, and F. Kitagawa, "Optical trapping – chemical analysis of single microparticles in solution," J. Photochem. Photobiol. C **4**, 227–247 (2003).

Chapter 8
GPC-Based Programmable Optical Micromanipulation

The realization that the radiation pressure force of laser light, although miniscule in absolute terms, can potentially generate quite large accelerations in microscopic particles, prompted Arthur Ashkin to experimentally demonstrate its manifestations [1]. This proved seminal in developing the field of optical trapping and manipulation. Small particles ranging from tens of micrometers down to atomic dimensions have since been optically trapped, thus encompassing lengthscales with multidisciplinary interest in biology, chemistry, medicine and physics. A single-beam optical trap finds utility in a wide range of inter-disciplinary research and is a practical tool for the measurement of interaction forces and manipulation of cells, sub-cellular structures and individual DNA-molecules [2, 3, 4], as well as in the assembly of microstructures on the micro- and nano-scale [5]. With all the applications of a single beam trap, it becomes exciting to envision a multi-beam system that can, for example, trap and manipulate an array of particles simultaneously yet independently. The multiple beams can drive processes in parallel or work in concert towards a common end such as in the assembly of micro-devices, or the synchronous actuation of a complex microstructure. This highlights the need to generate multiple optical traps where the shape, size, position and intensity of each trap can be controlled individually and preferably manipulated in real-time.

The previous chapters have established the capacity of the generalized phase contrast approach to arbitrarily shape light laterally. Thus, GPC can naturally enable optical trapping systems and imbibe them with *both* the ability to create independently controllable optical traps and the flexibility to simultaneously render, in real time, arbitrary dynamics for these traps. The main advantage of the GPC approach lies in its encoding simplicity, where each point in the trapping plane maps to a unique point in the programmable modulator. In this chapter, we illustrate various systems to showcase the flexibility and versatility facilitated by this point-wise mapping scheme in optical trapping and micromanipulation. It is remarkable that the generalized formulation of GPC, which allows phase contrast to go beyond its traditional small-scale phase assessment, can also become an enabling tool for interactive microscopy where the user not only passively observes a microscopic system but also can dynamically manipulate it.

8.1 Multiple-Beam GPC-Trapping for Two-Dimensional Manipulation of Particles with Various Properties

Reconfigurable light patterns synthesized by generalized phase contrast can be coupled through a microscope objective to illuminate a sample chamber where microparticles can be trapped and manipulated. Figure 8.1 shows the schematic of an interactive light-driven manipulation system as a simple optical attachment to a commercially available inverted microscope. Such a GPC-based optical micromanipulation system features a real-time user-feedback control that can simultaneously trap colloidal particles in two dimensions (2D). It also packs a unique interactive sorting capability and arbitrary patterning of microscopic particles. Straightforward phase encoding in a spatial light modulator, controlled by a computer, directly mimics the multiple beam patterns for manipulation of particles in the observation plane of a microscope. This eases the dynamic reconfiguration of the trapping patterns, which enables independent control of the position, size, shape and intensity of each beam. Efficient user-directed sorting of (polydisperse) microsphere mixtures of different sizes and colours using multiple optical traps are demonstrated.

The multi-trap system in Fig. 8.1 used a 200 mW diode laser beam (λ =830 nm), which was expanded and collimated to read out the computer-encoded phase patterns from a reflection type phase-only SLM (Hamamatsu Photonics). The phase-encoded light was efficiently transformed into intensity patterns at an intermediate image plane (IP) through a phase contrast imaging system, composed of lenses L1 and L2 with a $\theta = \pi$ phase contrast filter (PCF) at the Fourier plane. The PCF was fabricated by depositing a circular transparent photoresist (Shipley, Microposit S1818), 30 μm diameter, on an optical flat. Figure 8.1(a)–(d) show various high-contrast intensity distributions taken at an intermediate image plane, demonstrating the effective generation of regular and irregular geometries and arbitrary arrays. These patterns were relayed and demagnified to serve as multiple trapping beams onto the observation plane of the microscope using lens L3 and a 100× oil-immersion microscope objective, MO (NA = 1.4). Light was steered to the input pupil of the objective via a dichroic mirror (DM) through the fluorescence port of the microscope. The trapped particles were monitored with a CCD camera using the standard imaging functionality of the microscope.

A GPC-based trapping and micromanipulation system can easily work well with the fact that optical forces mutually depend on the particle and trap beam properties. The disc-shaped light in Fig. 8.1(a) can trap particles with higher refractive index than the surroundings. The 10 × 10 array of dark-cored beams in Fig. 8.1(c) can trap particles whose refractive indices are lower than that of the surrounding medium. The 15 × 15 trap array in Fig. 8.1(d) can trap particles with two distinct diameters. Figure 8.1b showcase the flexibility with which each trap can be sculpted. In this section we will explore the various trapping possibilities enabled by the reconfigurable generation of such flexible light structures.

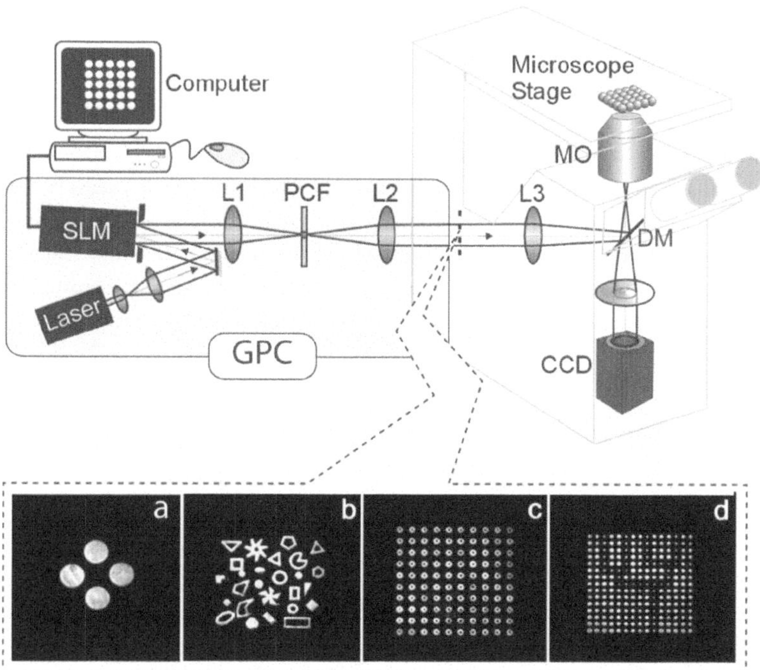

Fig. 8.1 Schematic diagram of an experimental setup for 2D interactive multi-particle optical manipulation. (a–d): Different geometries of multiple beams. The 10 × 10 array of annular beams in (c) can trap particles with refractive indices lower than the surrounding medium. The asymmetric array of 15 × 15 traps in (d) contains two distinct beam diameters. Input SLM phase modulations of 0 and π correspond to minimum and maximum output intensity, respectively.

In optical trapping experiments, radiation pressure can lift off polystyrene microspheres in an aqueous solution from the bottom surface of a glass chamber. The upward component of the radiation pressure force overcomes the particle weight to levitate the particle while the transverse component confines its path along the optical axis, subject to the drag force from the surrounding medium. Thus, in this two-dimensional geometry, trapped particles are normally held against the upper surface of the used glass chamber. Additional particles at the bottom of the chamber may also align underneath a previously trapped particle, allowing particles to be stacked on linear arrays along the beam axis [6, 7, 8]. Particle stacking can be undesirable when one intends to sort particles according to type, since particles with different properties can stack. Rapid pattern reconfigurability becomes a strong asset in this case, as illustrated by one solution depicted in Figure 8.2. Stacked polystyrene microspheres (Bangs Laboratories, Fishers, IN) were in this case distinguished from singly trapped particles by the image blurring from out of focus particles in the stack, as shown in Fig. 8.2(a). A computer's pointing device (e.g. mouse cursor) selected the beam trapping the stacked microspheres and displaced it to remove the excess particle (Fig. 8.2(b)–(c)) without disturbing the other trapped particles. A new trap was then immediately introduced to the 2 μm bead that was removed from the stack to independently trap the microspheres (Fig. 8.2(d)). In contrast to a collective mechanical shaking procedure (e.g. modulated transverse movement of the sample stage), this interactive solution localizes the disturbance to the problematic trap and avoids ejecting other particles from their traps. Stacks of particles were easily separated regardless of the size and the new position of the excess bead, again owing to rapid trap reconfigurability. Removing particle stacks does not compromise trapping performance and maintains the rest of the trapped particles undisturbed (Fig. 8.2(d)). This feature, in conjunction with the dynamic and parallel motion of the beams, makes the system a viable tool for user-interactive transverse particle sorting.

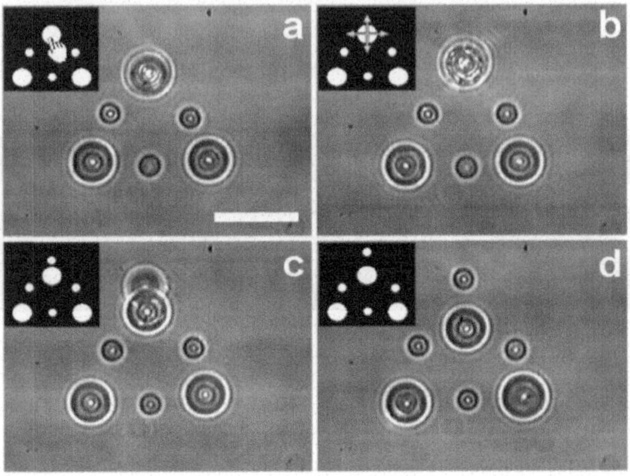

Fig. 8.2 "Click-and-drag" removal of particle stacking. (a) Blurred image of the topmost microsphere (5μm-diameter) indicates that another particle is trapped underneath. Inset: the problematic trap is selected by a computer "mouse" pointer on the computer graphics. (b) Motion of the selected graphic along the indicated directions displaces the selected trapping beam. (c) Introduction of an additional trap. Inset: the graphic corresponding to the new beam positioned at the site of the ejected particle (2-μm-diameter). (d) Final configuration with uniquely trapped distinguishable particles. Scale bar, 10 μm.

Upon separating stacked particles, sorting experiments can proceed by a "click-and-drag" interactive operation. Using the final configuration in Fig. 8.2 as starting point, Fig. 8.3 shows user manipulation of the inhomogeneous mixture of microspheres. The computer graphical user interface was used to directly manoeuvre the traps resulting in the subsequent sorting of the beads according to size. The image sequence show snapshots of intermediate configurations of the 2-μm and 5-μm polystyrene spheres, culminating with the formation of layers separating the two distinct sizes in the last frame.

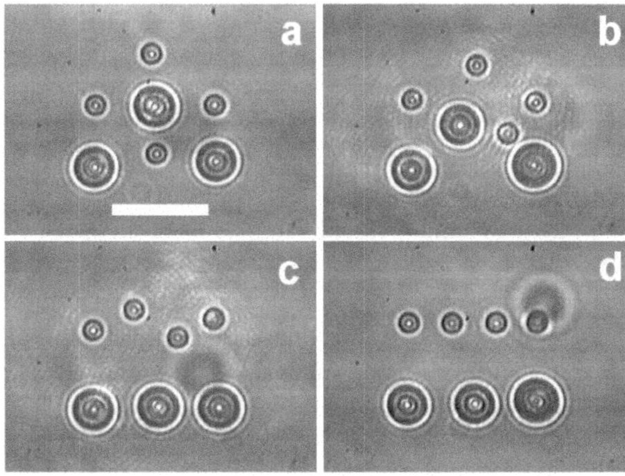

Fig. 8.3. Image sequences of trapping and sorting of inhomogeneous size-mixture of polystyrene beads in water solution with < 1% surfactant. (a) Dispersed beads with diameters 2 μm and 5 μm are first captured by corresponding trapping beams. The beads are held just below the upper surface of the glass chamber. The size of the beam used at each trapping site is proportional to the size of the trapped particle. (b)–(d) User-coordinated sorting of the beads according to size. Scale bar, 10 μm.

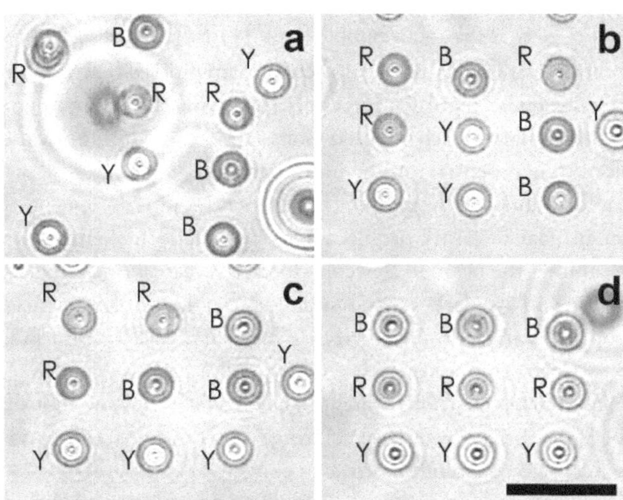

Fig. 8.4 Snapshots from interactive trapping and sorting of an inhomogeneous mixture of commercially dyed polystyrene beads (R-red, Y-yellow, B-blue). The beads are segregated according to colour and assembled into a 3 × 3 array. All beads have a diameter of 3 μm. Scale bar: 10 μm.

In practical applications, colour can be an important indicator such as, for example, when sorting on the basis of fluorescence emission of the particles. Figure 8.4, demonstrates colour-based sorting using multiple dyed polystyrene spheres (Polysciences), 3 μm nominal diameter. Randomly dispersed multicoloured microbeads were independently trapped and manoeuvred to arrive at the configuration in Fig. 8.4d where triplets of blue, red and yellow microspheres were sorted into three separate layers.

The GPC-based system shows inherent adaptability to create a variety of trapping beam patterns to potentially enhance, and even combine altogether, current optical micromanipulation schemes such as the rotation of an irregularly shaped object with a rectangle-profiled beam [9], trapping of low-index particles by hollow beams [10], and controlled trapping of a single particle surrounded by multiple beams [11]. In all cases, the advantage of a straightforward encoding approach cannot be overemphasized because it naturally lends itself to a graphical user interface for rapidly encoding phase patterns at modulator-limited rates. Multiple beams generated by the system can be utilized, not only for optically assisted synthesis of functional microstructures, but also for non-contact and parallel actuation of these devices. The illustrated GPC-system can thus provide flexible control over an assembly of microdevices, a crucial element for advanced light-powered lab-on-a-chip future applications.

As Arthur Ashkin originally observed in [1], a transparent dielectric microsphere suspended in water is readily pulled into the optical axis of a Gaussian laser beam where the intensity is strongest. This phenomenon is observed with particles having relative refractive index m greater than unity ($m = n/n_0$ where n and n_0 are the refractive indices of the particle and the suspending medium, respectively). On the other hand, Ashkin also noted at that time that for an air bubble ($m < 1$) in water the sign of the radial force due to the intensity gradient is reversed; hence, a low-index particle is repelled away from the beam axis. These repulsive forces also act on a low-index microsphere when illuminated by a stationary tightly focused Gaussian beam. Thus, optical tweezers does not provide a confining potential for low-index particles.

One solution for stable optical trapping of a low-index microscopic particle is to use a beam having an annular intensity profile. High-speed deflectable mirrors can trace out this profile by scanning the beam in a circular locus to create a ring of light that confines a low-index particle in its dark central spot [12]. A low-index particle can also be trapped in an optical vortex produced from a focused $TEM_{01}{}^*$ beam [13]. An optical vortex has been used to trap a low-index sphere and a high-index sphere, at the same time, in two neighbouring positions along the beam axis [14]. Low-index particles were also trapped between bright interference fringes produced at the focal plane of an objective lens where two coherent plane waves converge [15].

What if one needs to dynamically manipulate a large array of low-index particles where each particle can be independently steered? What about working with a heterogeneous mixture of both low- and high-index particles? If one can imagine trap shapes that suit each particle then, certainly, it becomes a strong indication for adopting a

GPC-based approach. The experiments discussed below demonstrate real-time user-interactive manipulation of a mixture of high- and low-index particles. For spherical particles, trapping beams with radial symmetry are utilized. High-index microspheres were efficiently trapped and manipulated using trapping beams with nearly top-hat transverse intensity profiles at the trapping plane. On the other hand, low-index particles are trapped using beams with annular transverse profiles. The experiments clearly demonstrate that, unlike other methods, the GPC approach readily provides *both* the ability to create independently controllable optical traps for high- and low-index particles, and the flexibilityf to render, in real time, arbitrary dynamics for these two types of particles simultaneously. This exceptional functionality may facilitate particle encapsulation in air-bubbles or in water-in-oil emulsions applied in petroleum, food, and drug processing.

Trapping and manipulation of a heterogeneous mixture of high- and low-index colloidal particles was achieved using the experimental setup shown in Fig. 8.5. The system used of a continuous wave (CW) Titanium:Sapphire (Ti:S) laser (wavelength-tuneable, 3900s; Spectra Physics) pumped with a CW frequency-doubled Neodymium:Yttrium Vanadate (Nd:YVO_4) laser (532 nm, Spectra Physics, Millennia V). The experiments were performed at $\lambda = 830$ nm using the built-in birefringent quartz filter plate in the Ti:S. which provided wavelength tuneability within the near infrared (NIR) spectrum from 700 to 850 nm. Pumping at 5.0 W with the Nd:YVO_4 yielded a maximum Ti:S output power of 1.5 W. The laser beam, expanded and collimated, read out phase information from a reflection-type phase-only SLM (Hamamatsu Photonics). The SLM phase elements, which utilize parallel-aligned nematic liquid crystals, were optically addressed using computer-controlled video projections from a VGA-resolution liquid crystal projector element. The SLM, located at the input plane of a 4f filtering system, imprinted programmable 2D binary phase patterns (0 or π phase levels) onto the readout 830 nm laser beam. The 4-f filtering system consisted of lenses L1 and L2 with a phase contrast filter (PCF) at their common focal plane (the Fourier plane). The PCF introduced a π phase-shift between low and high spatial frequency components of the phase-encoded beam. The diameters of the SLM iris (Ir) and the on-axis PCF respectively, were chosen to optimize light throughput and contrast of the generated output intensity distributions [16]. The resulting high-contrast intensity distributions generated at the image plane (IP) directly depicted the video output from the computer. The intensity distribution could be monitored through a pellicle beam splitter (~3% reflection) that projected a duplicate image onto a CCD camera. The main intensity patterns at the IP were scaled and relayed by lens L3 and microscope objective (MO) to a trapping plane within a sample chamber in an inverted microscope (Leica, DM-IRB). The beam entered through the microscope's fluorescence port where a dichroic mirror steered it to the back-focal plane of the MO. A CCD captured bright-field microscope images of the sample being manipulated.

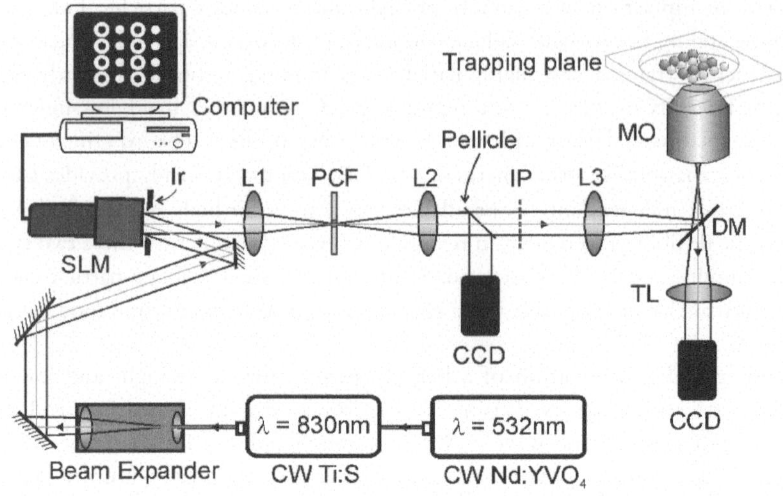

Fig. 8.5 Experimental setup for the simultaneous optical manipulation of high- and low-index particles. The expanded beam (λ = 830 nm) incident at the spatial light modulator (SLM) comes from a CW Ti:Sapphire (Ti:S) laser pumped by a visible CW $Nd:YVO_4$ laser. Under computer control, arbitrary 2D phase patterns are encoded onto the reflective SLM. A high-contrast intensity mapping of the phase pattern is formed at the image plane (IP) and is captured by a CCD camera via partial reflection from a pellicle. The intensity distribution is optically relayed to the trapping plane. Standard bright-field imaging is used to observe the trapped particles. PCF: phase contrast filter, Ir: iris diaphragm, L1, L2 and L3: lenses, MO: microscope objective, DM: dichroic mirror, TL: tube lens.

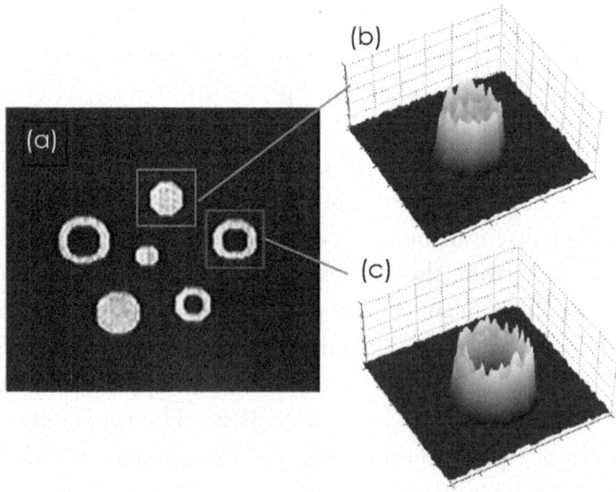

Fig. 8.6 (a) Measured high-contrast intensity pattern at the GPC output plane IP. Surface intensity plots are shown for representative (b) top-hat and (c) annular or doughnut trapping beams.

Images of intensity patterns captured from the intermediate image plane showed the quality of the GPC-generated patterns. A sample pattern is depicted in Fig. 8.6, showing different sizes of top-hat and annular transverse intensity profiles at different positions at the transverse (x-y) plane. These apply the optimal design conditions thoroughly described in the previous chapters. Optimum criterion calls for π-shifting PCF when using binary input phase conditions considered where intended bright regions are π-shifted with respect to the desired dark regions. The chosen operating diameters of the SLM iris and the PCF, corresponding to $K=1$, means π-shifting 25% of the input pixels yields a maximum output intensity of the trapping pattern is approximately four times the average intensity when directly transmitting the input beam.

A top-hat trapping beam profile provided a radially symmetric potential well for a high-index spherical particle as shown in Fig. 8.7(a). A high-index particle displaced away from the centre encountered restoring forces that pulled it back. Observations showed that a beam slightly larger than the particle provided better transverse confinement especially when the trapped particle was moved along the horizontal plane. In contrast, a top-hat beam acted as a potential barrier for a low-index particle with an unstable equilibrium at the beam centre. A low-index particle was repelled away from the centre as shown in Fig. 8.7(a). This was evident in the GPC-based experiments with spherical shells made of soda lime glass material (Polysciences) dispersed in de-ionized water. These air-filled hollow glass spheres have shell thickness of ~1 μm and outer diameters in the range of 2–20 μm. The hollow glass spheres with outer diameters greater than 5 μm effectively behave as low-index particles in water (n_0 = 1.33). Similar hollow glass spheres were found to have average density of ~0.2 g/mL and effective refractive index n_L = 1.2 [13].

Figure 8.8 shows a 6 μm hollow sphere in the presence of a top-hat beam. The images sequence shows optical manipulation without optical trapping. The low-index particle is displaced by optical fields as a result of its repulsion from the region of stronger light intensity.

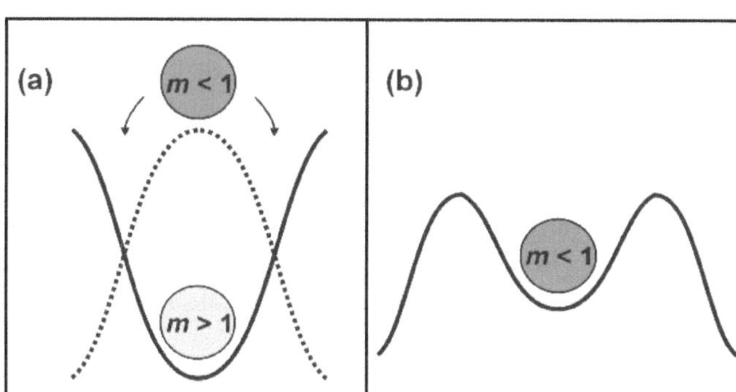

Fig. 8.7 Diagram of the optical potential (a) for a particle with high-index (solid curve) and a low-index (dashed curve) exposed to a top-hat beam; (b) for a low-index particle exposed to a beam with an annular intensity profile.

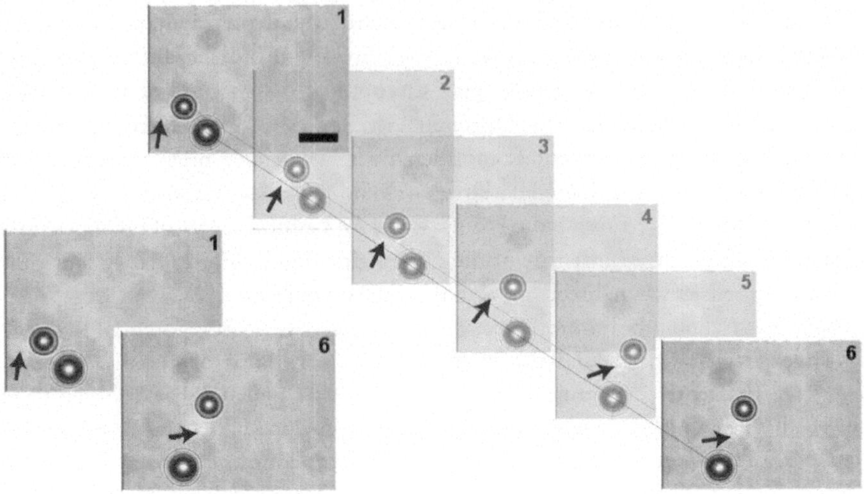

Fig. 8.8 Pushing a soda lime hollow glass sphere with a top-hat beam. An arrow indicates the user-controlled placement of the beam at each frame. Scale bar, 10 μm.

The optical manipulation that resulted from the repulsive forces exerted by intensity gradients can be exploited better by engineering suitable intensity patterns. One possibility is shown in Fig. 8.9, where a vertical line beam is used instead of a circular top-hat. The sample used contained a mixture of polystyrene microspheres (index n_H = 1.57, Bangs Laboratories) and the low-index hollow spheres in de-ionized water in a ~30 μm-thick glass cell. When mounted on a microscope stage, the polystyrene spheres (1.05 g/mL) settled to the bottom surface of the glass cell while the air-filled hollow glass spheres (0.2 g/mL) floated to the top portion. The image sequence in Fig. 8.9 illustrates an "optical raking" scheme where GPC synthesizes a scanning vertical line beam pattern to produce simultaneous deflection of low-index particles in the scan direction. This simple procedure can drag a number of low-index particles into the operating region where polystyrene spheres are found directly below. This non-mechanical scanning is facilitated by the simplified GPC phase encoding, which can be easily reconfigured to scan multiple parallel lines with adjustable spacing and duty cycle, if needed. The same effect can be demonstrated using a static pattern by lateral displacement of the microscope stage but programmable beam scanning can offer more flexibility.

To improve particle controllability, we shifted away from manipulation-without-trapping to achieve both trapping and manipulation. This required replacing the unstable circular flattop with one that enabled stable trapping. The dark centre of an annular intensity profile provided a minimum potential and stable equilibrium for a low-index particle, as shown in Fig. 8.7(b). However, the bright ring was an unstable equilibrium and nearby particles outside this ring would still be repelled. Thus, trapping low-index particles required the particle to be rapidly enclosed by the light ring, unlike high-index particles that could be slowly approached by a top-hat trap.

The GPC-based scheme fits well with trapping and manipulating low-index particles. The same computer video signal used to display graphic images on a computer monitor was channelled to an SLM that then "displayed" the images as phase patterns. Using customized software, doughnut traps could be created at desired locations and assigned independent dynamics such as gradual displacements, instant repositioning, or removal, as needed. Figure 8.10 demonstrate the steps for trapping low-index particles with doughnut optical traps. In the first frame, a doughnut trap was positioned next to a particle located almost outside the field of view. In the second frame, the trap was directly positioned on the particle and moved slightly to the centre of the observation region. In the third frame, a new trap was created by a mouseclick and brought to one of the freely moving particles in the next frame. The same procedure was repeated in the succeeding frames to trap all four particles as shown in the 15th frame. New traps could be created directly on the particles and the procedure simply illustrates that the traps can jump directly into a desired location as well as move along a chosen path. To illustrate independent manipulation of the trapped particles, they were brought into a diamond formation (20th frame) and then into a linear arrangement (25th frame). The doughnut traps were configured with appropriate diameters and thickness by a "click and draw" computer mouse sequence to match the particle diameters, which varied in the range 6–10 μm.

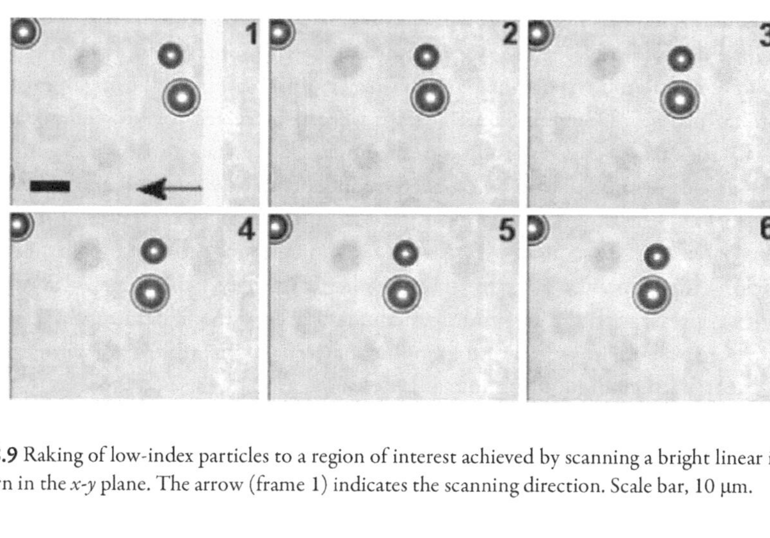

Fig. 8.9 Raking of low-index particles to a region of interest achieved by scanning a bright linear intensity pattern in the *x-y* plane. The arrow (frame 1) indicates the scanning direction. Scale bar, 10 μm.

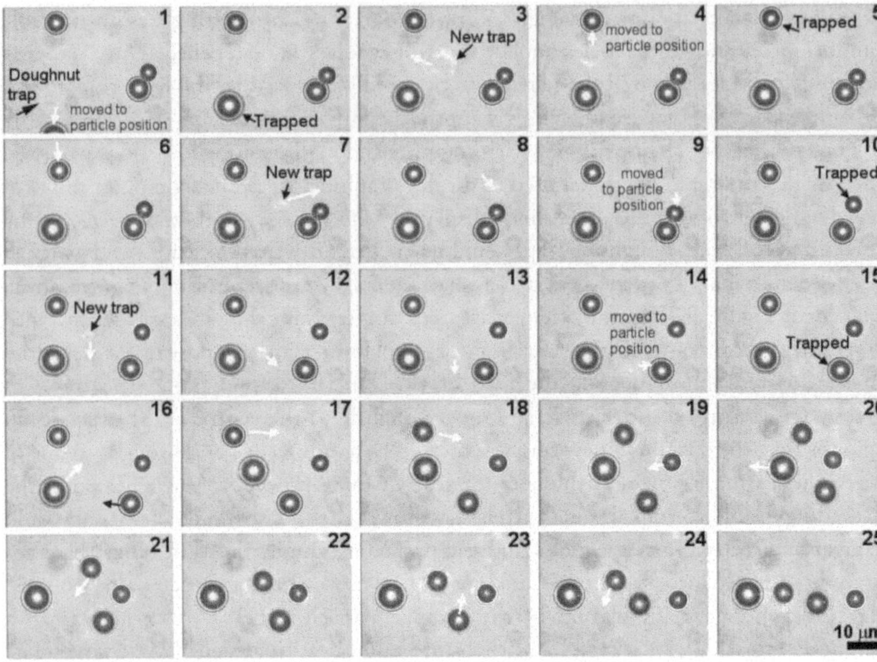

Fig. 8.10 User-interactive optical trapping and manipulation of different sizes of hollow glass spheres using doughnut optical traps.

A heterogeneous mixture of high-index polystyrene microspheres and hollow microspheres (effectively lower refractive index) will separate vertically due to buoyant forces. When manipulating this mixture, one may choose to first align the microscope to focus on the low-index particles at the upper surface of the sample cell. The high-index microspheres are then lifted off the bottom surface of the sample cell using optical traps with top-hat profiles. Doughnut optical traps were also created for low-index particles while high-index particles were brought to the upper glass surface by top-hat beams. The trapped particles in the mixture could then be independently manipulated. The image sequence in Fig. 8.11 shows controlled optical manipulation of a mixture of high- and low-index particles. The sequence starts with individually trapped particles in an irregular spatial distribution. The particles are individually displaced and sorted according to their index contrast with the suspending medium. This process illustrates the versatility of the GPC method in generating trapping patterns with arbitrary (symmetric or asymmetric) spatial configurations in real-time.

Generating predefined motions of trapped particles is definitely possible with a GPC-based system. This takes minimum computational power since the user can simply generate an animation of the desired motion and directly use the animation video to modulate the SLM without having to compute corresponding Fourier phase diffractive patterns. An example of such an experiment is illustrated in Fig. 8.12 where a row of trapped high-index polystyrene spheres and a row of low-index particles are simultaneously set into oscillatory motion. The minimal computational load in GPC means that the dynamics of the trapping beams are mainly limited only by the modulator refresh rate. The response time of the liquid crystal SLM used in the demonstrations is in the order of ~100 ms (i.e. time for one SLM pixel to change between two extreme states, e.g. phase delays 0 and π). When animating a trap as a sequence of images shifted one pixel at a time, the modulator refresh rate corresponds to a maximum lateral speed of ~2.5 μm/s for the scaling used in Fig. 8.5. Faster motion can be achieved by shifting by more pixels for each frame in the animation sequence, producing a coarser or more discrete motion of the traps. It is worth noting that the GPC method requires only binary phase objects to generate binary 2D intensity patterns with arbitrary symmetry. Thus, one may take advantage of SLM devices with faster refresh rates (e.g. ferroelectric liquid crystals or MEMS-based mirror arrays) to achieve faster, yet smooth, trap displacements. Assuming an SLM with high refresh rate, the maximum speed of a trapped particle depends on the stiffness of the optical trap and the hydrodynamic drag force from the surrounding medium. The trap stiffness for both the top-hat and annular beams may be improved by increasing the laser power.

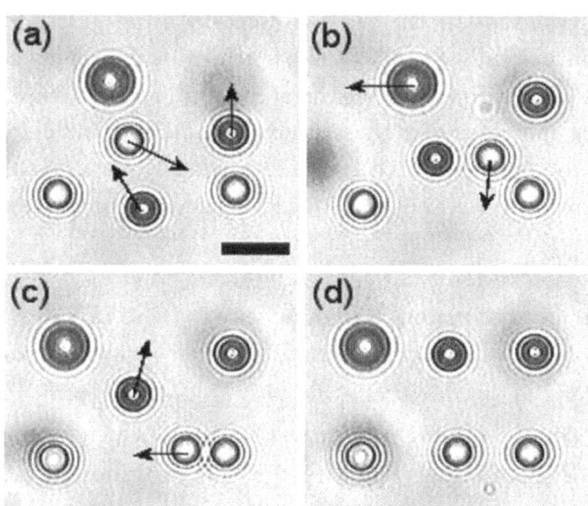

Fig. 8.11 Image sequence of trapping and user-interactive sorting of an inhomogeneous mixture of soda lime hollow glass spheres and polystyrene beads in water solution. (a) The particles are first captured by appropriate trapping beams and then (b-c) displaced one by one. The size of the beam used at each trapping site is proportional to the size of the corresponding particle. Arrows indicate the directions at which particles are transported. (d) Two separate rows of optically trapped high-index (lower row) and low-index particles (upper row). Scale bar, 10 μm.

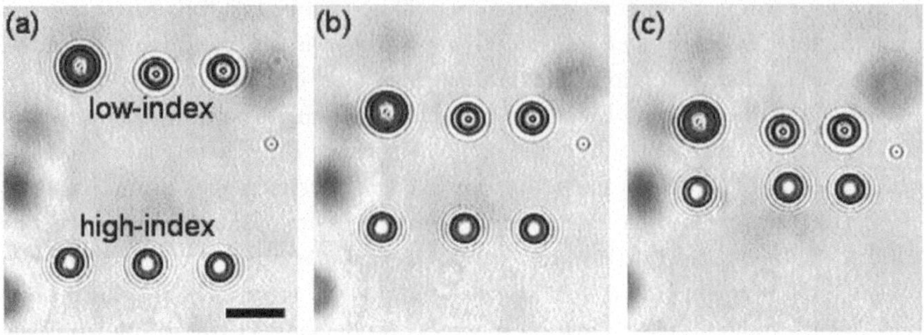

Fig. 8.12 Simultaneously transported high- and low-index particles confined in respective optical traps with pre-programmed dynamics. The time interval between adjacent frames is ~15 s. Scale bar, 10 μm.

8.2 Probing Growth Dynamics in Microbial Cultures of Mixed Yeast Species Using GPC-Based Optical Micromanipulation

The various demonstrations in the previous section using polystyrene microspheres illustrate the versatility of GPC-based multi-particle optical trapping and manipulation. This section demonstrates simultaneous trapping and manipulation of a plurality of living cells. This not only illustrates the capacity of GPC to handle multiple living cells but also shows the methods potential for investigating biological questions. Multi-cell trapping and manipulation may be used in microscale experiments to study issues that cannot be explicitly resolved in bulk experiments.

What happens when mixed yeast species interact? Microbiologists have previously observed, through bulk experiments at high density (>10^7 CFU/mL)†, that viable yeast cells of the species *Saccharomyces cerevisiae* cause growth arrest of two non-*Saccharomyces* yeast species in mixed cultures [17, 18]. Monospecies cultures of the two non-*Saccharomyces* do not exhibit early growth arrests, which seemed not caused by nutrient limitation. These findings kindle further investigations of cell growth in mixed yeast cultures. Is there a cell-cell contact mechanism between *S. cerivisiae* and non-*Saccharomyces* species at work? Microscale optical manipulation experiments with a traceable number of CFUs can offer discovery modes inaccessible to bulk experiments. This may provide clues to understanding the mechanism causing growth arrest in yeast.

A GPC-based optical trapping and manipulation system has been utilized in microbiological experiments to investigate whether one yeast species can impose a confine-

† CFU or Colony-Forming Units indicates the number of viable cells in a sample.

ment stress on another yeast species. Viable *S. cerevisiae* cells were trapped and positioned to surround non-*Saccharomyces* yeast cells, *Hanseniaspora uvarum*, in a mixed culture. Small aliquots of culture broth from 22-hour old single cultures of *H. uvarum* CBS 314 and *S. cerevisiae* (Saint Georges S101, Bio Springer, France), grown in yeast extract peptone dextrose (YPD) medium (9 g/L glucose, pH 5.6) with 140 rpm agitation at 25 °C, were transferred to a perfusion chamber (CoverWell, Sigma-Aldrich) with fresh YPD medium. The cells were allowed to settle on the glass coverslip. The chamber was sealed with grease and subsequently mounted on an inverted microscope (DM-IRB, Leica) with a non-immersion objective lens (×63, NA = 0.75; C PLAN, Leica).

The optical system was similar to previous trapping demonstrations (see Figs. 8.1 and 8.5). A GPC-based system synthesized and controlled multiple optical traps introduced through the microscope fluorescence port. The laser was operated at 830 nm wavelength, where water and biological molecules have very low absorption, to minimize photodamage to cells [3]. Since the illuminated cells have a higher refractive index relative to the surrounding YPD medium they were attracted to high intensity regions of the traps. The image sequence in Fig. 8.13 depicts real-time, interactive optical manipulation to trap *S. cerevisiae* cells and assemble them in a ring around a *H. uvarum* cell. Additional GPC-generated optical traps were also used to isolate the region of interest from other cells.

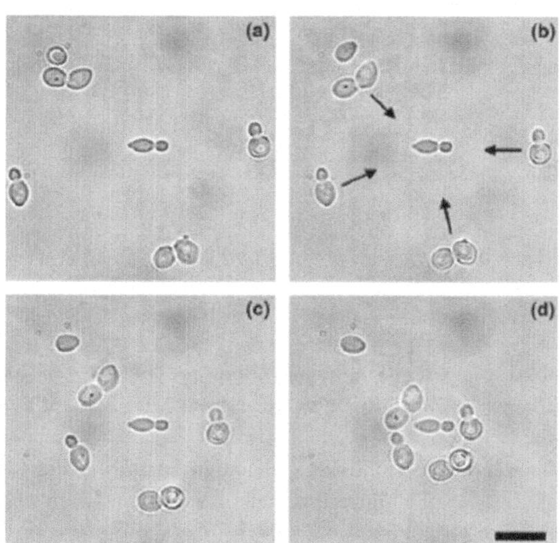

Fig. 8.13 (a)–(d) Snapshots from a monitored, microscale, mixed-culture fermentation using an interactive optical trapping system to arrange *Hanseniaspora uvarum* and *Saccharomyces cerevisiae* cells. Reconfigurable optical traps simultaneously move *S. cerevisiae* cells to surround a single *H. uvarum* cell. Arrows show the direction of movement. Scale bar, 10 μm.

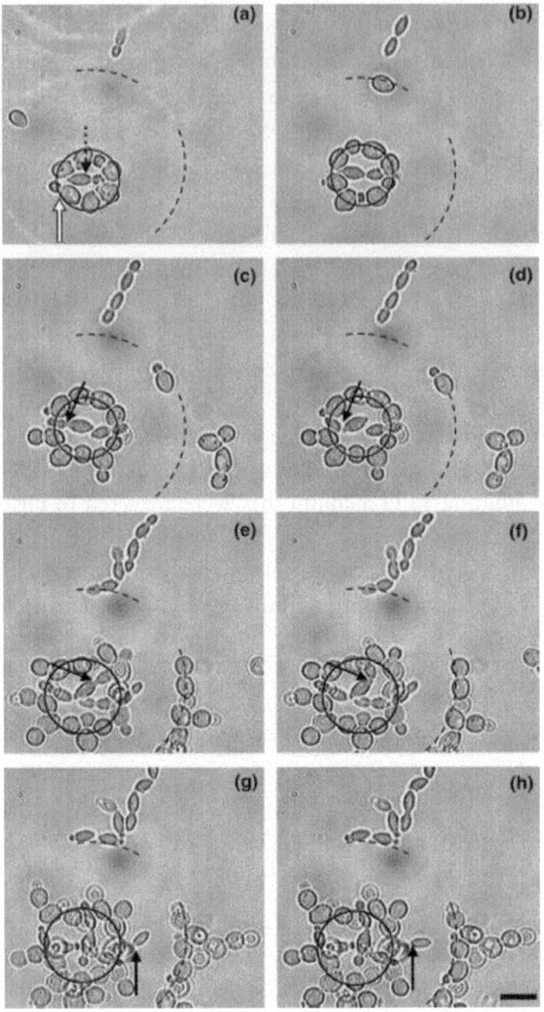

Fig. 8.14 Microscale, mixed-culture fermentation of *Hanseniaspora uvarum* and *Saccharomyces cerevisiae*. Images from one representative experiment were recorded at: (a) 5.0 min; (b) 51.5 min; (c) 114.5 min; (d) 116.2 min; (e) 184.3 min; (f) 186.0 min; (g) 217.5 min; (h) 219.2 min. *S. cerevisiae* cells (white arrow) were trapped in a ring-shaped optical beam pattern (full line circle), surrounding an individual *H. uvarum* cell (dashed black arrow). This region of interest was kept free from any interference with other cells by a larger ring-shaped optical beam pattern (dashed line circle). Black arrows show cytokinesis of individual, surrounded *H. uvarum* cells at 114.5–116.2 min ((c) and (d)), 184.3–186.0 min ((e) and (f)), and 217.5–219.2 min ((g) and (h)) of fermentation. Non-surrounded *H. uvarum* cells are located at the top centre of the images. Scale bar, 10 μm.

Yeast growth was monitored through video recordings and Fig. 8.14 show snapshots taken at 1.7 minute intervals. Cell division of *H. uvarum* was detected when cell pairs exhibited a distinct break during each cytokinesis (see e.g. Fig. 8.14(c)–(h)). The *generation time,* T_G, of a given cell is the time difference between two successive cytokinetic events (e.g. in Fig. 8.14, T_G = time[(f)] – time[(d)] = 186.0 min – 116.2 min = 69.8 min). Cell growth was only tracked for three generations as cell densities became too large for proper analysis afterwards. Data from twenty growth experiments were analyzed, consisting of more than 150 surrounded and non-surrounded *H. uvarum* cells. The average generation time for surrounded *H. uvarum* cells (91.7 min) was ~15% higher than that of non-surrounded cells (77.9 min). This indicates that the confinement imposed by viable *S. cerevisiae* cells on *H. uvarum* inhibits growth of the latter. The generation time of surrounded cells and non-surrounded cells correspond to growth rates of 0.45 and 0.53 per hour, respectively, which are comparable to population-averaged data reported in literature for aerobically grown *H. uvarum* [19].

8.3 Three-Dimensional Trapping and Manipulation in a GPC System

Microscopic particles can be optically trapped and manipulated stably in two-dimensions using moderately focused laser beams even in the presence of excess axial forces from radiation scattering by exploiting surface forcers from the sample chamber. Optical tweezers [20] use strongly focused laser beam using high-NA objective lenses to create large three-dimensional intensity gradients that can trap a microscopic particle within the focal volume. This has turned out to be a valuable tool in various fields that involve spatial confinement and control of mesoscopic particles observed under an optical microscope.

In his pioneering experiments in 1970, Ashkin used a moderately focused Gaussian laser beam and balanced the excess axial optical force with another beam, coaxially propagating in the opposite direction, to create a stable three-dimensional trap. The beam waists were slightly separated along the optical axis and the symmetry point [1] was a stable equilibrium point for a trapped high-index dielectric particle. Counter-propagating beams, known in the literature as a *counterpropagating-beam* or *dual-beam trap*, form a three-dimensional optical potential well for a dielectric particle [1, 21, 22, 23].

Controlled modifications on the relative strengths of the counter-propagating beams create a more general configuration – the *optical elevator*. An optical elevator dynamically controls the axial location of the potential minimum where the particle finds a stable equilibrium position. By using the GPC-based dynamic light shaping to setup multiple counter-propagating beams one can generate multiple independently translatable optical elevators to create independently controllable three-dimensional optical traps. This section outlines the main principles for designing a GPC-based system capable of independently trapping and manipulating multiple particles in three dimensions.

Dynamic and non-mechanical control of the axial position of trapped objects may be applied for the study of interaction forces between micro-organisms and functionally modified surfaces [24], and for experiments that require trapped particles to be kept away from sample surfaces. Here we will describe the design of an array of optical elevators with real-time reconfigurable beam patterns synthesized by the GPC method. We will also describe an experimental realization of such a GPC-based system for simultaneous trapping and manipulation of polystyrene spheres and yeast cells *S. cerevisiae* in three dimensions.

In a counter-propagating beam trap, particle displacement along the optical axis is due to an induced differential force that accompanies the change in axial location of the optical potential well minimum. For a GPC-based multi-trap system, this is accomplished through an energy-efficient polarization encoding scheme that varies the relative powers of the oppositely directed moderately focused beams. This extends the trapping functionality demonstrated in the previous section function to simultaneous 3D trapping and manipulation of particles, biological or otherwise. In the system described below, trapped particles can be independently steered along a transverse region covering almost the entire field of view while simultaneously being axially manoeuvred over a depth of tens of μm.

The general principle for realizing multiple optical elevators is illustrated in Fig. 8.15. In two-dimensional trapping, light patterns generated by GPC are scaled and relayed into a microscope as multiple traps in the trapping region. By splitting the beams during relay and coupling them into oppositely facing microscope objectives, one can setup an array of stable multiple counter-propagating beam traps that can be individually steered. To achieve independent axial control for each beam, the GPC output beams

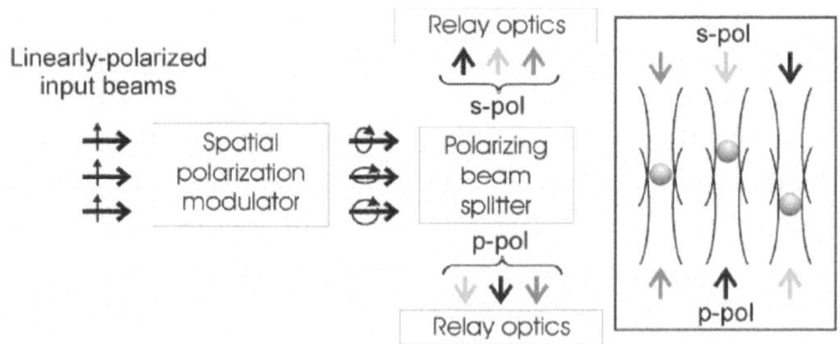

Fig. 8.15 Schematic diagram describing the generation of an array of optical elevators. A spatial polarization modulator converts a set of linearly polarized input beams into parallel beams with arbitrary elliptic polarization states. A polarizing beam splitter separates the s-polarized (s-pol) and the p-polarized (p-pol) components, whose strengths scale in proportion with the dark shade of the arrows. Pairs of orthogonally polarized beams are directed toward the sample volume to trap particles (inset).

are first independently encoded with polarization states using a spatial polarization modulator (SPM). An SPM can be realized with a phase-only SLM operating in polarization mode as described in the appendix. Using a polarizing beam splitter (PBS) and the same relay and scaling optics as before, the resulting counter-propagating beam traps have orthogonal s- and p-polarizations (labelled s-pol and p-pol in Fig. 8.15). The relative strengths of the beams in each trap, and hence the stable axial position, is easily adjusted by programming the SPM to independently encode appropriate s- and p-polarized components into each trap.

The concept outlined in Fig. 8.15 may be experimentally demonstrated using the setup illustrated in Fig. 8.16. A GPC system synthesizes linearly polarized input beams with high efficiency for subsequent polarization encoding. The GPC system employs a video graphics array (VGA)-addressed phase-only SLM (PO-SLM; Hamamatsu Photonics) to encode phase patterns that mimic the desired intensity patterns. This is essentially the same as the system used in two-dimensional trapping and offers the same user-interactive reconfigurability at video-frame rates, to specify the number of traps, alter their distribution and define trap sizes and geometrical shapes. The traps are polarization-encoded and split into s-pol and p-pol beams that propagate through microscope objective as counter-propagating beams in the sample cell to create an array of optical elevators.

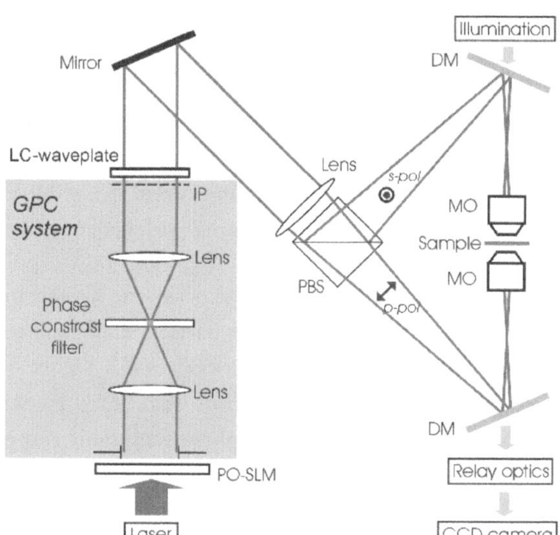

Fig. 8.16 Experimental setup. The generalized phase contrast (GPC) system converts PO-SLM-encoded phase patterns into high-contrast intensity pattern at plane IP. Relayed to the sample cell are two scaled versions of the trapping pattern with orthogonal polarizations (s-pol and p-pol) and with ~30 μm separation. Relative powers of s-pol and p-pol are governed by the voltage-dependent LC waveplate. Particles are monitored using standard bright-field detection via a microscope objective lens, MO (×60; NA = 0.85). PBS: polarizing beam splitter, DM: dichroic mirror.

To properly illustrate the concepts, we start with a proof-of-concept demonstration. The full implementation using spatial polarization modulation will be discussed in the next section. The present setup uses a voltage-controlled variable liquid crystal (LC) waveplate that encodes the same elliptic polarization to all beams. The LC-waveplate, aligned with its director axis at 45° with the vertically polarized incident beam, introduces a voltage-dependent phase retardation, $\varphi_g(V_{LC})$, where V_{LC} is the amplitude of the sinusoidal voltage driving the device. In an SPM implementation, each polarization-modulating pixel behaves similarly to this LC-waveplate. The output field through the LC-waveplate is described by a polarization vector showing the orthogonal polarization components

$$U' = \begin{bmatrix} \text{s-pol} \\ j\ \text{p-pol} \end{bmatrix} = \begin{bmatrix} \sin\left(\phi_g\left(V_{LC}\right)/2\right) \\ j\cos\left(\phi_g\left(V_{LC}\right)/2\right) \end{bmatrix} \tag{8.1}$$

It is obvious from Eq. (8.1) that one can vary the relative strengths of the orthogonal components (s-pol and p-pol) of the elliptically polarized output beam by changing $\varphi_g(V_{LC})$.

Results of quantitative measurements carried out to test the performance of the LC-waveplate are shown in Fig. 8.17. The power of s- and p-polarization components for different voltage settings, normalized by the incident power, denoted P_s and P_p respectively, are shown in Fig. 8.17(a). The average of the power sum, $(P_p + P_s)$, over the voltage range $2.0\ \text{V} \leq V_{LC} \leq 8.0\ \text{V}$ shows ~96% light efficiency of the polarization-based scheme. The inset plots a calibration curve for the differential power $(P_p - P_s)$ as a function of the driving voltage. The solid curve plots a least-squares fit to a decaying exponential. The two polarization components are equal at $V_{LC} = 2.9$ V.

Stable and controllable 3D positioning using voltage-controlled intensities of counter-propagating traps was demonstrated using fluid-borne polystyrene spheres (diameter = 3 μm; Polysciences). In a calibration experiment, a freely moving sphere was trapped in an optical elevator (beam diameter ≈ particle diameter) to observe how its axial position depended on the differential power $(P_p - P_s)$. Figure 8.17(b) shows the observed dependence of the sphere's axial position with $(P_p - P_s)$ within the applied voltage range $2.6\ \text{V} \leq V_{LC} \leq 3.2$ V. The microsphere was stably trapped in 3D within this voltage range. The axial positions are determined by referencing with bright-field images of an identical sphere taken at calibrated z-positions. Assigning $z = 0$ as the sphere's position of best-focus, the images enable unambiguous estimates of axial position in the range $-10\ \mu\text{m} \leq z \leq +10\ \mu\text{m}$ with ~1 μm precision. Axial position control was achieved within a 20 μm dynamic range. As will be discussed later, the operating range can be tremendously improved by using objective lenses with lower NA. Thus, the axial position control can be superior to optical tweezers that typically rely on wave-front modification techniques to axially shift the beam waist [25].

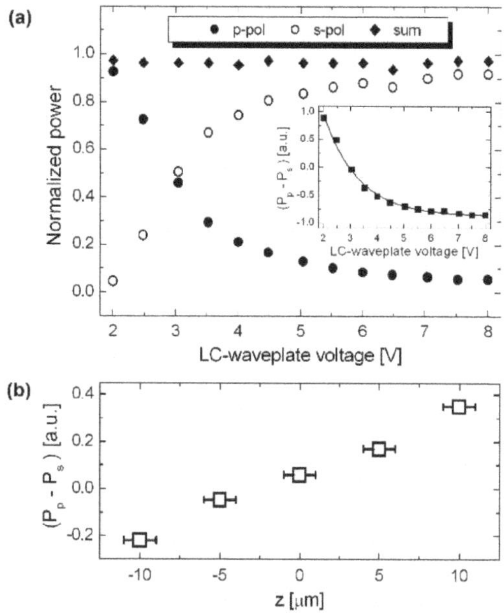

Fig. 8.17 (a) Measurement of relative powers of s-polarized (O) and p-polarized (●) components versus the LC-waveplate voltage. The sum (◆) of the two components is approximately unity. The inset plots the difference ($P_p - P_s$) versus the LC voltage. (b) Power difference ($P_p - P_s$) versus axial trap position of a 3 μm polystyrene sphere in an optical elevator with total power in the order of ~10^{-2} W.

The axial control demonstrated for a single particle can be extended to 3D trapping of a plurality of particles using a GPC-based implementation of multiple optical elevators. The setup employs a Titanium:Sapphire laser (Spectra Physics 3900S, 1.5 W max. power) operating at 830 nm wavelength. It can handle not only colloidal polymer spheres but also biological systems of viable yeast cells *S. cerevisae*. Figure 8.18(a)–(c) show snapshots of microspheres optically launched into circular orbit combined with simultaneous axial translation, which produces the observed defocusing. Snapshots for similar experiments with yeast cells are shown in Fig. 8.18(d)–(f), where four yeast cells orbit around another yeast cell. The laser power achieves tangential speeds of up to 5–10 μm/s (depending on *z*-position) while keeping both types of particles trapped. Yeast cells with elongated shapes tend to reorient and align with the optical axis.

Compatibility with biological systems is doubly supported by the biologically safe operating wavelength and the use of weakly focused beams, which minimizes local pressures on trapped cells. Observations indicate that the elevator traps do not significantly influence the measured generation time of budding yeast cells (see Sect. 8.2 for more experiments with yeast cells). The full implementation extending optical elevators to independent three-dimensional control of the multiple particles using a spatially addressable polarization modulator is described in the next section.

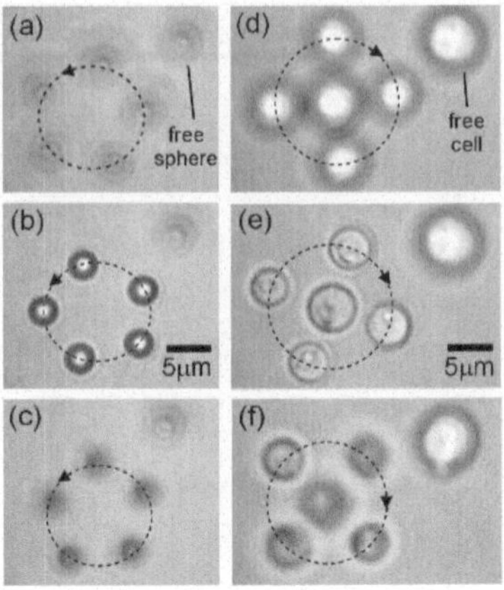

Fig. 8.18 Dynamic axial position control and simultaneous rotation of trapped particles in real-time reconfigurable arrays of optical elevators. Left and right columns show images of five microspheres set into orbit at various planes (a) $z = -10$ μm, (b) $z = 0$, (c) $z = +10$ μm, and five yeast cells in a rotating cross pattern at (d) $z = -5$ μm, (e) $z = 0$, (f) $z = +5$ μm, respectively. Dashed circles show the paths of the rotated particles with the arrows indicating the rotation direction.

8.4 Real-Time Autonomous 3D Control of Multiple Particles with Enhanced GPC Optical Micromanipulation System

The previous section outlined the principles for achieving three-dimensional trapping and manipulation of particles in a GPC approach. This was supported by proof-of-principle demonstrations showing direct and user-interactive trapping and manipulation of a colony of particles and living cells. A GPC-based system can independently trap and manipulate colloidal particles in real-time within an operating volume. As discussed previously, this requires a novel and light-efficient polarization scheme for independently controlling multiple counterpropagating beam traps. Spatial polarization modulation provides independent axial control for each trapped particle through lossless tuning of the relative powers of orthogonally polarized counterpropagating beams. This fits well with GPC-based synthesis of transverse trap positions and dynamics that are highly reconfigurable in real-time. The synergetic union of these two schemes enables true real-time user-interactive 3D optical manipulation of a plurality of simultaneously trapped microscopic particles.

In this section, we consider a system that fully realizes the design principles outlined in the schematic shown in Fig. 8.15 for real-time independent 3D control of multiple particles. The instrumentation for the optical trapping and manipulation system is schematically illustrated in Fig. 8.19. Light patterning by GPC is implemented, as before, using an expanded near-infrared laser beam to read encoded patterns information on a reflection-type phase-only SLM. Binary SLM phase corresponding to $\varphi = 0$ and $\varphi = \pi$, described by the spatial function $\phi(u,v)$, is directly programmed by a computer video signal that is also conveniently displayed on a computer monitor. This means that faster binary SLMs based on ferroelectric liquid crystals or MEMS-based deformable mirrors can be applied, if needed.

The GPC projection setup efficiently converts the uniform light carrying binary phase $\phi(u,v)$, into a high-contrast intensity pattern $I(u',v')$ that directly mimics the phase pattern that is also displayed on the monitor. For polarization encoding, the output intensity pattern is projected onto a spatial polarization modulator (SPM). The process is approximately described by:

$$\phi(u,v)=\pi\sum_{n=1}^{N} f_n(u-u_n,v-v_n) \xrightarrow{\text{GPC}} I(u',v')\approx I_0\sum_{n=1}^{N} f_n(u'-u_n',v'-v_n')\ , \qquad (8.2)$$

where $f_n(u,v)$ describes the desired spatial features of a designated nth trap , assuming traps do not overlap. For a GPC system with magnification M, we may write $f_n(u,v)=f_n(u'/M,v'/M)$. Neglecting the diffractive blurring of the traps, we may describe the trapping intensity pattern using a scaled form of $I(u',v')$. Circular traps with top-hat intensity profiles are described by $f_n(u,v)=\text{circ}(\sqrt{u^2+v^2}/a_n)$, where a_n is an adjustable trap radius. Following GPC optimization, described in previous chapters, high photon efficiency and optimal peak irradiance ($\sim I_0$) and visibility is achieved in generating $I(u',v')$.

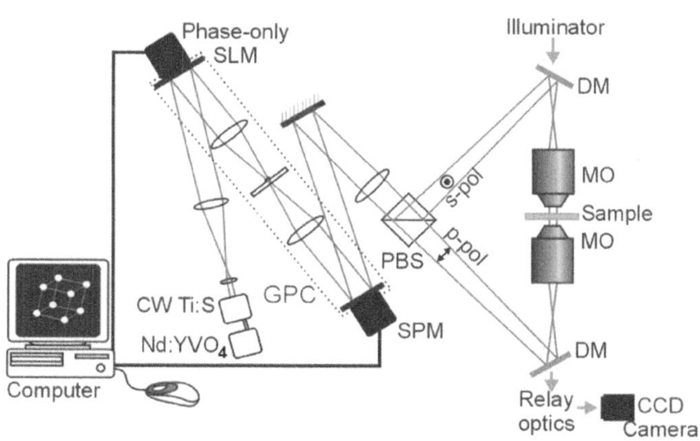

Fig. 8.19 Setup for implementing 4D (or real-time 3D) optical manipulation. CW Ti:S, continuous-wave titanium:sapphire (wavelength = 830 nm; maximum power = 1.5 W); Nd:YVO_4, neodymium:yttrium-vanadate (wavelength = 532 nm); GPC, generalized phase contrast system; SLM, spatial light modulator; SPM, spatial polarization modulator; PBS, polarizing beam splitter; MO, microscope objective (×60; NA = 0.85); DM, dichroic mirror.

To understand the operating modes of the system, let us consider the spatial polarization modulation (SPM). The SPM modulates a vertically polarized incident field, $E(u',v')\,\hat{\mathbf{e}}_v$, corresponding to the GPC output intensity pattern, $I(u',v')$. The SPM encodes a computer-controlled phase function, $\psi(u',v')$, onto this field. With the SPM aligned with its axis at 45° with the incident polarization, the reflected intensity pattern $I_r(u',v';t)$ is described by a complex field given by (see Appendix),

$$\vec{\mathbf{E}}_{\text{reflected}}(u',v') = E(u',v')\{-i\sin(\psi(u',v';t)/2)\hat{\mathbf{e}}_{\mathbf{u}} + \cos(\psi(u',v';t)/2)\hat{\mathbf{e}}_v\} \tag{8.3}$$

The bracketed terms of Eq. (8.3) denotes a computer-controlled polarization landscape that, in general, can be programmed both in space and time. The operating modes for real-time 3D optical manipulation is determined by the spatiotemporal dynamics of the input phase, $\phi(u',v';t)$, and the SPM phase, $\psi(u',v';t)$. The input phase controls the intensity $I(u',v';t)$ and the SPM phase independently controls the intensity ratio in each of the counterpropagating beam traps. A polarizing beam splitter (PBS) decomposes the orthogonal polarization components that are then scaled and projected from opposite sides of a sample along a common optical axis. The two projections are imaged onto respective planes axially separated by ~30 μm. The relative strengths of the polarization components are locally controllable at any given time, as seen from Eq. (8.3). Thus, when trapping N particles in a colloidal system, each can be independently controlled and manipulated with unique dynamics. The transverse position is controlled by the input spatial phase modulation while the axial position is controlled by the spatial polarization modulation.

As an initial demonstration of independent 3D control of multiple particles, Fig. 8.20 shows an operating mode where polystyrene microbeads are set along an imaginary "roller coaster" full-space trajectory. First, three particles are trapped and positioned at the vertices of an imaginary triangle. Then, the SPM is programmed with a fixed polarization landscape to define the "roller coaster track" and the particles assume their respective positions on this track. When the GPC input phase is modified to rotate the triangular trap configuration, the microparticles move along the imaginary track.

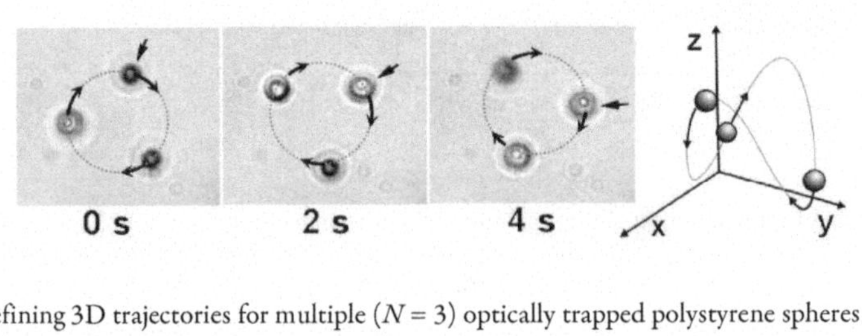

Fig. 8.20 Defining 3D trajectories for multiple ($N = 3$) optically trapped polystyrene spheres (diameter = 3 μm). The trace of the 3D path is shown on the right.

In its most general operating mode, the axial trap position is also treated as an active control parameter, like the transverse trap position, and is no longer confined by a fixed polarization landscape, as in the previous demonstration. This entails tracking each trap and independently programming polarization modulation on each of them. In this case, the polarization encoding is mathematically described as

$$\psi(u',v';t)=\pi\sum_{n=1}^{N}\gamma_n(t)\,f_n\left(u'-u_n'(t),v'-v_n'(t)\right), \tag{8.4}$$

where the spatial profile now follows the input phase, $\phi(u,v)$, where each trap has an associated user-defined amplitude factor, $0\leq\gamma_n(t)\leq 1$. Since $\psi(u',v')$ determines the relative strengths of the orthogonal polarization components, the trap-specific factor, $\gamma_n(t)$, enables independent axial control for each trap.

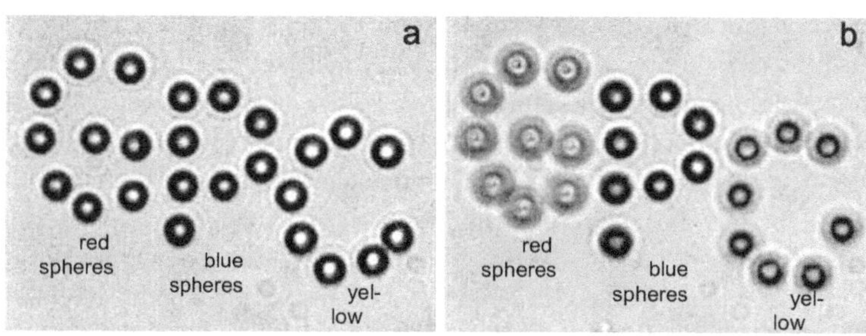

Fig. 8.21 User-coordinated patterning of commercially dyed polystyrene spheres (diameter = 3 μm) in (a) 2D; and (b) 3D, (N = 25), forming the text "GPC", using a distinct sphere-colour for each character. In **b**, virtual planes z = −5 μm, z = 0, and z = +5 μm intersect the centres-of-mass of the red, blue, and yellow-dyed spheres, respectively.

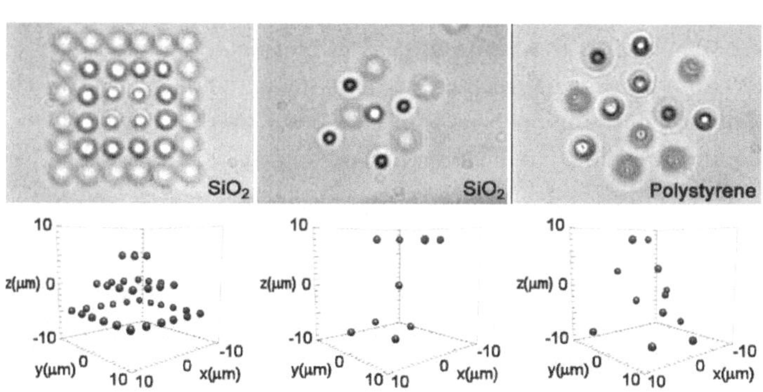

Fig. 8.22 Optical trapping (N_{max} = 36) and manipulation into 3D constellations of colloidal microspheres (SiO_2, diameter = 2.25 μm; polystyrene, diameter = 3 μm), top. 3D rendered view of the spheres' relative positions, bottom.

The various operation modes enable genuine real-time interactive 3D optical trapping and manipulation of large arrays of particles (see Fig. 8.21 and Fig. 8.22). The capacity for interactively forming colloidal constellations and assigning user-defined dynamics can have several profound implications. A GPC-based system allows access into exciting new experiments involving arbitrarily patterned or dynamically driven systems of colloidal particles, such as colloidal arrays created to model systems that are unmanageable or inaccessible within atomic and molecular domains [26]. It can serve as a non-invasive tool for microbiologists, such as for manipulating spatial cell configurations to investigate interaction that may affect developmental features (see Sect. 8.2).

A GPC-based system for manipulating multiple particles in 3D offers several advantages over other alternatives. Real-time user-interactive manipulation can easily support scaling to larger particle arrays since GPC allows high photon efficiency without computational overhead. In contrast, similar scaling in approaches based on holographic or diffractive optics (DO) [27, 28] can suffer computational setbacks from the associated iterative optimization algorithms. In a GPC-based system, the traps can be stretched out along the entire microscope field of view. Photon efficiency in a DO based system depends on the spatial trap locations, and may be compromised by several factors, such as: (1) the limited space-bandwidth product (SBP) of the SLM; (2) deleterious zero-order and higher-orders and; (3) Potentially strong spherical aberrations along the depth dimension associated with the use of high-NA objective lens focusing. The particular implementation of a GPC-based real-time 3D manipulation system described here uses two key devices: (1) an SLM to modulate phase at the GPC input and (2) an SPM to modulate polarization of light at the GPC output. It is intuitively easy to appreciate that the GPC approach, which distributes transverse and axial trap control to the two devices, sets more reasonable device requirements over a DO-based method that demands synthesis of a 3D light distribution on a single 2D phase-encoding device. Given practical device constraints, better performance can thus be expected from the distributed approach. Moreover, the simplicity of the binary spatial phase modulation and the spatial polarization modulation schemes in a GPC-based system implies the future use of low-cost liquid crystal display technology. Finally, the counterpropagating geometry enables stable trapping with lower NA objectives, which opens a much more liberal working volume for the sample. The implications of this final point are explored in the next section.

8.5 GPC-Based Optical Micromanipulation of Particles in Three Dimensions with Simultaneous Imaging in Two Orthogonal Planes

As demonstrated in the previous section, GPC used in tandem with an ingenious polarization modulation scheme, creates an optical system for real-time interactive assembly of multiple microparticles into 3D configurations using counterpropagating-beam traps. A GPC-based system traps particles on a transverse plane using gradient forces and uses the polarization-modulated counterpropagating beams to achieve equilibrium on axial scattering forces [1, 23]. A trapping beam that matches the size of a given particle can be generated using microscope objectives with different numerical apertures (NA) by either changing the lens before the objective or adjusting the size of the SLM patterns to compensate. This grants flexibility for choosing microscope objectives to suit experimental needs. Since moderately focused counterpropagating beams can form stable traps, we can choose to implement a GPC based 3D trapping system using low numerical aperture, non-immersion, objective lenses. The low-NA implementation of a GPC trapping system offers a wider manipulation region and covers a larger imaging field of view and allows for a long working distance (>10 mm) between the two opposing objective lenses. A large working distance can be vital as it removes restrictions on sample chamber size and allows accessory optical systems to be implemented along the orthogonal axis. To illustrate this possibility, we can implement an imaging system along this orthogonal axis for simultaneous monitoring of the trapped particles on orthogonal observation planes.

The large axial positioning range in GPC-based counterpropagating-beam traps means that particles can be positioned well beyond the plane of best focus in the microscope. This can result in highly blurred and hardly visible images, which makes observations difficult. We can solve this problem of highly defocused particles by exploiting the flexibility to choose between different NA values to select low-NA objectives with long working distances (WD). This increases the WD by more than one order of magnitude, from less than 0.5 mm in previous implementations to 10.6 mm. This free region from objective lens to the trapping region cannot be obtained in conventional optical tweezers and it conveniently removes previous constraints on the sample chamber dimensions. This is a desirable feature for microfluidics-related applications, especially those that require accessories such as space-consuming tube connectors for the flow system.

As a sample illustration of the utility of having a wide working space, let us consider an accessory imaging system fitted along the orthogonal axis. A schematic showing the setup and the workflow associated with optical trapping is illustrated in Fig. 8.23. The GPC-based trapping system is essentially the same as in the previous section, but uses two ×50, NA = 0.55 objectives (Olympus LMPLFL), both with 10.6 mm WD. Side-view images are captured with a CCD camera through a $f = 150$mm lens and a ×20, NA = 0.42 objective (Mitutoyo MPlanApo) with 20.0 mm WD. An easily assembled and disposable sample chamber is custom-designed to contain a window for side-view access.

The accessory system makes it possible to obtain volumetric particle information by observing simultaneously on two orthogonal planes to get top- and side-view images. This makes it easy to monitor particle dynamics in the whole trapping volume by direct observation. Streaming images acquired from cameras looking from the top and side can be integrated in a customized graphical user-interface. This information provides essential feedback that can help a user to make decisions or, alternately, be computer-processed and analyzed in an automated particle manipulation scheme. This setup gives a view from the side into the sample volume and provides a frame of reference by showing a perspective view of the sample bottom surface.

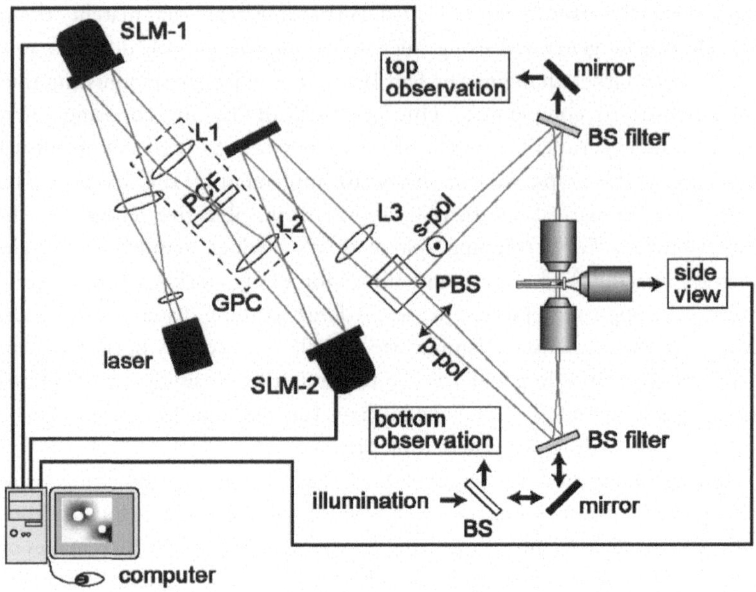

Fig. 8.23 Schematic diagram of the optical setup. L1, L2, and L3 are achromatic lenses. Identical lenses L1 and L2 have a focal length of 300 mm, while L3 has 400 mm. Top and bottom objectives are both ×50 and a ×20 objective is used for side view. The computer acquires video input from two of the cameras and controls SLM-1 and SLM-2 based on user interaction.

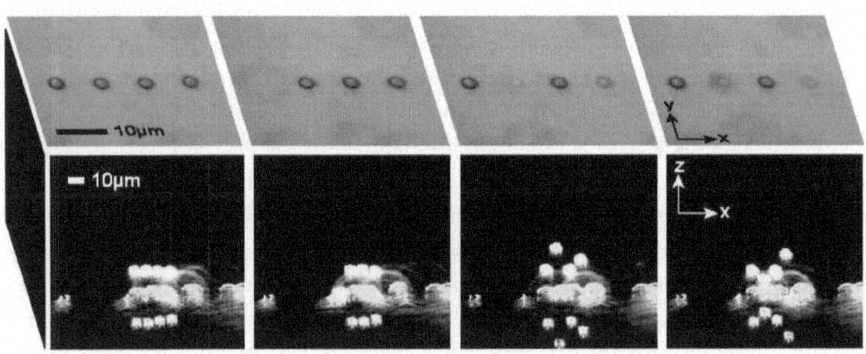

Fig. 8.24 Composite image from snapshots of concurrent top- and side-view imaging showing four 3 μm polystyrene beads optically trapped and manipulated in a GPC-based system. All beads are initially 15 μm above the bottom surface. The beads were manipulated through user-interactive control with ~30 μm axial dynamic range. The side-view (*x-z*) imaging system also captured reflections of the beads from the lower glass surface. The top-view (*x-y*) imaging system sharply images a plane 15 μm above the bottom surface; out-of-focus beads are easily identified with the auxiliary side-view system.

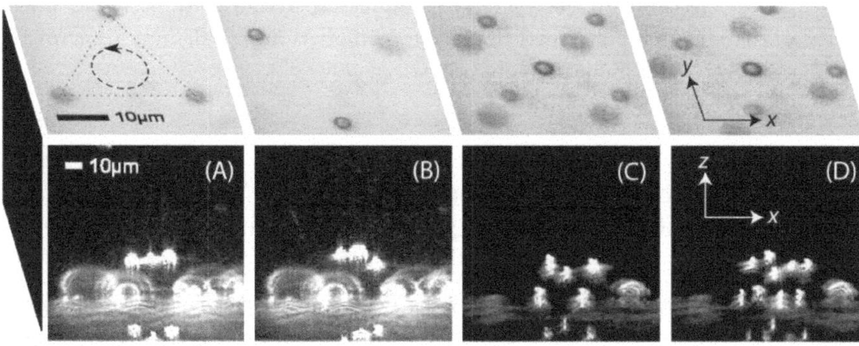

Fig. 8.25 Composite image from snapshots of concurrent top- and side-view imaging of optically trapped and manipulated microparticles in a GPC-based system. **A–B**: Three 3-µm-sized beads trapped at the vertices of an imaginary triangle coast along an imaginary "roller-coaster track" where each particle encounters an assigned depth-stroke during each revolution. **C–D**: Nine 3-µm-sized beads are arranged in a configuration resembling a body-centred cubic unit cell and the whole structure is rotated about the *z*-axis.

A sample chamber with viewing windows at the top, bottom and the side was constructed using microscope cover glass of varying sizes glued together by UV-curable adhesive. A dilute aqueous solution of 3.0 µm polystyrene beads (Polysciences) was introduced through capillary forces and the channel was sealed with UV adhesive. The laser provided 5mW at each counterpropagating-beam trap at 830 nm wavelength. Each trap approximately matched the size of the beads. Concurrent side- and top-view images of optically trapped and dynamically manipulated particles acquired from the setup are shown in Figs. 8.24 and 8.25.

The counterpropagating beams in a GPC system do not require tight focusing and rely on opposing scattering forces to achieve stable axial trapping. This reduces the light intensity needed to trap particles, as compared to standard optical tweezers, and grants flexibility in choosing the numerical aperture of the microscope objectives. Thus, one may choose a large field of view and volume of manipulation, if needed. We have seen that a GPC-based trapping architecture achieves previously unseen working distance by utilizing low-NA objectives. This, in turn, offers enhanced data acquisition possibilities, such as the demonstrated concurrent top- and side-view observations. The gathered axial data may also be obtained by confocal microscopy [29], but dual imaging allows simultaneous orthogonal observation in real-time. Full volumetric particle position information may be used for calibrating and fine-tuning the counterpropagating beams and for observing dynamics of particles located beyond the imaging region of a top view microscope. Side-view imaging with multi-particle trapping can be used to study various particle interactions, such as axial optical binding [30], as well as other inter-particle interactions that can arise in counterpropagating fields. These may lead to interesting novel schemes for constructing denser 3D colloidal crystals. Finally, the large working distance gives a freedom to observe and manipulate particles in devices (e.g. microfluidics and lab-on-a-chip systems) that are too cumbersome to fit into a microscope with high-NA oil/water immersion objectives.

8.6 All-GPC Scheme for Three-Dimensional Multi-Particle Manipulation Using a Single Spatial Light Modulator

Minute radiation pressure forces from laser beams are normally strong enough to influence the motion of microscopic and nanoscopic particles. A great deal of progress has been achieved in optical trapping techniques and applications since the pioneering demonstrations by Ashkin. For example, optical trapping and manipulation of ensembles of microparticles, which opens for promising themes of studies within colloid science and microbiology, are now viable using reconfigurable patterns of optical fields [31, 32, 33, 34]. Beam modulation techniques employing computer-programmable spatial light modulators (SLM) can produce reconfigurable optical potential landscapes.

The generalized contrast uses SLM technology to synthesize reconfigurable light patterns that, coupled through microscope objectives, can be used to trap microscopic particles and cellular organisms in real-time. In the previous sections, we discussed a GPC-based system that generates multiple independently steerable counterpropagating-beam (CB) traps for three-dimensional (3D) interactive manipulation [31, 32]. The scheme employed two spatially addressable electro optic devices operated in complementary modes: (1) as a phase-only modulator (SPhM) that provides reconfigurable input in GPC-based light shaping for transverse particle control and (2) as spatial polarization modulator (SPoM) that controls the power ratio in each CB trap for axial positioning. The counterpropagating trap geometry provides ample space for the sample chamber that is otherwise impossible using optical tweezers that relies on tight beam focusing with high NA objectives.

In this section we consider a full-GPC implementation of a multi-particle 3D trapping and manipulation system. The current scheme uses a single phase-only SLM and does not need a polarization modulator. In the earlier system, GPC created an on-axis set of trapping patterns that was duplicated by beam splitting to produce counterpropagating beams. The current system uses GPC to simultaneously create two sets of trapping patterns that are then independently steered to serve as counterpropagating beams in the trapping volume. The transverse intensity profile governing transverse confinement, as well as the power ratios of counterpropagating beams for controlling axial position, is fully controlled by the single SLM at the GPC input. In the following, we will outline theoretical and experimental considerations for implementing GPC with a dual-beam illumination. We will also describe the single-SLM, full-GPC micromanipulation system and demonstrate its use for multiple-particle 3D trapping experiments. This exploits developments in spatial light modulators to boost the degree of control in particle manipulation. Contemporary modulators with resolutions such as 1920×1080 pixels can address more pixels than a 4×2 arrangement of older SLMs with only 480×480 pixels.

8.6.1 GPC system with Two Parallel Input Beams

There are several options for using GPC to generate two sets of light patterns that can be introduced as counterpropagating beam traps in a sample chamber. Using a circular input aperture and a circular PCF, one may encode duplicate phase patterns on each half-circle. It is also possible to use the full rectangular clear aperture of an SLM together with a rectangular PCF and encode duplicate phase patterns on each half-side of the SLM. However, considering that the trapping pattern will eventually be projected within the circular field of view of a microscope, it makes sense to implement a GPC scheme with two parallel circular input beams with a circular PCF. We will focus our attention on the third scheme since the first two schemes correspond to standard GPC configurations already described in previous chapters.

Consider a GPC setup illuminated by two off-axis circular beams. In this case, the input field, $e(x,y)$, and phase contrast filter (PCF) transfer function, $H(f_x,f_y)$, are defined respectively by

$$\begin{aligned} e(x,y) = {} & \mathrm{circ}\left(\sqrt{(x+x_0)^2+y^2}/\Delta r\right)\exp\left(j\phi_1(x+x_0,y)\right) \\ & +\mathrm{circ}\left(\sqrt{(x-x_0)^2+y^2}/\Delta r\right)\exp\left(j\phi_2(x-x_0,y)\right) \end{aligned} \tag{8.5}$$

$$H(f_x,f_y)=\left[1+\left(\exp(j\theta)-1\right)\mathrm{circ}\left(\sqrt{f_x^2+f_y^2}/\Delta f_r\right)\right]. \tag{8.6}$$

The circ-functions in Eq. (8.5) represent two off-axis top-hat beams (with symmetric lateral shifts, x_0 and identical radii, Δr) that read out input phase patterns, ϕ_1 and ϕ_2, written on regions R_1 and R_2 of the SLM, respectively, as shown in Fig. 8.26. The axially centred Fourier filter introduces a phase shift, θ, to spatial frequency components within a radius Δf_r.

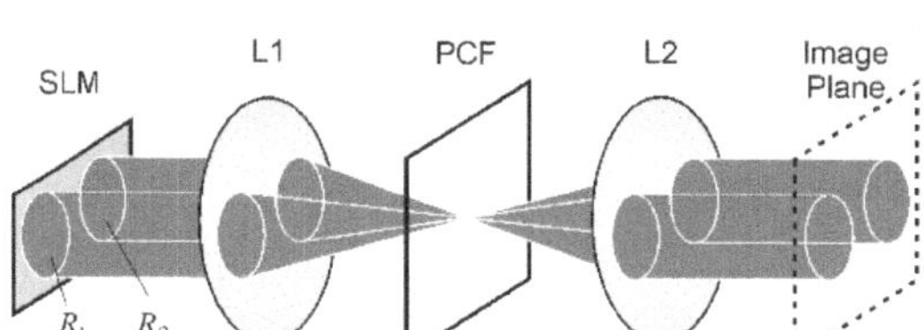

Fig. 8.26 $4f$ setup for implementing the GPC method with dual-beam illumination onto circular regions R_1 and R_2 of the spatial light modulator; PCF, phase contrast filter; L1 and L2, lenses (focal length = 300 mm).

A GPC system with binary encoded inputs (0 and π) and PCF filter shift, $\theta = \pi$, generates an output given by (neglecting space-inversion of the $x'y'$-coordinate system)

$$\begin{aligned} o(x',y') &= \Im^{-1}\{H(f_x,f_y)\Im\{e(x,y)\}\} \\ &= \mathrm{circ}\left(\sqrt{(x'+x_0)^2+y'^2}/\Delta r\right)\exp\left(j\phi_1(x'+x_0,y')\right) \\ &\quad -2\bar{\alpha}_1\Im^{-1}\left\{\mathrm{circ}\left(\sqrt{f_x^2+f_y^2}/\Delta f_r\right)\Im\left\{\mathrm{circ}\left(\sqrt{x^2+y^2}/\Delta r\right)\right\}\exp\left(j2\pi f_x x_0\right)\right\} \\ &\quad +\mathrm{circ}\left(\sqrt{(x'-x_0)^2+y'^2}/\Delta r\right)\exp\left(j\phi_2(x'-x_0,y')\right) \\ &\quad -2\bar{\alpha}_2\Im^{-1}\left\{\mathrm{circ}\left(\sqrt{f_x^2+f_y^2}/\Delta f_r\right)\Im\left\{\mathrm{circ}\left(\sqrt{x^2+y^2}/\Delta r\right)\right\}\exp\left(-j2\pi f_x x_0\right)\right\} \end{aligned} \tag{8.7}$$

where

$$\bar{\alpha}_i = \left[\pi(\Delta r)^2\right]^{-1}\int_{R_i}\exp\left(j\phi_i\right)\mathrm{d}x\mathrm{d}y\ ;\ \text{for } i = 1, 2 \tag{8.8}$$

In single-beam GPC, the term $\Im^{-1}\left\{\mathrm{circ}\left(\sqrt{f_x^2+f_y^2}/\Delta f_r\right)\Im\left\{\mathrm{circ}\left(\sqrt{x^2+y^2}/\Delta r\right)\right\}\right\}$ is interpreted as a so-called synthetic reference wave (SRW) and is suitably described by a circularly symmetric function $g(r')$:

$$\begin{aligned} g(r') &= \Im^{-1}\left\{\mathrm{circ}\left(f_r/\Delta f_r\right)\Im\left\{\mathrm{circ}\left(r/\Delta r\right)\right\}\right\} \\ &= 2\pi\Delta r\int_0^{\Delta f_r} J_1\left(2\pi\Delta r f_r\right)J_0\left(2\pi r' f_r\right)\mathrm{d}f_r \end{aligned} \tag{8.9}$$

where $r' = \sqrt{x'^2+y'^2}$ and $f_r = \sqrt{f_x^2+f_y^2}$. Similar Fourier transform terms in Eq. (8.7) may also be interpreted as SRW, where by using the Shift Theorem [35], linear phase factors in the Fourier plane provide the necessary lateral displacements at the output plane so that they interfere with their respective off-axis signals. Thus, we can have a simple picture for describing the system by writing the output intensity $I(x',y')$ of the dual-illuminated GPC setup as

$$I(x',y') = \left|o_1(x',y') + o_2(x',y')\right|^2, \tag{8.10}$$

where

$$o_1(x',y') = \mathrm{circ}\left(\frac{\sqrt{(x'+x_0)^2+y'^2}}{\Delta r}\right)\exp\left(j\phi_1(x'+x_0,y')\right) - 2\bar{\alpha}_1 g\left(\sqrt{(x'+x_0)^2+y'^2}\right) \tag{8.11}$$

and

$$o_2(x',y') = \mathrm{circ}\left(\frac{\sqrt{(x'-x_0)^2 + y'^2}}{\Delta r}\right)\exp\left(j\phi_2(x'-x_0,y')\right) - 2\bar{\alpha}_2 g\left(\sqrt{(x'-x_0)^2 + y'^2}\right) \tag{8.12}$$

Any significant overlap between $o_1(x,y)$ and $o_2(x',y')$ can be avoided by choosing a sufficiently large shift value x_0. Under this condition, we can interpret the dual-beam system as being two independent GPC systems compactly implemented to share the same PCF.

Figure 8.27 shows that experimental results are in good agreement with the theoretical results for the two-beam input GPC illustrated in Fig. 8.26. Without a PCF, Fig. 8.26 becomes a 4*f* imaging system and simply reproduces the two circular input illuminations, as shown in Fig. 8.27(a). Figures 8.27(b) and 8.27(c) show high-contrast intensity patterns obtained upon introducing a PCF. The PCF is a glass optical flat with a tiny cylindrical pit with depth chosen to produce $\sim\pi$ phase shift and radius of ~7.5 μm chosen to match the input aperture [16]. The intensity linescans in Fig. 8.27 show that the output intensity can reach up to about four times the input intensity level, as in single-beam GPC. The output shown in Fig. 8.27(b) uses a binary input phase with π-phase discs on a 0-phase background. The dual outputs can be used to form counter-propagating-beam traps with matching power. The intensity patterns in Fig. 8.27(c),

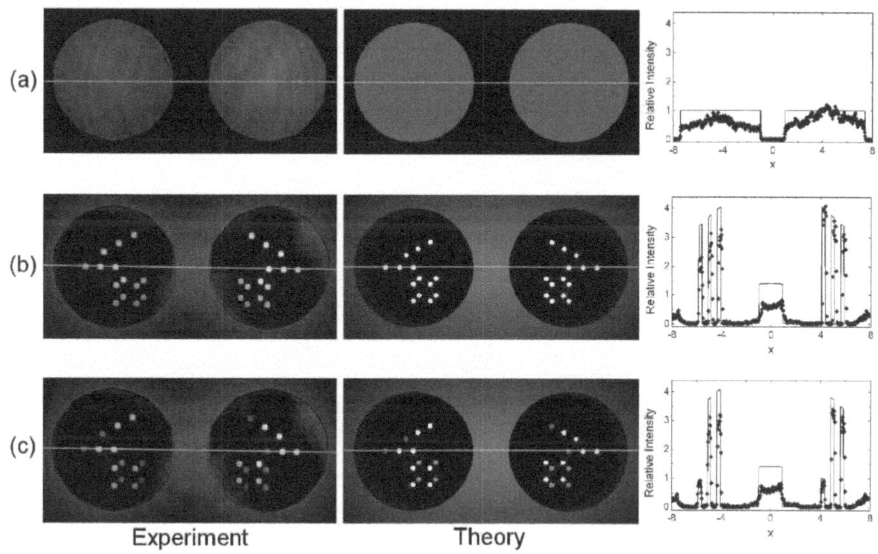

Fig. 8.27 Comparison of theoretically (solid curve in the line-scan) and experimentally obtained intensity patterns at the image plane of a GPC 4*f* setup with two adjacent input beams (modelled with top-hat intensity profiles) (a) in the absence of the PCF, (b) with an aligned PCF and a binary phase-dot array input, and (c) with an aligned PCF and a multilevel phase-dot array input. Line-scans are taken along the green lines.

which uses multilevel input phase encoding on the SLM, can be used to form CB traps with different power ratios. The intensity of individual light discs is controlled by the encoded phase on the corresponding input phase disc on the SLM. Sweeping the encoded phase from zero to π sweeps the corresponding output intensity from minimum to its maximum level.

8.6.2 Single-SLM Full-GPC Optical Trapping System

The dual output of a GPC system with dual-beam illumination can be scaled and relayed as counterpropagating beam traps as a full-GPC alternative to the GPC-SPM (spatial polarization modulator) optical trapping system previously described [31, 32]. Figure 8.28 shows a schematic of an optical trapping system with a folded GPC for a more compact implementation. To create two readout beams, the two-beam illumination system (not shown) employs a beam splitter and steering optics to create duplicate images of a circular iris illuminated by an expanded laser beam (Nd:YVO_4 laser, Laser Quantum Excel; wavelength = 532 nm; maximum power = 1.5 W). The two sets of trapping intensity patterns from the dual-beam GPC setup, separated by a right-angle mirror, are separately scaled and relayed into the sample through identical long-working-distance low-NA objective lenses (Olympus LMPLFL) to form the array of CB

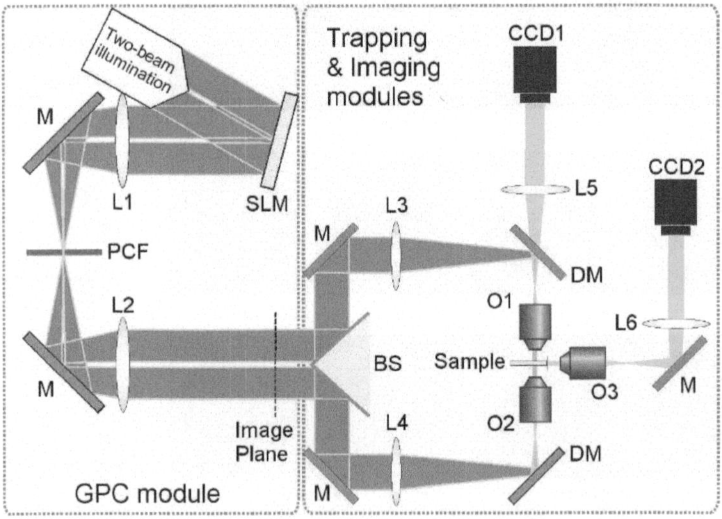

Fig. 8.28 Schematic diagram of the proposed optical micromanipulation system. SLM, spatial light modulator; PCF, phase contrast filter; M, mirror; L1 and L2, achromats (focal length = 300 mm); L3 and L4, achromats (focal length = 400 mm); L5 and L6, singlets (focal length = 300 mm and 200 mm, respectively); BS, beam splitter, DM, dichroic mirror; O1 and O2, trapping objective lenses (x50, NA = 0.55); O3, *yz*-view objective lens (×50, NA = 0.55); CCD1, *xy*-view camera; CCD2, *yz*-view camera.

traps. A graphical user-interface (GUI) developed using LabVIEW displays microscope images of the sample and allows user-directed light synthesis of trapping patterns. The GUI provides a user-friendly point-and-click and drag-and-drop features to interactively select and position CB traps along the *xy*-plane with real-time video overlay of streaming microscope images from the sample. The GUI can also used to set the power ratios of individual CB traps using the multilevel phase encoding scheme described in the previously.

One unique feature of a GPC-based trapping system is its capability to monitor the trapping volume in two perpendicular views [36]. The use of low-NA (air immersion) objective lenses provides sufficient working space, as shown by the earlier demonstration of this imaging modality, where lateral (*xy*-plane) bright-field imaging and side (*yz*-plane) laser scatter imaging of the trapped particles were accomplished. The available space allows various imaging modalities to be incorporated along the orthogonal axis.

Figure 8.29 shows *xy* and *yz* brightfield images concurrently obtained from 3D-assemblies of microparticles held by CB traps. These images, displayed on the GUI, are vital for a user who uses the same GUI to control the lateral trap profiles and adjust the power ratio of each CB trap to achieve desired 3D control. The three microspheres displayed in Fig. 8.29(a) are stably trapped at distinct positions with the lowermost and topmost spheres axially separated by ~30 μm. The beads appear diagonally aligned on both views after shifting the *xy* positions of the two outer particles, as shown in Fig. 8.29(b). Figure 8.29(c) shows trapped microparticles that have been interactively assembled into a rhomboid structure using eight CB traps.

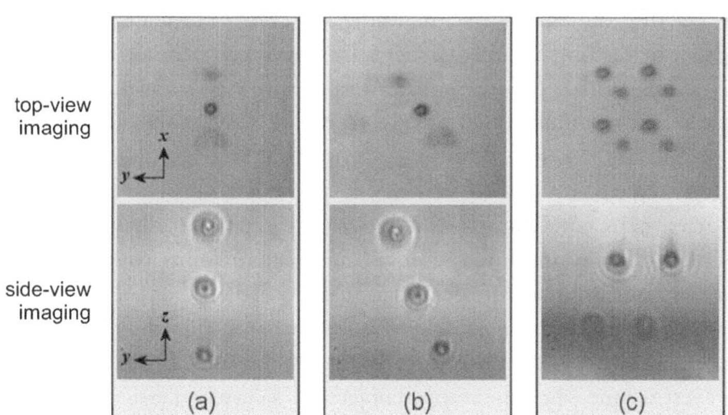

Fig. 8.29 Optically assembled arrays of 3-μm diameter polystyrene spheres in 3D simultaneously viewed in *xy* (top frame) and *yz* planes (bottom frame). (a)–(b) Two of three spheres are translated in the *xy*-plane. (c) Eight spheres optically positioned and stably kept in the corners of a virtual parallelepiped.

A GPC-system can trap more than one particle along the axial direction by axially stacking particles, if required. The axial particle spacing may be controlled by exploiting the so-called optical binding effect (particle interaction due to light-scattering). Holographic optical tweezers can create axial traps for two or three beads with particle spacing constrained by particle shadowing and the spherical aberrations and short confocal range of a high-NA focusing objective [33, 34].

These results, together with the earlier theoretical analysis, illustrate that a single SLM may be used in a GPC setup to implement a 3D real-time micromanipulation system that uses a plurality of independently controllable counterpropagating-beam traps. By utilizing counterpropagating-beam traps, the accessible volume for particle manipulation in a GPC-based system can be tuned using a wide selection of objectives with different NAs. Other trapping systems based on tightly focused beam gradient-traps (e.g. holographic trapping [33, 34, 37, 38]), have a limited manipulation range due to the necessary use of oil or water immersion high-NA objectives. The flexibility in choosing microscope NA can be utilized to implement an accessory bright-field side-view imaging module. This can be used to acquire orthogonal bright-field imaging of the sample volume to achieve a genuine visual inspection and measurement of optically assembled microparticles in 3D. The flexibility in adding an accessory system can be employed in future extensions to include other useful spectroscopic and imaging modalities (e.g. fluorescence, Raman).

8.7 GPC-Based Optical Actuation of Microfabricated Tools

Optical forces permit the non-contact handling and controlled manipulation of microscopic objects. Spherically symmetric particles are readily trapped in a three-dimensional potential well created either by a pair of counterpropagating Gaussian beams [1] or by a single-beam gradient trap known as optical tweezers [20]. Microfabricated objects with anisotropic geometries may also be actuated by optical fields. One of the key motivations for fabricating custom-shaped structures is to realize miniaturized machines that can mimic, or perhaps even outperform, their macroscopic analogues. It may also be possible to realize micromachines without any macroscopic counterpart. Optical tweezers have been utilized both for translational [39, 40] and angular [40] control of tools such as a micromechanical mass-spring system, and hinged submicron tweezers and needles. Optical tweezers can transfer linear momentum to an illuminated object and this can exert torque when the trapped object has a custom-fabricated shape [41, 42, 43]. Micromachined birefringent materials align with the polarization direction of the optical tweezers [44] and can be rotated about its axis at rates up to hundreds of Hertz using circularly polarized light [45, 46].

Earlier discussions of GPC-based multitrap systems focused on an operating mode where each trap controls an independent particle. Here we describe another operating mode where multiple traps work synchronously to control a single particle, like fingers grabbing a single tool for better manoeuvrability. Its journal publication marked the first demonstration of a combined use of *multiple* real-time reconfigurable optical traps coordinated to efficiently manipulate microfabricated objects. This represented a departure from the prevailing approach at the time, where a *single* laser trap actuates one microtool to achieve light-driven translational and rotational control of microfabricated elements. Manipulation of shape-defined structures with multiple traps has not been previously exhibited either with time-shared, holographic optical trap arrays.

This section highlights the value of a GPC-based array of counterpropagating-beam traps in actuating microstructures. Synchronized multiple counterpropagating beam traps are generated using the GPC-SPM technique described in Sect. 8.4. A GPC-synthesized high-contrast intensity pattern is spatially polarization-modulated and decomposed into orthogonally polarized beams that are then subsequently scaled and projected from opposite sides of a sample chamber to form multiple counterpropagating-beam traps [31, 32, 36]. A GPC-based method can realize various actuation schemes (i.e. translational, rotational and angular control) with its plurality of optical manipulators. The illustrative demonstrations are shown using a fabricated microstructure having flat coin-like handles. These specific structures may be highly difficult to manipulate in the same manner using holographic or time-shared tightly focused optical traps since coin-like microdisks are known to align their longest diagonal with a laser tweezers' propagation axis, found in both theoretical [47] and experimental [48] considerations.

8.7.1 Design and Fabrication of Micromachine Elements

The microstructures were fabricated by etching elements from a uniformly deposited thin layer of SiO_2 material using masks with lithographically defined shapes. The substrate was prepared by uniformly depositing 1.0 μm-thick layer of SiO_2 on both sides of a standard silicon wafer (100). The mask was prepared on another wafer where a 1.5 μm-thick photoresist was spin-coated, followed by a 30-minute HMDS (hexamethyldisilazane) treatment in an oven. Photoresist masks for the desired microtool shapes were patterned by standard UV lithography. Developed photoresist masks were transferred to the SiO_2 layer by anisotropic reactive-ion etching and remaining resist residue was removed with acetone. The microtools were under-etched from the wafer and released into an isotropic silicon etch consisting of 1:2 HF and HNO_3. The solution containing the released microtools was then passed through a mechanical filter followed by a water rinse to stop further etching of the tools and to remove remaining HF and HNO_3. The water solution containing the microtools was dried by slow heating in a Petri dish and finally examined in a microscope.

SEM images of sample microtool structures are shown in Fig. 8.30. The prototype microtool designs were based on simple polygons equipped with multiple coin-like disks as handles for the counterpropagating-beam traps. The adopted fabrication procedure can synthesize flat structures with defined lateral shapes and uniform thickness. The photolithographic procedure is similar to semiconductor fabrication and can employ similar technologies for efficient mass production. Photolithography offers vast parallelism due to the small size of the individual microstructures, giving literally millions of structures from a single four-inch silicon wafer. Multilayer structures can be fabricated by multiple electron beam lithography steps and even more intricate 3D structures with fine features down to an order of 100 nm could be synthesized using two-photon polymerization techniques [39, 40].

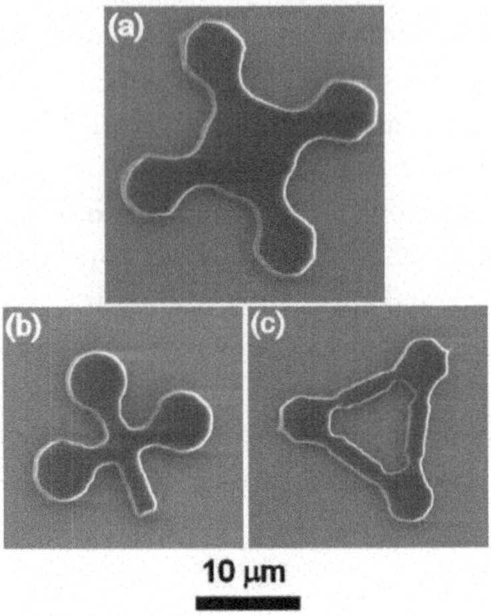

Fig. 8.30. SEM images of SiO_2 microfabricated structures.

8.7.2 *Actuation of Microtools by Multiple Counterpropagating-Beam Traps*

The microtools were optically actuated using reconfigurable arrays of counterpropagating-beam traps, which may be synthesized by the alternate methods described in Sects. 8.4 and 8.6. The multiple traps were coaxially projected from opposite sides of a sample through two identical objective lenses to create a collection of three-dimensional (3D) optical traps in the sample region. The original demonstration used the GPC-SPM method described in Sect. 8.4, where opposing beams have orthogonal polarizations and the power ratio in each pair is adjustable while preserving the power sum. These features enabled 3D particle control where the user can easily specify the number of traps, the shape, size and 3D position of each trap. These features were essential in GPC-based 3D optical manipulation of large ensembles of independent colloidal microspheres [32]. These same features are also vital when powering and actuating specially fabricated and shaped glass structures with micrometer dimensions and submicron features.

Optical manipulation of functionally shaped micro/nanofabricated structures can have future applications that can complement and potentially surpass those achievable with colloidal spheres. Figure 8.31 depicts one of the fabricated prototype microstructures to illustrate available control modes in a GPC-based multitrap system for controlled optical manipulation of microtools. Figure 8.31(a) shows a reference orientation where user-defined optical traps hold a microstructure by their rounded handles. The microstructure is oriented perpendicular to the optical axis (z-axis) using counterpropagating beams with identical power. Figures 8.31(b)–(d) show available rotational and translational control.

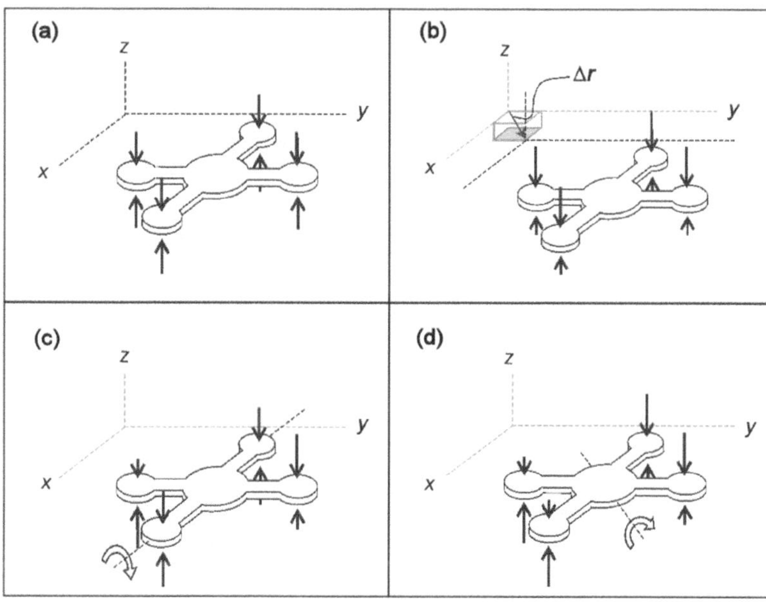

Fig. 8.31 Explanatory illustration of an optically actuated four-limbed microfabricated structure. (a) Respective counterpropagating-beam traps, formed by s-polarized (down-pointed arrows) and p-polarized (up-pointed arrows) beams, are positioned at all four circular handles. Arrow length is proportional to beam power. (b) Purely translational displacement $\Delta \boldsymbol{r}$ of the structure in 3D. (c) Angular orientation of the structure about a line parallel to the x-axis. (d) Angular orientation about a line at 45° with x-axis and y-axis.

Figure 8.31(b) shows the whole microstructure displaced in 3D without any angular displacement. The transverse component of the microstructure displacement is controlled by coordinated transverse displacement of the traps while the axial displacement is governed by the synchronized change in the power ratios of the counterpropagating traps. The transverse and axial displacements may be implemented sequentially or simultaneously. Tilting or rotating the structure entails coupled axial and transverse displacement. Figures 8.31(c) and 8.31(d) show rotation or tilt about different horizontal axes (lying on the *xy*-plane) where the arrow lengths depict the relative power of the counterpropagating beams. Combined with structure rotation about the optical axis (not shown), these control modes form a basis set that can be combined to achieve a desired particle orientation.

Dry microstructures were extracted from a Petri dish by repeated discharge and suction of 100 μL distilled water using a manual pipette. The solution was introduced by capillary action into a sample chamber made from bonded pieces of microscope coverglass. The sample was mounted on a motorized *xyz*-translation stage for high-precision positioning between two opposing objective lenses. Digital video microscopy monitored and grabbed images of the structures through one of the objective lenses.

Figure 8.32 shows images acquired from experiments on controlled manipulation of a four-limbed microstructure, which demonstrates different actuation schemes. Applying the procedures described in Fig. 8.31 yields the following controlled manipulation: Fig. 8.32(a) shows the structure being tilted; Fig. 8.32(b) shows the structure being lifted along the *z*-axis, making its image out of focus. In Fig. 8.32(c), the structure can be rotated about its centre (at typical rates of ~4.5 rpm) while keeping it lie on the *xy*-plane. This is done by setting the four counterpropagating-beam traps in synchronized circular orbit to maintain them in square-configuration. The rotation can be set clockwise or anticlockwise, as desired.

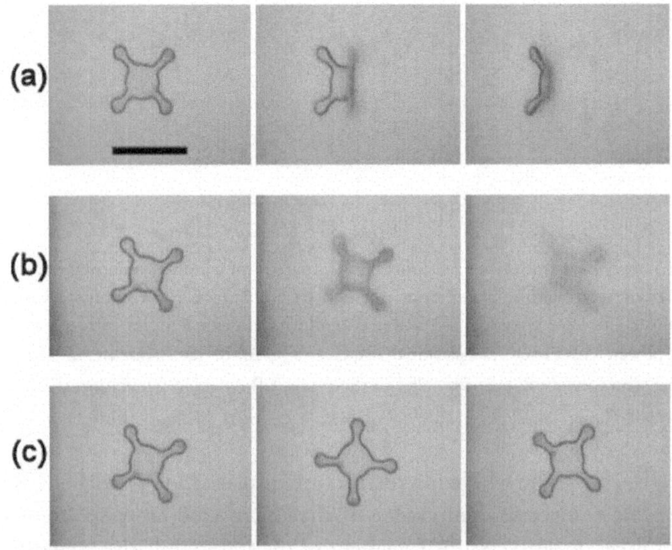

Fig. 8.32 Experimental results showing optical actuation of a microstructure. (a) Angular reorientation showing approximately 80° maximum angular deflection, which allows visual inspection of the structure's flatness. (b) Axial displacement of the structure. (c) Rotation of the structure about a normal axis through its centre. Scale bar, 20 μm.

These results show the capacity of a GPC-based trapping system to create multiple counterpropagating-beam traps for real-time and computer-controlled actuation of microfabricated structures. The synchronized operation of optical traps enabled us to translate, rotate, or angularly tilt microtool-prototypes in 3D. The ability of a GPC-based system to rapidly specify the number of traps and reconfigure them based on specific needs can be a vital element in a future optical workbench where micro-elements may be integrated and assembled by light into functional upper-hierarchy micromachines. Such integration is expected to require a larger number of trapping beams and thus requires high-power laser sources. The power tolerance of the liquid-crystal-based SLMs being used in GPC may likely become a constraint. However, a next-generation GPC-based optical trapping system with a relatively high input laser power tolerance is currently being developed [49].

Trapping stability and actuation control can be further improved by optimizing the microstructures, such as fabricating spherical handles and/or employing higher refractive index materials such as polymers. Furthermore, the ability to precisely customize the shape of microtools and functionalize them towards specific applications is expected to be a key issue for unlocking their full potential.

8.8 Autonomous Cell Handling by GPC in a Microfluidic Flow

The use of optical forces has, over the past decades, been proven efficient and versatile for non-invasive manipulation of mesoscopic objects. Recent advances on specially tailored structures of light have brought about more versatile and general manipulation of particles and cell colonies, such as in organizing small particles, including microorganisms, in desired patterns and to sorting samples of particles according to their attributes, to name a few applications [37, 38, 50]. Meanwhile, parallel developments in microfluidic technologies have opened a new world of possibilities for the bio/chemical community and hybrid approaches have been proposed. Earlier demonstrations of laser assisted microfluidic systems have revealed exciting degrees of freedom, such as the ability to create an arbitrary environment for one or more selected cells, or even the simulation of *in vivo* conditions [51, 52, 53].

Next-generation smart platforms that integrate these unique technologies into fully parallel and real-time controlled micro- or optofluidic systems will rely heavily on automation to overcome the slow and serial nature of human interfacing. The generalized phase contrast (GPC) method shows promise in this arena due to its capacity for *true* real-time and parallel manipulation of a plurality of particles in a 3D environment [32]. This capacity, rooted on its straightforward design of phase inputs, has been instrumental in facilitating the human interface in earlier demonstrations. In this section we will now consider a computer-automated GPC-based system that works with multiple particles in parallel within a rapidly changing micro-environment, where cells and suspension media are continuously replenished at high flow rates. Key advantages arise when combining the GPC approach for optical trapping with a microfluidic system through two simple applications.

8.8.1 Experimental Setup

Optical System

A possible implementation of a microfluidic system featuring a GPC-based system for automated particle manipulation is depicted in Fig. 8.33. Multiple counterpropagating beams traps for handling particles in the microfluidic chamber may be synthesized using alternate GPC implementations described in Sects. 8.4 and 8.6. The original demonstration [54] used two 50×, NA=0.55 IR objectives (Olympus LMPL) with a 12mm working distance between the barrel tips. The light synthesis using the spatial light modulator and data acquisition from the CCD camera (JAI CV-M4+CL monochrome camera with 2/3" sensor) was controlled using a LabVIEW program using standard image acquisition (IMAQ) modules.

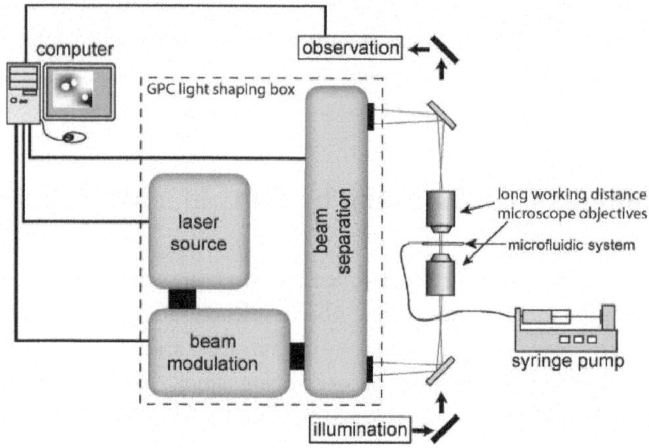

Fig. 8.33 Schematic diagram of the experimental setup. The long working distance between the objective lenses significantly eases the insertion of a microfluidic system. The computer undertakes multiple tasks such as receiving feedback from an observation module, processing the acquired data and lastly generating control signals used for addressing the spatial light modulation module.

Microfluidic System

The microfluidic system was comprised of a custom-designed rectangular channel made using ~170 μm thick microscope glass cover slips assembled using UV adhesive (Norland), as illustrated in Fig. 8.34. Two 60 mm × 24 mm pieces formed the top and base and two 24 mm × 10 mm and one 24 mm × 4mm were arranged to form a channel with two inlets (see Fig. 8.34). The final channel measured 170 μm × 340 μm × 15 mm. The width was chosen to reduce the ratio between pump rate and resulting flow velocity and thereby maximizing the range of experimental pump rates. Two clinical needles served as interface between the channel and the feeding tubes. The pump needle (0.8 mm external diameter) securely coupled to a short silicon tube (inner diameter of 0.5 mm) while a slightly smaller sample inlet needle (0.6 mm external diameter) was used to ease the insertion/removal of a small disposable syringe for sample injection. A 15 cm PTFE tube (inner diameter 0.25 mm) connected the microfluidic system to a syringe pump (Harvard model 11 plus). A selection of glass syringes (Hamilton) ranging from 25 μl to 1 ml was utilized to optimize the pump stroke and achieve the desired flow rates.

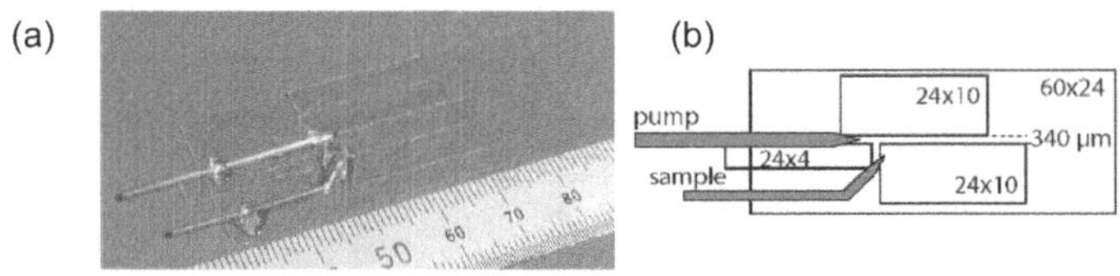

Fig. 8.34 (a) An image of one of the microfluidic systems and (b) the corresponding schematic showing the layout of component needles and microscope cover slips.

Sample Preparation

Yeast cells (*Saccharomyces cerevisiae*) were diluted in purified water to cell densities of approximately 1 to 50 cells per 100 μm x 100 μm area within the trapping region. Prior to introducing the yeast solution, the fluidic system (feeding tube and microsystem) was first primed with purified water, which removed air from the system. Fluid was then pumped at a rate in the order of 10 μl/hour through the pump needle using a glass syringe (typically 25 μl). The filled channel required minor optical realignment to compensate for variations between the different microfluidic systems – this included fine adjustments in the computer program to ensure pixel correlation between the SLM and the CCD. The yeast solution was then injected through the sample needle, taking care to keep any incidental air bubbles away from the flow.

The experiments demonstrated the key advantages of how an automated real-time trapping system can assist in the handling of a microbiological experiment, particularly when the amount of cells carried by the continuous flow might be overwhelming for a human operator.

8.8.2 Experimental Demonstration

Initial experiments were performed to calibrate the trap stiffness relative to the fluid drag force. This exploited the parabolic flow profile in laminar flow at very low Reynolds numbers, which drags particles at different velocities depending on their distance from the surface. Pumping at a volume flow rate of 10 μl/hour corresponded to an average flow velocity of ~50 μm/s for the fabricated channel (170 μm × 340 μm cross section) with a peak velocity of ~75 μm/s at the centre of the channel. Particles tended to settle at the channel floor where they crept at a slower rate. For trap stiffness calibration, particles at the bottom of the channel were first laterally centred to operate on the peak region of the lateral velocity profile.

Particle Lift and Escape

To evaluate the axial dependency of the trap stiffness, the top beam was disabled to lift particles from the bottom as illustrated in Fig. 8.35. The fluid drag force increased as the particle was elevated away from the channel floor into a region with higher fluid velocity. At some point, the drag force prevailed over the optical trapping force and the cell was carried downstream with a fairly constant velocity (see Fig. 8.35 right). The escape velocities of the yeast cells for a single beam trap (diameter at focal point: 5 μm, measured power: 10 mW) depend on cell size and lie between 60 to 80 μm/s.

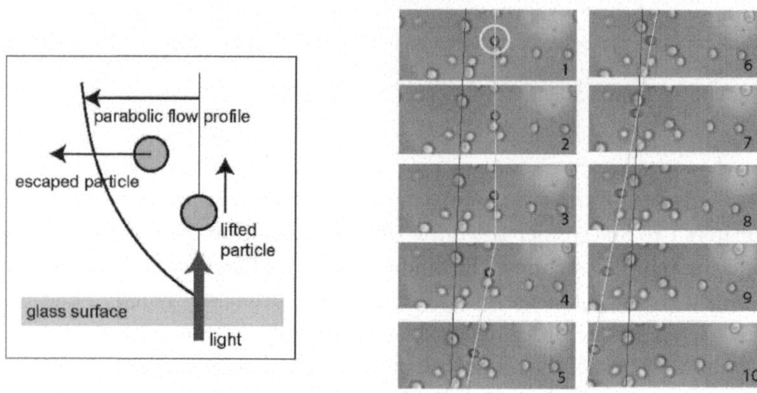

Fig. 8.35 *Left* Illustrating the particle dynamics in a lift-and-escape experiment. Forward movement is temporarily stopped while the optical trap lifts the particle toward the fast flow region. *Right* Image sequence showing yeast cells in a flow, 200 ms between successive frames. The black line tracks a reference free flowing cell. The grey line tracks a trapped cell, which exits the trap between frames 3 and 4 and eventually passes the other particle at frame 7. The exit velocity of the lifted yeast cell exceeds 60 μm/s, approximately 6 times that of a free flowing cell.

The escape velocity was determined by two opposing factors: trap strength and fluid drag force. Measurements using colloidal dispersions of polystyrene microspheres showed that the escape velocity exhibits dependence on the volume flow rate (see Fig. 8.36). This result points to the inadequacy of a collimated beam model for an optical trap. In a collimated trap, the lateral trap strength is uniform along the axial direction and the lateral escape velocity would be independent of the volume flow rate. The particle would always break free of the trap upon reaching an axial position where the local fluid velocity causes drag forces that exceed the fixed trap strength, independent of the flow rate. A more sophisticated propagation model for a GPC-based trap is discussed in Sect. 8.10, which shows that trapping strength varies with particle position along the beam propagation direction. The coupled effects of a volume flow rate-dependent velocity profile and the axial variation of the trap strength determine the variation observed in Fig. 8.36.

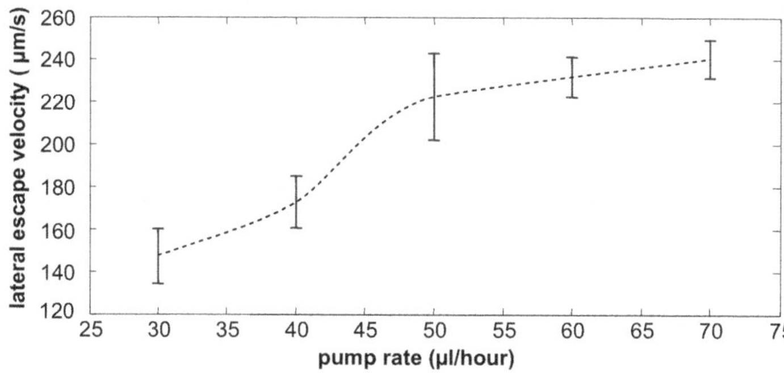

Fig. 8.36 Escape velocities of polystyrene beads as a function of pump rate (bead size 5.68 μm). Mean values determined using 50–100 data points per pump rate; error bars indicate the standard deviation, mainly caused by syringe pump pulsation. The microbeads are more strongly trapped and have significantly larger escape velocity than the yeast cells. The dashed line is drawn as visual guide.

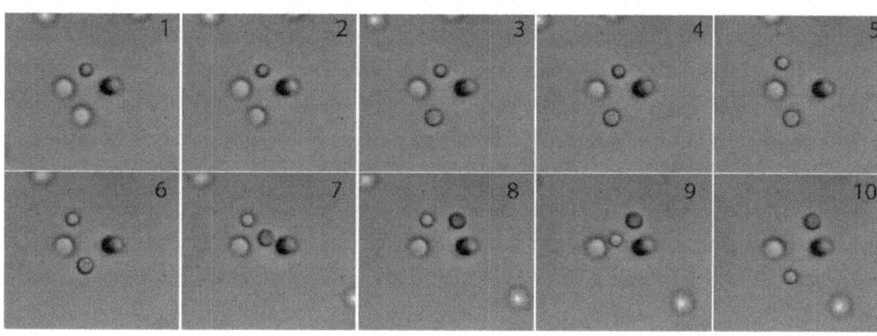

Fig. 8.37 Realtime interactive manipulation of yeast cells in a microfluidic system (0.75 s between frames). Free moving cells are out of focus and creeping from left to right along the lower surface at ~10 μm/s. Five yeast cells are trapped (rightmost trap contains two cells). The yeast cells that are lifted into focus enter a region with the flow velocity exceeding 50 μm/s (estimated by turning off traps). Frames 1–3: the lower cell is lifted. Frames 4–10: the upper and lower cells are repositioned by the user via computer mouse control.

Three-Dimensional Manipulation in a Flow

The optical traps lifted yeast cells off the bottom surface and manipulated them in the presence of a surrounding fluid flow. Working at a given volume flow rate, the surrounding fluid velocity could range from 10 μm/s to more than 60 μm/s as the cells navigated through the parabolic velocity profile.

A human operator can cope with moderate fluid flows to perform real-time, interactive optical manipulation of multiple living yeast cells in 3D within the microfluidic flow shown in Fig. 8.37. The figure indicates that trapped cells were kept close together in a group to minimize the risk of collision with incoming cells. Given the mismatch between a serial computer mouse interface and the parallel nature of the task, a human operator could easily be overwhelmed by incoming cells carried through the trapping volume by the fluid.

Automated Cell Detection and Locking

A major advantage of a GPC-based optical manipulation system is that the patterns to be encoded into the spatial light modulator require no computation since they merely mimic the desired light configuration. Thus, available computational power can be dedicated to value-added features such as process automation. The direct mapping from the SLM pattern to the light intensity distribution in the sample volume allows trapping patterns to be rapidly reconfigured with ease.

Process automation features were activated by extending the customized LabVIEW control program employed in earlier experiments. A detection area was defined in the image together with a set of detection parameters (e.g. minimum and maximum size of incoming cells) for identifying targets to trap. For proof-of-principle experiments, a simple detection and trapping subroutine was developed with the following core steps: 1. Acquire image from camera. 2. Detect cells. 3. Discard those already having a trap. 4. Set trap at free cell position. 5. Save image file for reference. 6. When more than a preset number of cells are trapped, discard traps and wait for next activation. The extended program could provide the same full control to a user to design, move, add and remove traps, if needed.

Figure 8.38 shows snapshots from an experiment using a system equipped with automated cell detection and trapping. The automated system successfully detected and captured cells with high efficiency, even at cell concentrations and flow rates that could overwhelm a human operator. Upon enabling the detection algorithm, all detected cells in the designated area were trapped in place. Demonstrations showed that incoming cells could stack at the upstream edge of the detection area. Automating the removal particle stacking, described in Sect. 8.1, could yield faster and more reliable results as compared to a manual approach. This could be applied to address the bunching of passing particles in the microfluidic system if it is problematic for the target application.

These results show the successful integration of a microfluidic system with a GPC-based optical trapping system. Its journal publication was the first demonstration of autonomous and 3D real-time multi-cell laser-manipulation in a microfluidic environment [54]. Automated detection and trapping of cells was achieved at rates beyond the capacity of a human operator using moderate computing power. Furthermore, the system can easily add more functionality in cell manipulation, e.g. lifting trapped cells into the faster flow stream, through program extensions. The cells can be maintained at a given position after trapping until a more elaborate experiment can take place. With the aid of a microfluidic system, this can involve the replacement of the sample fluid with one carrying growth media or other chemical substances that imitate special environments for microbiological experiments. Computational power can generate other value-added features such as automated image analysis of the trapped cells to measure cell size or the circularity of the cell as an indication of cell growth. It can also automate simple yet tedious tasks, such as running biological experiments spanning several hours or even days with limited user intervention. The bottom line is that a GPC-based system for controlling microfluidic systems grants sufficient flexibility for incorporating vital automation features suited for a specific application.

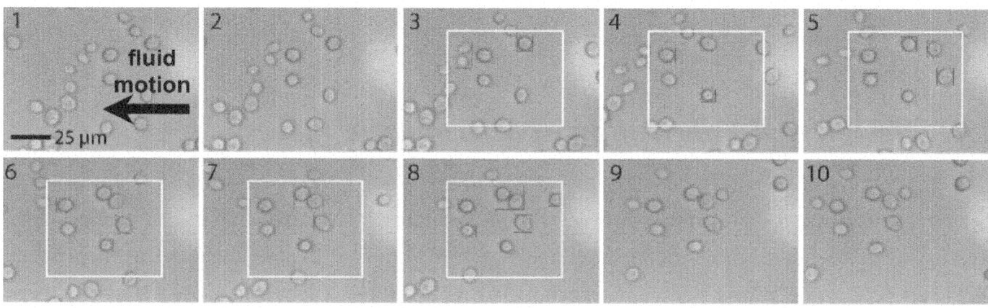

Fig. 8.38 Snapshots showing automated cell detection and trapping (0.75 s between successive frames). The square marks the detection/trapping area. Detection is disabled and the cells are released in frame 9 where the square is off. The flow is set to 20 μl/hour, giving a cell velocity of approximately 15 μm/s.

8.9 Autonomous Assembly of Micropuzzles Using GPC

In micromechanics, the assembly of minute components particularly with dimensions of 1 – 100 μm still remains a challenge [55]. Search for new micro-assembly methods is therefore important as it may lead to the potential development of novel micro-(opto)electromechanical systems [55, 56, 57]. Fluidic micro-assembly, for instance, is increasingly pursued due to several reasons including the reduced undesirable effects of van der Waals forces and electrostatic interactions in liquid host environments. Most of the current methods of fluidic-based micro-integration rely on the probabilistic nature of random assembly processes [56, 57]. At present, microscale random assembly in liquid is constrained by the trade-off between positioning accuracy on the bonding of micro-elements (normally to receptor sites on a template) and bonding yield. Possible yet cumbersome solutions for achieving reasonable yield are sample recirculation and template agitation. Bonding selectivity has nevertheless been enhanced by suitably matching the geometrical shapes of the tiny building blocks and the receptor sites [57].

In this section we demonstrate an all-optical, directed micro-assembly scheme in liquid by tiling a plurality of microscopic structural elements on a planar substrate using real-time reconfigurable optical traps from a variant of the optical setup described in our previous work [58, 59]. The number of traps, their intensity profiles and spatial locations can be controlled either interactively or in an automated way by a computer. The system demonstrates the capability for fully autonomous search-and-collect routines without any user-intervention [54]. Results show that optical traps of a few milliwatts are able to achieve good positional and rotational control of the microstructures. We also make use of shape complementarity among the micropuzzle pieces. The puzzle pieces have identical geometrical shapes and in-plane rotational symmetry. Furthermore, the puzzle pieces have an elongated aspect ratio so that orientations are readily discerned by an image analysis subroutine and optically controlled by likewise elongated traps. The microfabrication of the puzzle pieces via femtosecond laser two-photon polymerization technique [60, 52] is also described.

8.9.1 Design and Fabrication of Micropuzzle Pieces

As proof-of-concept demonstration of micro-assembly by optical forces, we use a GPC-based system to perform a microtessellation assembly. The task entails tiling micropuzzle pieces to cover a plane without overlapping and without leaving gaps. The prototype micropuzzle pattern adopted, shown in Fig. 8.39(a), falls under the "p4g" category [61] among different plane symmetry groups (also called "wallpaper groups"). The target puzzle pattern exhibits a characteristic checkerboard arrangement of horizontally and vertically symmetric tiles. Each micropuzzle piece can be specified by using intersecting circles as depicted in Fig. 8.39(b), which can guide the microfabrication process. The pattern specification also helps design the trap geometry and determine their relative positions in a packed cluster, which serve as useful inputs in developing the control software for automated optical manipulation.

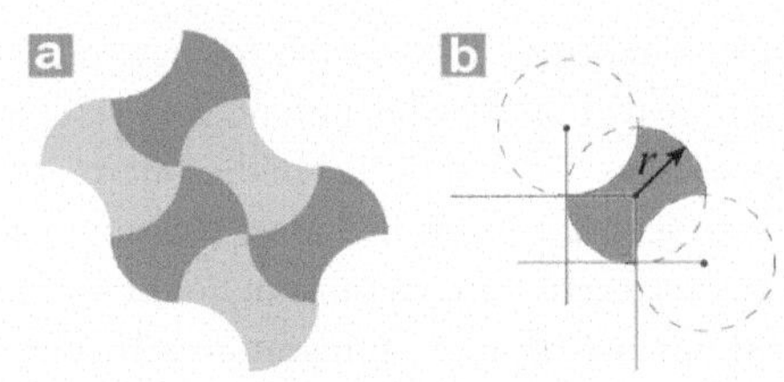

Fig. 8.39 Illustration of (a) the desired tessellation to be optically assembled and (b) a single micropuzzle element, which is a symmetric cutout of a circle of radius r.

We have previously established the system's capacity to optically manipulate SiO_2 microstructures fabricated using semiconductor processing technology in Sect. 8.7. The present demonstration considers polymer micropuzzle structures fabricated by two-photon polymerization (2PP), which is also expected to be utilized for more general 3D structures. The optical and mechanical setup is as described in ref. [52]. The process starts out by spin-coating ~15-μm layer of an epoxy-based SU8 negative photoresist resin (Michrochem, Newton, MA, USA) onto a glass microscope coverslip. A 100-fs pulse laser (790nm, 80MHz) is focused to a diffraction-limited spot in the resin, about 5 μm from the glass substrate, by a 100× oil-immersion microscope objective. Photopolymerization takes place only within a focal volume with the highest photon concentration – making 2PP a powerful high-resolution 3D microfabrication technique. The resin is translated relative to the beam with nanometre precision to polymerize the along the path shown in Fig. 8.40. The polymerized lines merge during this process, thus realizing a continuous structure. Viable structures are fabricated with a laser power (before the objective) of ~3 mW and 5 μm/sec translation speed. The finished structures are held above the glass surface by a thin layer of unsolidified resin.

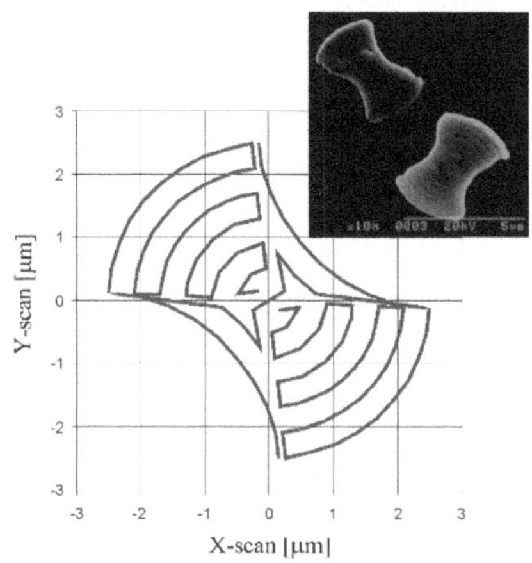

Fig. 8.40 Selected trajectories for the focused femtosecond laser beam in the 2PP fabrication of the puzzle pieces with characteristic radius r = 2.5 μm. Inset shows the corresponding SEM micrographs of the 2PP-fabricated structures.

After completing the laser-pattern-writing, the sample is baked at 95 °C for 2-minutes and then cooled down. The portion of substrate containing the polymerized puzzle pieces is placed, SU8 layer face up, into a reservoir. Unsolidified SU8 is removed by gently adding 100 μL of a developer solution (Micro Resist Technology GmbH, Berlin, Germany) into the reservoir. In the process, puzzle pieces slowly drift and eventually settle on the glass substrate. The developer is replaced with another fresh 100-μL solution without causing further positional drift. Then the developer is removed and the reservoir is gently rinsed with ethanol and dried. To prevent puzzle pieces from sticking together in prepared samples, a few microlitres of 5% surfactant solution (Tween 20, Sigma-Aldrich) are first added to the reservoir. Then the puzzle pieces are detached and extracted under a low-magnification microscope using a glass capillary tube connected by silicone rubber tubing to a microsyringe. The free end of the capillary tube has approximately 30-μm diameter opening and is positioned by a motorized 3D manipulator arm. An SEM image of typical micropuzzle pieces is shown in the inset of Fig. 8.40. The experimental parameters yield an estimated 400–500 nm lateral resolution of the 2PP method, which produces micropuzzle pieces with acceptable morphology and uniformity. The thickness of the structures is estimated to be 1.0 ± 0.2 μm.

8.9.2 Optical Assembly of Micropuzzle Pieces

Initial experiments characterized the achievable spatial and angular confinement for a single trapped puzzle piece. A puzzle element is positioned at the centre of the imaging field of view using a pair of circular counterpropagating optical traps (Fig. 8.41 inset). Microscope images, acquired at 10 frames/sec, are processed by an image analysis subroutine to measure the angular orientation, θ, of the long axis and the centroid coordinates (x, y) of the micro-object. Figure 8.41 plots the time-evolution of the angular position, $\theta(t)$, and the centroid radial distance $\rho(t)= x(t)^2 + y(t)^2$. Translational control of the microstructure can be achieved with submicron resolution using ~6 mW total power in the composite of trap. Angular position is controlled to within a few degrees. The standard deviations of θ and ρ over a ~95 s trapping interval are 39.6 nm and 3.41°, respectively. The puzzle piece wanders away with unstable angular orientation due to Brownian motion when the trap is removed.

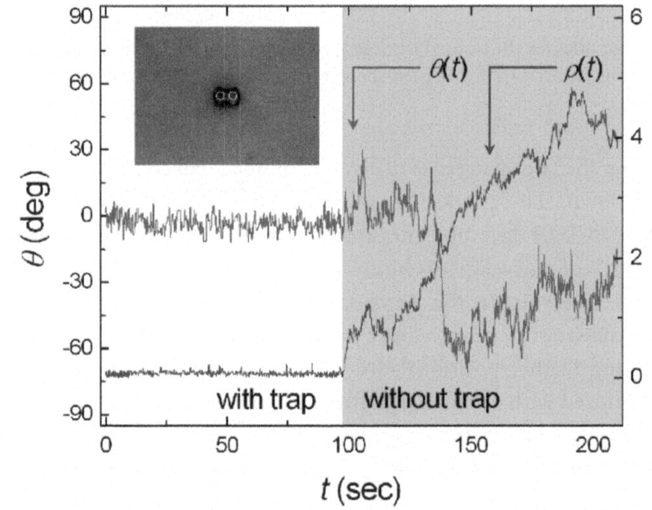

Fig. 8.41 Plot of the angular orientation and radial position of a puzzle piece as a function of time in the presence and absence of paired trapping beams. Relative positions of the paired traps are indicated by the two circular markers overlaid with the image of the trapped and horizontally oriented micro-object (inset).

The puzzle pieces may be interactively manipulated and assembled in real-time using a drag-and-drop user interface. A user manually tiles the microstructures sequentially and controls the motorized sample stage to search for additional pieces (Fig. 8.42). This operation mode allows the user to intelligently react to unexpected setbacks that might arise. This rather slow process (it took 4 min to assemble the 4×4 structure) helps identify potential problems that an automated system must cope with.

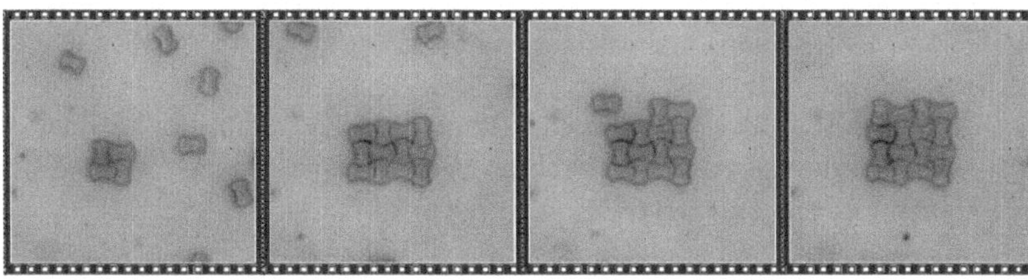

Fig. 8.42 Snapshots from an interactive optical manipulation and assembly of micropuzzle pieces into a 4×4 tiling. The linear and angular speeds at which the pieces are translated and rotated are ~2 μm/sec and ~20 deg/sec, respectively.

Automated assembly can rapidly accomplish the same task. Here, a user simply activates the process and the trapping system automatically moves all constituent pieces simultaneously into their proper places. Starting with enough pieces within the field of view, a 4×4 tessellation (see Fig. 8.43) was completed within a few seconds. A sample with lower density may be automated by also automating the sample stage control. Once assembled, the tiled structure can be translated and rotated as one entity, using coordinated motion of the trap array. Using ~6 mW total trapping power for each puzzle piece, linear and angular speeds of approximately 1.5 μm/s and $10° \, s^{-1}$, respectively, were achieved without compromising structural integrity. Thus, the sample stage may be moved at similar speeds and maintain the structure to automatically "hunt and collect" free pieces to build a bigger tessellation.

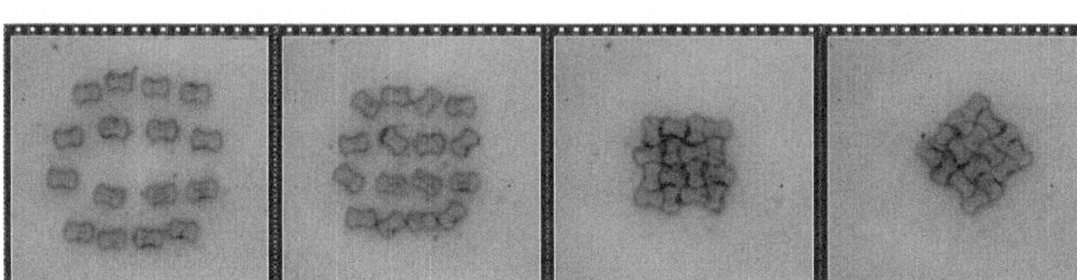

Fig. 8.43 Snapshots from parallel optical assembly of 16 micropuzzle pieces into a 4×4 tessellated configuration. Once assembled (~7 sec from 1st to 3rd frame), adjacent elements remain intact while the superstructure is displaced and rotated.

Figure 8.44 shows snapshots from a fully automated operation where puzzle pieces are autonomously gathered and assembled without user intervention. Automation was accomplished using basic image analysis packages available in LabVIEW. During operation, the sample stage moves at a constant speed to bring free pieces into the field of view. An image analysis subroutine determines the position and orientation of micro-puzzle pieces within a designated detection region (marked rectangular area in Fig. 8.44). Another subroutine launches composite traps on each detected piece and guides them to their correct place and orientation in the tessellation. Fully autonomous operation eliminates tedious manual tasks, minimizes user involvement and accomplishes the task faster.

These results indicate that GPC-based light-based manipulation can be a potential micro-assembly method, particularly for objects with dimensions much below 100 μm. A few milliwatts of trapping power can hold microstructures with only tens of nanometres and a few degrees variations in translational and angular positions, respectively. This can be enhanced by increasing the trapping power. Fully autonomous search-and-collect capability is especially significant when working with low-density distributions of microparticles.

An automated microstructure gathering and positioning system can form part of a bigger micro-assembly process. For example, another sub-process may engage in permanently bonding the micro-assemblies. One way to do this is to use a polymerizable liquid medium and a focused laser to locally fuse micro-elements while they are held in place by optical traps [62]. The micro-elements may also be attached together by surface functionalization or critical point drying methods. Microcomponents with complementary shapes, accurately fabricated by the two-photon polymerization technique, may serve as microscale building blocks for the optical assembly of more intricate geometries.

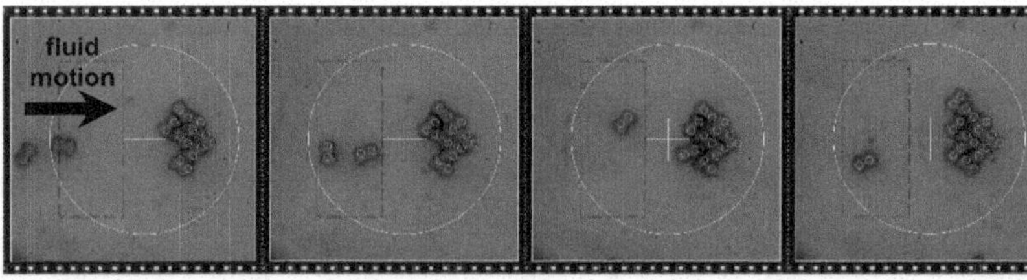

Fig. 8.44 Snapshots taken from a computer-automated "hunt-and-collect" procedure for tiling micropuzzle pieces. The dashed rectangle highlights the selected detection area where pieces coming from the left due to constant-speed sample stage movement are automatically detected. Once detected, trapping beams with appropriate target trajectories are immediately assigned.

8.10 Optical Forces in Three-Dimensional GPC-Trapping

The preceding sections have provided a glimpse on various possibilities for exploiting GPC-based optical manipulation. The versatility of GPC-based counterpropagating beam traps derives from its ability to interactively manipulate a large number of microspheres or cells, and to actuate microfabricated structures in 3D [31, 32, 36, 63]. Genuine real-time interactive micromanipulation stems from its straightforward beam modulation technique that does not require iterative computation to generate arbitrary trapping patterns. In this section we will calculate the optical forces in counterpropagating-beam traps of variable power ratios to provide theoretical support for GPC-based optical trapping and manipulation. Ashkin's pioneering demonstration of stable optical trapping in three dimensions (3D) utilized mildly focused counterpropagating Gaussian beams [1]. These beams required a positive separation, s, between their waists to form a stable optical potential well at the midpoint of separation. Positive beam waist separation means that both beams are diverging when they overlap in the trapping region. A similar situation is qualitatively observed in counterpropagating-beam (CB) traps from two coaxially aligned optical fibres [21, 64].

In the following, we will derive a rigorous model by analyzing the optical force on a single sphere in a GPC-based CB trap and study how the trap stability depends on s. The model points to a minimum critical separation, s_c, below which the potential well becomes unstable. We will study the value of s_c using a balanced symmetric GPC-based trap. We will also consider how the trap efficiency varies with other parameters such as the microsphere's refractive index and radius for these symmetric traps. The axial force curves for unbalanced counterpropagating beams provide an estimate for the dynamic range of axial position control. These considerations allow us to find conditions for producing stable counterpropagating beam traps, which are vital for designing efficient GPC-based systems.

8.10.1 Optical Forces on a Particle Illuminated by Counterpropagating Beams

Let us consider a purely dielectric microsphere with diameter = $2a$ illuminated by two orthogonally polarized optical fields counterpropagating coaxially along the z-axis (Fig. 8.45a). The fields have associated intensity distributions, $I^{+}(x,y,z)$ and $I^{-}(x,y,z)$, where + and –denote propagation towards the $+z$-axis and $-z$-axis, respectively. Let us begin with the optical force induced by $I^{+}(x,y,z)$ on the particle. Referring to Fig. 8.45b., the force arising from an ray incident on an infinitesimal surface element dS may be decomposed into two components:

$$\mathrm{d}\mathbf{F}_{\parallel}^{+} = \hat{\mathbf{e}}_{\parallel}\frac{n_1}{c}q_{\parallel}\mathrm{d}P, \tag{8.13a}$$

$$\mathrm{d}\mathbf{F}_{\perp}^{+} = \hat{\mathbf{e}}_{\perp}\frac{n_1}{c}q_{\perp}\mathrm{d}P, \tag{8.13b}$$

where n_1 is the refractive index of the host medium, dP is the differential power of the ray, c is the speed of light in vacuum, and $\hat{\mathbf{e}}_{\parallel}$ and $\hat{\mathbf{e}}_{\perp}$ are unit vectors parallel and perpendicular to the incident ray, respectively. The orientations of $\hat{\mathbf{e}}_{\parallel}$ and $\hat{\mathbf{e}}_{\perp}$ with surface normal $\hat{\mathbf{n}}$ are shown in Fig. 8.45b together with the angles of incidence α_i and refraction α_r. Factors $q_{\parallel}$ and $q_{\perp}$ denote momentum components, transferred to the particle by the ray, along $\hat{\mathbf{e}}_{\parallel}$ and $\hat{\mathbf{e}}_{\perp}$, respectively. They are related to α_i and α_r by [23]

$$q_{\parallel} = 1 + R\cos 2\alpha_i - T^2\frac{\cos(2\alpha_i - 2\alpha_r) + R\cos 2\alpha_i}{1 + R^2 + 2R\cos 2\alpha_r}, \tag{8.14a}$$

$$q_{\perp} = -R\sin 2\alpha_i + T^2\frac{\sin(2\alpha_i - 2\alpha_r) + R\sin 2\alpha_i}{1 + R^2 + 2R\cos 2\alpha_r}, \tag{8.14b}$$

where the surface reflectance R and transmittance T depend on the incident polarization and angle of incidence. The optical power dP through the surface dS is given by

$$\mathrm{d}P = I^{+}(x_p, y_p, z_p)\cos\alpha_i \mathrm{d}S, \tag{8.15}$$

where (x_p, y_p, z_p) are the coordinates of the point of incidence P on the unprimed Cartesian coordinate system in Fig. 8.46. The components of the total axial and transverse force on the microsphere are obtained by integrating the scalar products of $\mathrm{d}\mathbf{F}_{\parallel}^{+} + \mathrm{d}\mathbf{F}_{\perp}^{+}$ with unit vectors $\hat{\mathbf{x}}$, $\hat{\mathbf{y}}$, and $\hat{\mathbf{z}}$ over the illuminated sphere surface.

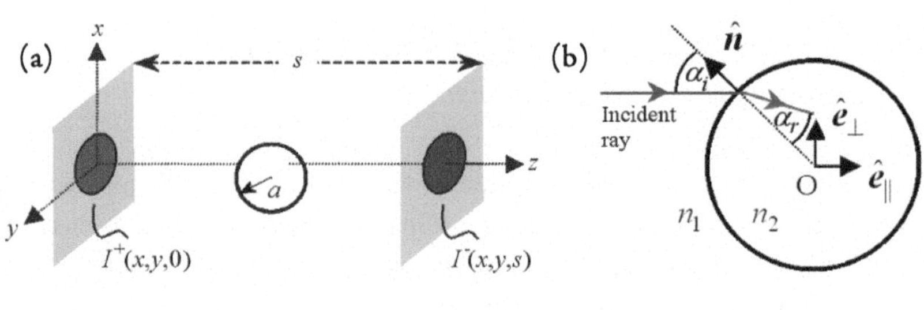

Fig 8.45 (a) Graphical illustration of a microsphere centred at (x_o,y_o,z_o) illuminated by two counter-propagating beams I^+ and I^-. (b) The microsphere's cross-section in the plane of incidence associated with a ray hitting the surface at an angle α_i measured from the normal.

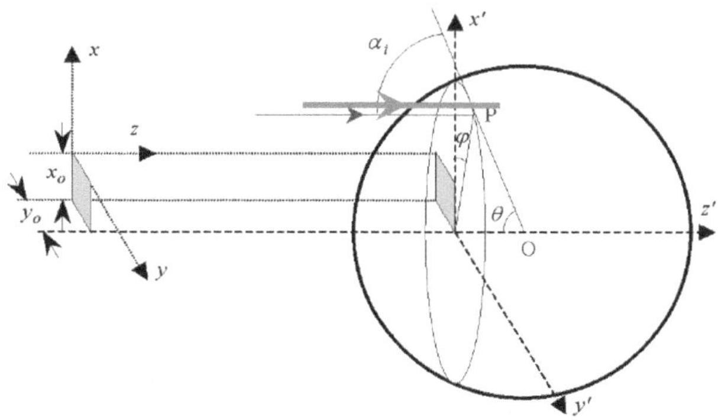

Fig. 8.46 3D illustration of an incident ray (grey arrow) impinging at point P on the surface of a microsphere centred at point O, which is located at (x_o,y_o,z_o) of the unprimed Cartesian coordinate system.

The geometry makes it more convenient to use spherical coordinate system in the analysis. The unit vectors are related by $\hat{\mathbf{e}}_{\|} = \hat{\mathbf{z}}$ and $\hat{\mathbf{e}}_{\perp} = \hat{\mathbf{x}}\cos\varphi + \hat{\mathbf{y}}\sin\varphi$ while the angle of incidence is simply the zenith angle of P (i.e. $\alpha_i = \theta$). The optical force and its components are then obtained as

$$\mathbf{F}^{+} = \frac{n_1 a^2}{c}\int_0^{2\pi} \mathrm{d}\varphi \int_0^{\pi/2} \mathrm{d}\theta (\hat{\mathbf{e}}_{\|} q_{\|} + \hat{\mathbf{e}}_{\perp} q_{\perp}) \sin\theta\cos\theta I^{+}(x_p, y_p, z_p), \tag{8.16a}$$

$$F_x^{+} = \frac{n_1 a^2}{2c}\int_0^{2\pi} \mathrm{d}\varphi \int_0^{\pi/2} d\theta \cos\varphi \, \sin 2\theta \, q_{\perp}(\theta) I^{+}(x_p, y_p, z_p), \tag{8.16b}$$

$$F_y^{+} = \frac{n_1 a^2}{2c}\int_0^{2\pi} \mathrm{d}\varphi \int_0^{\pi/2} d\theta \sin\varphi \, \sin 2\theta \, q_{\perp}(\theta) I^{+}(x_p, y_p, z_p), \tag{8.16c}$$

$$F_z^{+} = \frac{n_1 a^2}{2c}\int_0^{2\pi} \mathrm{d}\varphi \int_0^{\pi/2} d\theta \, \sin 2\theta \, q_{\|}(\theta) \, I^{+}(x_p, y_p, z_p), \tag{8.16d}$$

where the intensity is calculated at each point of incidence on the sphere, $x_p = x_o + a\sin\theta\cos\varphi$, $y_p = y_o + a\sin\theta\sin\varphi$, and $z_p = z_o - a\cos\theta$. The force components are determined through numerical integration of Eq. (8.16b) through Eq. (8.16d). Assuming a linearly polarized incident beam, the TE and TM components must be determined for each illuminated point on the surface. When using linear polarization along the x-axis, one must include a multiplicative weighting factor, $\sin^2\varphi$ or $\cos^2\varphi$, in the integrands of Eqs. (8.16) to account for the power fraction carried by the TE or TM component. Furthermore, the intensity $I^{+}(x_p, y_p, z_p)$ must account for the Fresnel propagation of the field, $e^{+}(x, y, 0)$, from its image plane (the objective focal plane, $z = 0$) to the incidence point (x_p, y_p, z_p).

8.10.2 Top-Hat Field Distribution and Propagation

A GPC-based counterpropagating-beam trap commonly generates a top-hat transverse field distribution at $z=0$, which is conveniently described as

$$e^{+}(x,y,0)=\begin{cases}\left(P^{+}/\pi R^{2}\right)^{1/2}, & \text{for } \left(x^{2}+y^{2}\right)^{1/2}\leq R \\ 0, & \text{otherwise}\end{cases}, \tag{8.17}$$

where the beam power P^{+} is uniformly distributed over a circular area of radius R. The volume intensity distribution $I^{+}(x,y,z)$ for $z>0$ may be determined by numerically propagating the initial field through the host medium of refractive index n_1 using a Fresnel integral [35].

Figure 8.47 shows an axial section of the calculated volume diffraction pattern in the *xz*-plane. The results apply to a trap (R = 1.5 μm, P^{+} = 5 mW and λ = 830 nm) propagating through water (n_1=1.33), which are typical parameters in past experiments. The global intensity maximum in Fig. 8.47 (about four times the intensity at $z=0$) occurs at z ~ 3.8 μm, which is consistent with $z_{I,\max}=n_1R^2/\lambda$ for the given beam parameters. The transverse plane, $z=z_{I,\max}$, represents a soft boundary for the Fraunhofer diffraction region where the central lobe monotonically broadens and decreases in peak intensity with subsequent propagation. In the following section, we will use these observations to explain the obtained dependence of axial and transverse forces on the position of the particle within the CB trap.

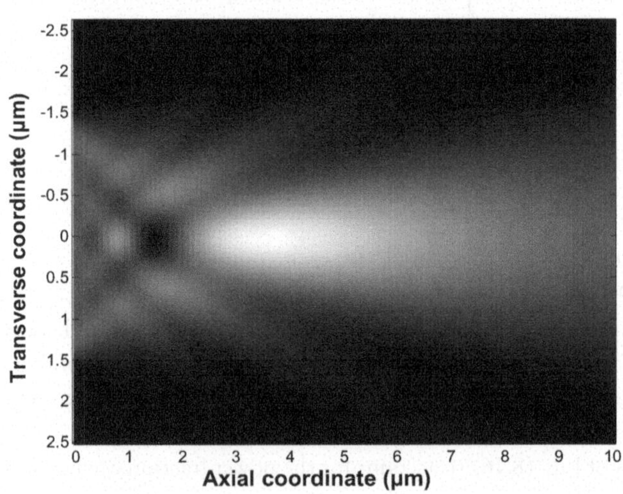

Fig. 8.47 Axial section (xz-plane) intensity distribution for a typical GPC-generated beam directed towards the positive z-axis. At z = 0, the beam has a top-hat field profile extending from −1.5 μm to 1.5 μm of the x-axis.

8.10.3 Numerical Calculation of Force Curves

Typical parameters in GPC-based trapping experiments will be used in simulations using the mathematical analysis outlined above. At power regimes where gravity may be neglected, the stability of a counterpropagating trap depends only on the power ratio of the beams and not on absolute power values. Thus, given a total power, $P_t = P^+ + P^-$, we will adopt normalized parameters to study the trap stability:

$$Q_{xt} = \frac{c}{n_1 P_t}\left(F_x^+ + F_x^-\right), \tag{8.18a}$$

$$Q_{zt} = \frac{c}{n_1 P_t}\left(F_z^+ - F_z^-\right), \tag{8.18b}$$

where the ± superscripts specify the propagation direction (±z-axis) (see Fig. 8.45a). From symmetry, forces from the opposite beam determined similar to Eqs. (8.16) and (8.17) with the power P^+ replaced by P^- and the axial position, z_o, replaced by $s - z_o$. The beam waist separation s is defined as the axial distance between the top-hat images, which are located at the focal planes of the two opposing microscope objectives.

Let us examine the case of identical counterpropagating beams (top-hat with radius R = 1.5 μm) and a polystyrene microsphere (refractive index n_2 = 1.59, radius a = 1.5 μm) with water as the host medium. Figures 8.48a and 8.48b plot the axial and transverse force components for respective axial and transverse positions of the microsphere, about the midpoint, for different focal separations, s. The transverse force curves in Fig. 8.48b indicate restoring forces towards the equilibrium point at the axis. Similar results are generally observed for the axial force curves in Fig. 8.48a, except for s = 20 μm. For s = 20 μm the midpoint is an unstable axial equilibrium point – a small axial perturbation will eject the particle along the perturbation direction where it can be trapped by two stable points located on either side. In general, stable traps can form at the overlapping Fraunhofer regions of the counterpropagating beams. The fluctuations of the axial force curve tails in Fig. 8.48a arise from cross-sectional intensity variations in the Fresnel region of each beam (see Fig. 8.47).

For displacements from the midpoint comparable with the particle radius, the traps may be considered as harmonic potentials. The normalized transverse and axial restoring forces may be, respectively, written as $Q_{xt} \approx -f_x x_o$ and $Q_{zt} \approx -f_z\left(z_o - s/2\right)$, which quantifies the trap stiffness using the force constants, f_x and f_z. The dependence of the trap stiffness on s is examined using the plots in Fig. 8.49. The radial force constant f_x decreases with s, but remains positive-valued, for the considered range of s. This occurs since at larger s the particle encounters smaller transverse intensity gradients from the broader and dimmer central lobes of the counter propagating beams (see Fig. 8.47).

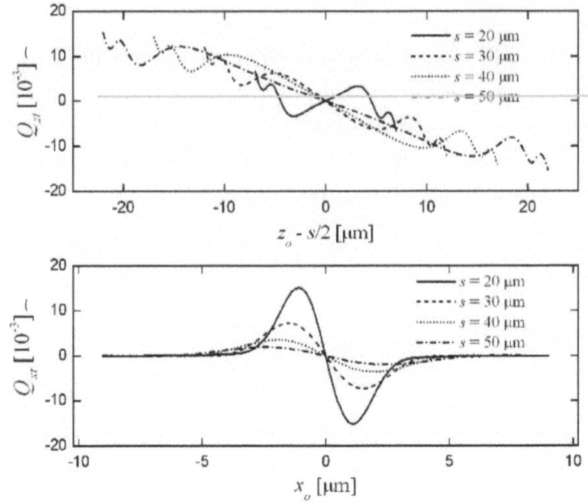

Fig. 8.48 Typical dependence of the (a) axial and (b) transverse force components on the respective axial and transverse distances of the microsphere from the midpoint between the two beam waists for different waist separation distances.

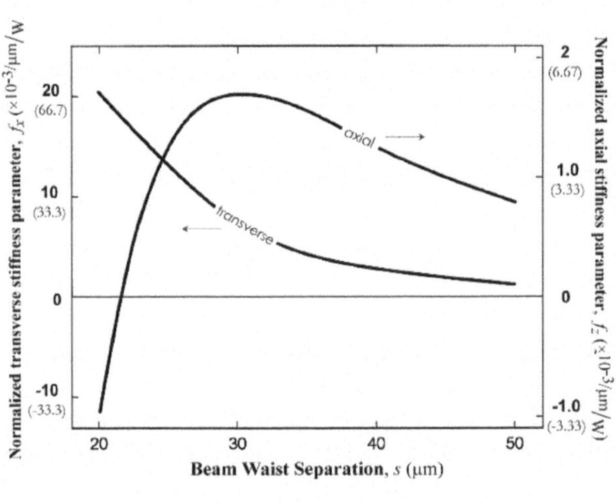

Fig. 8.49 Normalized transverse and axial stiffness parameters for a GPC-based CB trap as a function of beam waist separation. An alternate scale (numbers in parentheses) shows the trapping force constant in pN/μm per watt of total power in the counterpropagating beams.

The $f_z(s)$ plot in Fig. 8.49 shows a zero-crossing at $s = s_c$. This represents a critical separation value between $s > s_c$, which result in axially stable traps (i.e. f_z >0), and $s < s_c$, which generate unstable traps (i.e. f_z >0). The $f_z(s)$-plot can also be used to identify an optimum separation, s_{op}, that yields maximum axial force constant, $f_{z,\max}$. The optimal separation in Fig. 8.49, s_{op} = 30 μm, matches values estimated from previous experiments with the same set of parameters [31, 32]. The decreasing f_x with s can also factor into the eventual choice of focal separation. For instance, the intersection of the $f_z(s)$ and $f_x(s)$ plots in Fig. 8.49 can represent another optimal separation that forms a trap with matching axial and transverse stiffness.

Other parameters, such as the size of the largest particle (for size-varied colloids) and the axial dynamic range of manipulation, also need to be considered when deciding on an operational separation. Figures 8.50(a) and 8.50(b) illustrate the axial and transverse force curves for different particle indices n_2 for a focal separation (s = 40 μm) sufficiently larger than the critical value. The axial and transverse stiffness both increase with higher refractive index contrast between particle and host medium due to the increased momentum change when light crosses the n_1-n_2 interface.

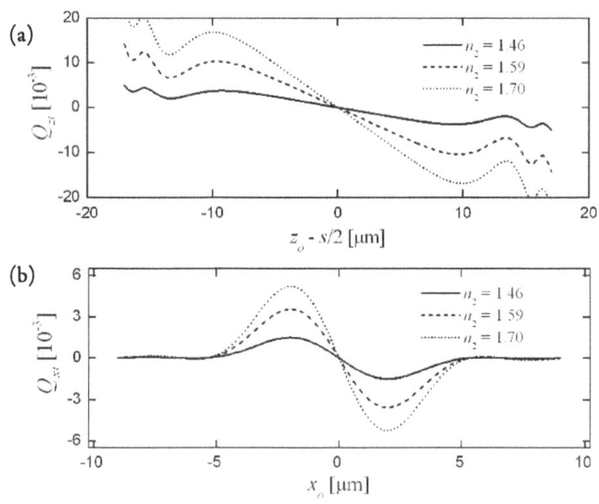

Fig. 8.50 Typical dependence of the (a) axial and (b) transverse force components on the respective axial and transverse distances of the microsphere from the midpoint between the two beam waists (s = 40 μm) for different refractive indices n_2. Spheres with radius a = 1.5 μm and equal counterpropagating top-hat beams of size R = 1.5 μm are considered.

To examine the role of particle size in finding an operational beam waist separation, let's consider, for example, a trapping experiment with a size-polydisperse sample (radii a = 1.0 μm, 1.5 μm, and 2.5 μm). The $f_z(s)$-curves in Fig. 8.51 shows that simultaneously trapping the three particles requires implementing s_{op} > 35 μm. Using a sufficiently large separation, s_{op}=40 μm, Figs. 8.52(a) and 8.52(b) plot the force curves for different microsphere radii. Smaller spheres encounter traps with weaker transverse stiffness since they capture less beam power. On the other hand, larger spheres may have smaller axial manipulation range, as suggested by the axial force plots in Fig. 8.52(a).

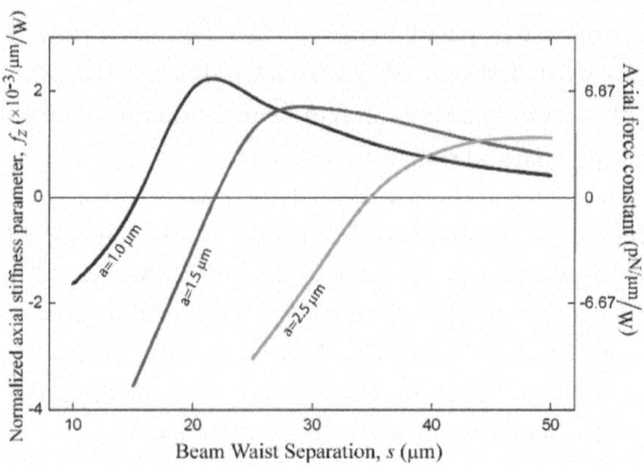

Fig. 8.51 Dependence of GPC-based CB trap's axial stiffness on beam waist separation for different microsphere radii. The trapping force constant in pN/μm per watt of total power in the counterpropagating beams can be read from the secondary axis. The zero-crossings denote critical separation s_c, which increases for larger particles.

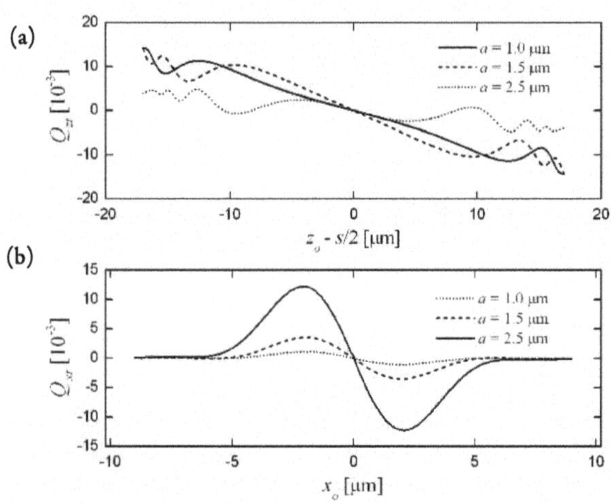

Fig. 8.52 Typical dependence of the (a) axial and (b) transverse force components on the respective axial and transverse distances of the microsphere from the midpoint between the two beam waists for different refractive indices n_2. Polystyrene spheres with refractive index $n_2 = 1.59$ and similar counterpropagating top-hat beams are considered (size R = 1.5 μm and separation s = 40 μm).

In practice, axial manipulation is achieved by varying the power ratio of the two beams. Figures 8.53a and 8.53b show the axial and transverse force curves for varied differential powers $P^+ - P^-$. The simulations are modelled from experimental GPC setups using the polarization scheme [31, 32, 36, 63], which maintain a constant total power while the differential power is varied. Starting with balanced beams, the differential power, $P^+ - P^-$, is tuned to shift the equilibrium point up to a defined cutoff axial offset, Δz, where the axial force constant has decreased by 50% (inset shows the trap stiffness at the different equilibrium positions). The region $-\Delta z \leq z \leq \Delta z$ defines a competitively large dynamic range for axial control of the particle. The transverse force curves in Fig. 8.53(b) were determined at respective axial equilibrium positions.

The theoretical modelling and results presented above is expected to play vital role in future optical micromanipulation systems that utilize GPC-technology. A comprehensive theoretical understanding of how trapping parameters affect trapping performance will help design, implement, calibrate, optimize and troubleshoot these systems. The

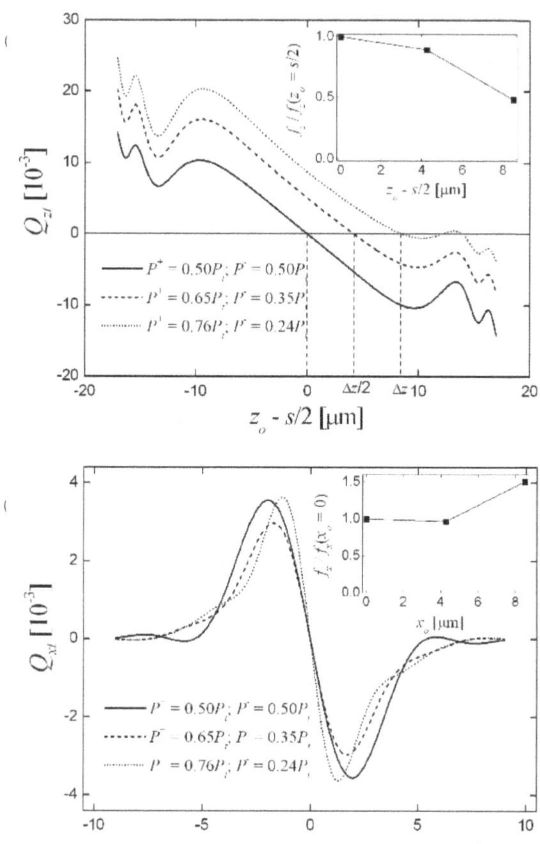

Fig. 8.53 Plot of the (a) axial and (b) transverse force curves for varied differential powers $P^+ - P^-$. Insets show the change in (a) axial and (b) transverse trap stiffness at three different points of stable equilibrium. The obtained dynamic range for axial position control is $\pm\Delta z$ corresponding to a depth of ~17.0 μm, in good comparison with experimentally obtained ~20.0 μm [31]. Polystyrene spheres (radius a = 1.5 μm and index n_2 = 1.59) and counterpropagating top-hat beams (size R = 1.5 μm and separation s = 40 μm) are considered.

model aids in the accurate understanding of optical forces in GPC-based CB traps, which is vital for developing advanced applications. Among the parameters affecting the trap forces and stiffness are beam waist separation, particle dimension and particle-host index contrast. The model allows us to find an optimal beam waist separation to achieve stable potentials, which are harmonic over a significant range within the trapping volume. We can also find optimal separation to tune the axial and transverse trap stiffness. Along with the inherently large transverse range for manipulation, the already large axial dynamic range can be further extended by selecting larger beam waist separation. The outlined formulations outlined can serve as a framework for modelling CB traps with complex intensity and/or polarization spatial profiles.

8.11 Summary and Links

In this chapter we have demonstrated a classic example of how GPC-based synthesis of customized light distributions can be carried further and exploited. The high information capacity in GPC-based light synthesis, established in Chapter 6, turns into desirable encoding simplicity and high-speed reconfigurability, which perfectly tie in with the demands of programmable optical trapping and manipulation. This was manifested in the various experimental demonstrations where GPC successfully provided the diverse traps needed for handling biological and synthetic microparticles possessing assorted geometry and optical properties. Requiring minimal computations, much of the available computational power in GPC-based trapping may be devoted to process automation, a highly practical feature in multiple cell handling and assembly of microstructures. We have also provided a theoretical analysis of optical forces to provide solid support and optimization possibilities for the counterpropagating beam (CB) traps employed in typical GPC trapping systems. The CB trap geometry allows the creation of stiff optical traps using microscope objectives having a long working distance. This leaves ample space for incorporating accessory systems orthogonally, such as the side-view microscope that we discussed. We presented options for creating counterpropagating beams– one that combines a GPC projector in tandem with a polarization modulator and another based on an alternative GPC implementation that uses two parallel input phase pattern arrangements. Other alternative GPC implementations and their practical advantages will be discussed in Chapter 9.

References

1. A. Ashkin, "Acceleration and trapping of particles by radiation pressure," Phys. Rev. Lett. 24, 156-159 (1970).
2. K. Svoboda, C. F. Schmidt, B. J. Schnapp, and S. M. Block, "Direct observation of kinesin stepping by optical trapping interferometry," Nature **365**, 721-727 (1993).
3. K. Svoboda and S. M. Block, "Biological application of optical forces," Annu. Rev. Biophys. Biomol. Struct. **23**, 247-285 (1994).
4. M. D. Wang, H. Yun, R. Landick, J. Gelles, and S. M. Block, "Stretching DNA with optical tweezers," Biophys. J. **72**, 1335-1346 (1997).
5. R. E. Holmlin, M. Schiavoni, C. Y. Chen, S. P. Smith, M. G. Prentiss, and G. M. Whitesides, "Light-driven microfabrication: Assembly of multi-component, three-dimensional structures by using optical tweezers," Angew. Chem. Int. Ed. Engl. **39**, 3503 (2000).
6. M. P. MacDonald, L. Paterson, K. Volke-Sepulveda, J. Arlt, W. Sibbett, and K. Dholakia, "Creation and manipulation of three-dimensional optically trapped structures," Science **296**, 1101-1103 (2002).
7. P. Zemanek, A. Jonas, L. Sramek, and M. Liska, "Optical trapping of nanoparticles and microparticles by a Gaussian standing wave," Opt. Lett. **24**, 1448-1450 (1999).
8. J. Arlt, V. Garces-Chavez, W. Sibbett, and K. Dholakia, "Optical micromanipulation using a Bessel light beam," Opt. Commun. **197**, 239-245 (2001).
9. A. T. O'Neil and M. J. Padgett, "Rotational control within optical tweezers by use of a rotating aperture," Opt. Lett. **27**, 743-745 (2002).
10. K. T. Gahagan and G. A. Swartzlander, "Optical vortex trapping of particles," Opt. Lett. **21**, 827-829 (1996).
11. Y. Ogura, K. Kagawa, and J. Tanida, "Optical manipulation of microscopic objects by means of vertical-cavity surface-emitting laser array sources," Appl. Opt. **40**, 5430-5435(2001).
12. K. Sasaki, M. Koshioka, H. Misawa, N. Kitamura, and H. Masuhara, "Optical trapping of a metal-particle and a water droplet by a scanning laser-beam," Appl. Phys. Lett. **60**, 807-809 (1992).
13. K. T. Gahagan and G. A. Swartzlander, "Trapping of low-index microparticles in an optical vortex," J. Opt. Soc. Am. B. **15**, 524-534 (1998).
14. K. T. Gahagan and G. A. Swartzlander, "Simultaneous trapping of low-index and high-index microparticles observed with an optical-vortex trap," J. Opt. Soc. Am. B **16**, 533-537 (1999).
15. M. P. MacDonald, L. Paterson, W. Sibbett, K. Dholakia, and P. E. Bryant, "Trapping and manipulation of low-index particles in a two-dimensional interferometric optical trap," Opt. Lett. **26**, 863-865 (2001).
16. J. Glückstad and P. C. Mogensen, "Optimal phase contrast in common-path interferometry," Appl. Opt. **40**, 268-282 (2001).

17. P. Nissen, D. Nielsen, and N. Arneborg, "Viable Saccharomyces cerevisiae cells at high concentrations cause early growth arrest of non-Saccharomyces yeasts in mixed cultures by a cell-cell contact-mediated mechanism," Yeast **20**, 331-341 (2003).
18. P. Nissen and N. Arneborg, "Characterization of early deaths of non-Saccharomyces yeasts in mixed cultures with Saccharomyces cerevisiae," Arch. Microbiol. **180**, 257-263 (2003).
19. C. Venturin, H. Boze, G. Moulin, and P. Galzy, "Glucose metabolism, enzymatic analysis and product formation in chemostat culture of Hanseniaspora uvarum," Yeast **11**, 327–336 (1995).
20. A. Ashkin, J. M. Dziedzic, J. E. Bjorkholm, and S. Chu, "Observation of a single-beam gradient force optical trap for dielectric particles," Opt. Lett. **11**, 288-290 (1986).
21. A. Constable, J. Kim, J. Mervis, F. Zarinetchi, and M. Prentiss, "Demonstration of a fiber-optical light-force trap," Opt. Lett. **18**, 1867-1869 (1993).
22. S. C. Grover, A.G. Skirtach, R. C. Gauthier, and C. P. Grover, "Automated single-cell sorting system based on optical trapping," J. Biomed. Opt. **6**, 14-22 (2001).
23. E. Sidick, S. D. Collins, and A. Knoesen, "Trapping forces in a multiple-beam fiber-optic trap," Appl. Opt. **36**, 6423-6433 (1997).
24. M. N. Liang, S. P. Smith, S. J. Metallo, I. S. Choi, M. Prentiss, and G. M. Whitesides," Measuring the forces involved in polyvalent adhesion of uropathogenic Escherichia coli to mannose-presenting surfaces," Proc. Natl. Acad. Sci. U.S.A. **97**, 13092-13096 (2000).
25. G. Sinclair, P. Jordan, J. Leach, M. J. Padgett, and J. Cooper, "Defining the trapping limits of holographical optical tweezers," J. Mod. Opt. **51**, 409-414 (2004).
26. M. M. Burns, J. -M. Fournier, and J. A. Golovchenko, "Optical matter - crystallization and binding in intense optical-fields," Science **249**, 749-754 (1990).
27. J. Leach, G. Sinclair, P. Jordan, J. Courtial, M. J. Padgett, J. Cooper, and Z. J. Laczik, "3D manipulation of particles into crystal structures using holographic optical tweezers," Opt. Express **12**, 220-226 (2004).
28. J. E. Curtis, B. A. Koss, and D. G. Grier, "Dynamic holographic optical tweezers," Opt. Commun. **207**, 169-175 (2002).
29. D. L. J.Vossen, A. van der Horst, M. Dogterom and A. van Blaaderen, "Optical tweezers and confocal microscopy for simultaneous three-dimensional manipulation and imaging in concentrated colloidal dispersions," Rev. Sci. Instrum. **75**, 2960-2970 (2004).
30. S. A. Tatarkova, A. E. Carruthers, and K. Dholakia, "One-dimensional optically bound arrays of microscopic particles," Phys. Rev. Lett. **89**, 283901 (2002).
31. P. J. Rodrigo, V. R. Daria, and J. Glückstad, "Real-time three-dimensional optical micromanipulation of multiple particles and living cells," Opt. Lett. **29**, 2270-2272 (2004).
32. P. J. Rodrigo, V. R. Daria, and J. Glückstad, "Four-dimensional optical manipulation of colloidal particles," Appl. Phys. Lett. **86**, 074103 (2005).

33. M. Reicherter, T. Haist, E.U. Wagemann, and H.J. Tiziani, "Optical particle trapping with computer-generated holograms written on a liquid-crystal display," Opt. Lett. **24,** 608-610 (1999).
34. G. Sinclair, P. Jordan, J. Courtial, M. Padgett, J. Cooper, and Z. J. Laczik, "Assembly of 3-dimensional structures using programmable holographic optical tweezers," Opt. Express **12,** 5475-5480 (2004).
35. J. W. Goodman, Introduction to Fourier Optics, Second Edition (McGraw-Hill, New York, 1996).
36. I. R. Perch-Nielsen, P. J. Rodrigo, and J. Glückstad, "Real-time interactive 3D manipulation of particles viewed in two orthogonal observation planes," Opt. Express **18,** 2852-2857 (2005),
37. D. G. Grier, "A revolution in optical manipulation," Nature **424**, 810-816 (2003).
38. K. Dholakia and P. Reece, "Optical micromanipulation takes hold," Nano Today **1**, 18-27 (2006).
39. S. Kawata, H. B. Sun, T. Tanaka, and K. Takada, "Finer features for functional microdevices - Micromachines can be created with higher resolution using two-photon absorption," Nature **412**, 697-698 (2001).
40. S. Maruo, K. Ikuta, and H. Korogi, "Submicron manipulation tools driven by light in a liquid," Appl. Phys. Lett. **82**, 133-135 (2003).
41. E. Higurashi, H. Ukita, H. Tanaka, and O. Ohguchi, "Optically induce rotation of anisotropic micro-objects fabricated by surface micromachining," Appl. Phys. Lett. **64**, 2209-2210 (1994).
42. P. Galajda and P. Ormos, "Complex micromachines produced and driven by light," Appl. Phys. Lett. **78**, 249-251 (2001).
43. P. Galajda and P. Ormos, "Rotors produced and driven in laser tweezers with reversed direction of rotation," Appl. Phys. Lett. **80**, 4653-4655 (2002).
44. E. Higurashi, R. Sawada, and T. Ito, "Optically driven angular alignment of micro-components made of in-plane birefringent polyimide film based on optical angular momentum transfer," J. Micromech. Microeng. **11**, 140-145 (2001).
45. M. E. J. Friese, T. A. Nieminen, N. R. Heckenberg, and H. Rubinsztein-Dunlop, "Optical alignment and spinning of laser-trapped microscopic particles," Nature **394**, 348-350 (1998).
46. Later in the peer-review process, we were made aware of a recent article [S.L. Neale, M. P. MacDonald, K. Dholakia and T. F. Krauss, "All-optical control of microfluidic components using form birefringence", Nat. Mat. **4**, 530-533 (2005)] that shows rotation of a microfabricated structure in a circularly polarized light due the object's form birefringence.
47. R. C. Gauthier, "Theoretical investigation of the optical trapping force and torque on cylindrical micro-objects," J. Opt. Soc. Am. B **14** 3323-3333 (1997).
48. Z. Cheng, P. M. Chaikin, and T. G. Mason, "Light streak tracking of optically trapped thin microdisks," Phys. Rev. Lett. **89**, 108303 (2002).
49. J. Glückstad, I. R. Perch-Nielsen, and P. J. Rodrigo, in preparation.
50. J. Glückstad, "Sorting particles with light" Nature Materials **3**, 9-10 (2004).

51. A. Terray, J. Oakey, and D. W. M. Marr, “Microfluidic control using colloidal devices,” Science **296**, 1841 (2002).
52. L. Kelemen, S. Valkai and P. Ormos, “Integrated optical motor,” Appl. Opt. **45**, 2777-2780 (2006).
53. J. Enger, M. Goksör, K. Ramser, P. Hagberg, and D. Hanstorp, “Optical tweezers applied to a microfluidic system”, Lab Chip, **4**, 196-200 (2004).
54. I. R. Perch-Nielsen, P. J. Rodrigo, C. A. Alonzo, and J. Glückstad, “Autonomous and 3D real-time multi-beam manipulation in a microfluidic environment,” Opt. Express **14**, 12199-12205 (2006).
55. M. Gauthier, D. Heriban, D. Gendreau, S. Regnier, P. Lutz and N. Chaillet, “Microfactory for submerged assembly: interests and architectures,” Proc. 5th Int. Workshop on Microfactories (2006).
56. J. J. Talghader, J. K. Tu and J. S. Smith, “Integration of fluidically self-assembled optoelectronic devices using silicon-based process,” IEEE Photon. Technol. Lett. **7**, 1321-1323 (1995).
57. K. Hosokawa, I. Shimoyama and H. Miura, “Two-dimensional micro-self-assembly using the surface tension of water,” Sens. Actuators A **57**, 117-125 (1996).
58. R. L. Eriksen, V. R. Daria, and J. Glückstad, "Fully dynamic multiple-beam optical tweezers," Opt. Express **10**, 597-602 (2002).
59. P. J. Rodrigo, R. L. Eriksen, V. R. Daria, and J. Glückstad, "Interactive light-driven and parallel manipulation of inhomogeneous particles," Opt. Express **10**, 1550-1556 (2002).
60. S. Maruo, O. Nakamura, and S. Kawata, “Three-dimensional microfabrication with two-photon-absorbed photopolymerization,” Opt. Lett. **22**, 132-134 (1997).
61. http://mathworld.wolfram.com/WallpaperGroups.html
62. A. Terray, J. Oakey and D.W.M. Marr, “Fabrication of linear colloidal structures for microfluidic applications,” Appl. Phys. Lett. **81**, 1555-1557 (2002).
63. P. J. Rodrigo, L. Gammelgaard, P. Bøggild, I. R. Perch-Nielsen, and J. Glückstad, "Actuation of microfabricated tools using multiple GPC-based counterpropagating-beam traps," Opt. Express **13**, 6899-6904 (2005).
64. E. R. Lyons and G. J. Sonek, “Confinement and bistability in a tapered hemispherically lensed optical fiber trap,” Appl. Phys. Lett. **66**, 1584-1586 (1995).

Chapter 9
Alternative GPC Schemes

The previous chapters presented the theoretical foundations of the generalized phase contrast (GPC) and established its advantages in quantitative wavefront sensing and arbitrary light shaping. For shaping light, the main advantage of the GPC approach lies in its encoding simplicity, where each point at the output plane maps to a unique point at the input plane. The typical GPC setup, shown in Fig. 9.1, uses a $4f$ image processing setup where input phase modulation is converted to high-contrast output intensity patterns using two Fourier lenses with a static phase contrast filter (PCF) located at the common focal plane between them. The input phase modulation could be an unknown wavefront that a user wishes to visualize, or a user-defined pattern encoded with a reconfigurable phase-only spatial light modulator (SLM) to synthesize desired dynamic sequences of light patterns.

Various modifications of the GPC setup can further enhance its flexibility, versatility and general performance. Adaptive operation can be achieved with a reconfigurable PCF that can be modified to match the input or to shift between different operation

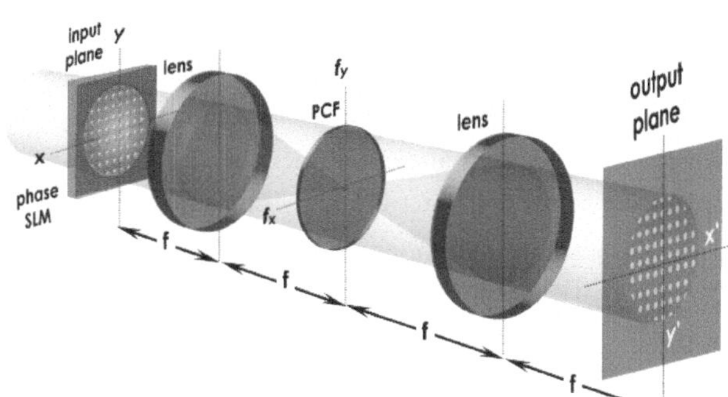

Fig. 9.1 Typical optical $4f$-setup used to synthesize patterned light using the generalized phase contrast method.

modes. Adaptive and self-alignment operations can be achieved for a PCF based on a nonlinear material that automatically sets up a light-induced phase shift. High-power operation can be enabled using multiple beams to illuminate the input from different angles. Higher output peak intensities can be achieved by incorporating a matched filter into the PCF. A robust and compact operation can be achieved in a planar micro-optics GPC implementation. We will illustrate these different enhancements in this chapter.

9.1 GPC Using a Light-Induced Spatial Phase Filter

The theoretical framework presented in earlier chapters allows one to find optimal experimental parameters. As with many optical techniques, these optimal parameters require matching optimal alignment for achieving optimal results. The numerous experimental demonstrations show that GPC have reasonable alignment requirements. Nevertheless, a nonlinear refracting medium situated at the spatial frequency domain can generate an adaptive and self-aligning PCF. In the following, we will present a mathematical model for GPC that uses a general Kerr-type nonlinear PCF for generating structured light. Structured light has applications in machine vision, particularly in systems involving 3D spatial measurements. We will also describe proof-of-concept experiments using bacteriorhodopsin, which is a highly nonlinear optical material. Results indicate that the scheme can also be used for dynamic contrast reversal of input amplitude objects.

The setup is based on the $4f$ architecture depicted in Fig. 9.1 but with a thin film Kerr medium located at the spatial Fourier domain instead of a prefabricated PCF. We consider a Gaussian-illuminated periodic array of rectangular phase shifting regions as the GPC input for generating the structured light illumination:

$$a(x,y)=\frac{\sqrt{2P}}{r_o}\exp\left(\frac{-\pi r^2}{r_o^2}\right)\mathrm{rect}\left(\frac{x}{\Delta x_A},\frac{y}{\Delta y_A}\right)\exp\left(i\phi_0\sum_{mn}\mathrm{rect}\left(\frac{x-m\Delta x}{\Delta x_c},\frac{y-n\Delta y}{\Delta y_c}\right)\right) \quad (9.1)$$

where $\Delta x_A \Delta y_A$ is the area of the phase mask aperture, $\Delta x_c \Delta y_c$ is the area of each cell that shifts the phase by ϕ, Δx and Δy are the spatial periods of the phase array, and P is the power of the laser beam. It is also convenient to define an encoding fill factor, $F=\Delta x_c \Delta y_c/\Delta x \Delta y$ to describe what fraction of the input area is phase-encoded by ϕ_0. Assuming the Gaussian beam is predominantly transmitted by the mask aperture, the field distribution in the spatial frequency domain is

$$A(f_x,f_y)=\frac{\sqrt{2P}r_0}{\lambda f}\left[\exp\left(-\pi r_o^2 f_r^2\right)+\frac{\Delta x_c \Delta x_c}{\Delta x \Delta y}\left[\exp(i\phi_0)-1\right]\mathrm{sinc}\left(\Delta x_c f_x,\Delta y_c f_y\right)\right.$$
$$\left.\times\sum_{mn}\exp\left(-\pi r_o^2\left[\left(f_x-\frac{m}{\Delta x}\right)^2+\left(f_y-\frac{n}{\Delta y}\right)^2\right]\right)\right] \quad (9.2)$$

where $\bar{\alpha} = 1 + F\left[\exp(i\phi_0) - 1\right]$ is a normalized zero-order amplitude, as in the standard GPC analysis, and $f_{\{x,y,r\}}$ refer to the Cartesian and radial spatial frequency coordinates. The input periodicity generates periodicity at the optical Fourier domain as is evident from Eq. (9.2). Each Fourier-order is modulated by a narrow Gaussian envelope, which isolates the different Fourier-orders from each other (i.e. $\Delta x, \Delta y << r_o$). The sinc-function imposes an overall intensity roll-off.

9.1.1 Self-Induced PCF on a Kerr Medium

The Fourier field illuminates a thin film Kerr medium situated at the spatial frequency domain. The film's refractive index changes according to the impinging light intensity due to the self-induced optical Kerr effect (a third order nonlinearity) and the induced refractive index change, Δn, is proportional to the intensity

$$\Delta n = n_2 |A|^2 \tag{9.3}$$

where n_2 is a material-dependent Kerr coefficient.

The refractive index change imposes a self-induced phase shift on the propagating light. This effectively creates a self-induced phase contrast filter when the zero-order region is shifted with respect to the higher orders. A localized phase shift is a good approximation in most cases since the sinc-rolloff strongly dampens the higher orders compared to the zero-order. If necessary, the maximum Kerr medium size may be restricted to $\sim(\lambda f)^2/(\Delta x \Delta y)$ to avoid higher orders that might have sufficiently high intensity to also create self-induced spatial phase changes.

The self-induced PCF is described by an intensity-dependent zero-order transfer function,

$$T_0\left(f_x, f_y\right) = \exp\left(i\frac{2\pi L}{\lambda} n_2 \left|A_0\left(f_x, f_y\right)\right|^2\right), \tag{9.4}$$

where L is the film thickness. Since the sinc envelope is ~ 1 around the zero-order, the intensity around the zero-order is

$$\left|A_0\left(f_x, f_y\right)\right|^2 = \frac{2\mathrm{P}r_o^2}{\lambda^2 f^2} \exp\left(-2\pi r_o^2 f_r^2\right) |\bar{\alpha}|^2 \tag{9.5}$$

Substituting the zero-order intensity into the transfer function, we can express T_0 by a series expansion [1]:

$$T_0\left(f_x, f_y\right) = \sum_{k=0}^{\infty} \frac{(i\theta_0)^k}{k!} \exp\left(-2k\pi r_o^2 f_r^2\right) \tag{9.6}$$

where the zero-order phase parameter is given by

$$\begin{aligned}\theta_0 &= \frac{4\pi L n_2 \Pr_o^2}{\lambda^3 f^2}|\bar{\alpha}|^2 \\ &= \frac{4\pi L n_2 \Pr_o^2}{\lambda^3 f^2}\left[1+2F(F-1)(1-\cos\phi_0)\right]\end{aligned} \tag{9.7}$$

Equation (9.7) contains, as a special case, the expected result that no self phase modulation occurs when there is no zero order beam ($\theta_0 = 0$ when $\bar{\alpha} = 0$). This occurs when half of the input area is phase shifted by an odd multiple of $\pm\pi$, which corresponds to $\phi_0 = \pm(1+2p)\pi$ (where $p = 0, 1, 2\ldots$) and $\Delta x_c \Delta y_c / \Delta x \Delta y = 1/2$.

The field distribution immediately after the film is

$$O(f_x, f_y) = A(f_x, f_y) + \frac{\sqrt{2P}r_o}{\lambda f}\left[\sum_{k=0}^{\infty}\frac{(i\theta_0)^k}{k!}\exp\left(-2k\pi r_o^2 f_r^2\right) - 1\right]\bar{\alpha}\exp\left(-\pi r_o^2 f_r^2\right), \tag{9.8}$$

where the zero-order has been separated from the higher order terms. The second Fourier lens in the $4f$ system then converts this to an output intensity given by

$$\left|o(x', y')\right|^2 = \left|a(x', y') + \bar{\alpha}\frac{2P}{r_o^2}\sum_{k=1}^{\infty}\frac{(i\theta_0)^k}{k!(1+2k)}\exp\left(-\frac{\pi r'^2}{r_o^2(1+2k)}\right)\right|^2. \tag{9.9}$$

As in the standard GPC, this output may be interpreted as the interference between the first term, $a(x',y')$, which is a direct image of the phase input, and the second term, which is visualized as a synthetic reference wave (SRW). Plotting this expression reveals an efficient and robust array illumination. Intensity levels can reach up to almost four times stronger than the input intensity level when using optimal parameters, such as $\psi = \pm(1+2p)\pi$ and $\Delta x_c \Delta y_c / \Delta x \Delta y \to 0^+$.

Figure 9.2 illustrates how the output intensity evolves as a function of increasing power levels (increasing zero-order self-phase modulation). Surprisingly, the spot array does not manifest any "oscillations" in the intensity level for increasing input power levels over a certain region, thereby increasing the robustness of the scheme. Actually, for all $|\theta_0|$ above $\sim\pi$ the spot array illumination is at an almost constant level. This is a very important feature for practical implementations since the scheme will then be robust both with respect to alignment (self-induced phase filter) and with respect to oscillations in input power. However, a given minimum power level is obviously required in order to perturb the nonlinear thin film into this robust operating regime. According to Eq. (9.7) a small area ratio value is preferable with respect to low minimum power requirements. In the next section we will discuss what happens when the area ratio value approaches the opposite extreme 1^-.

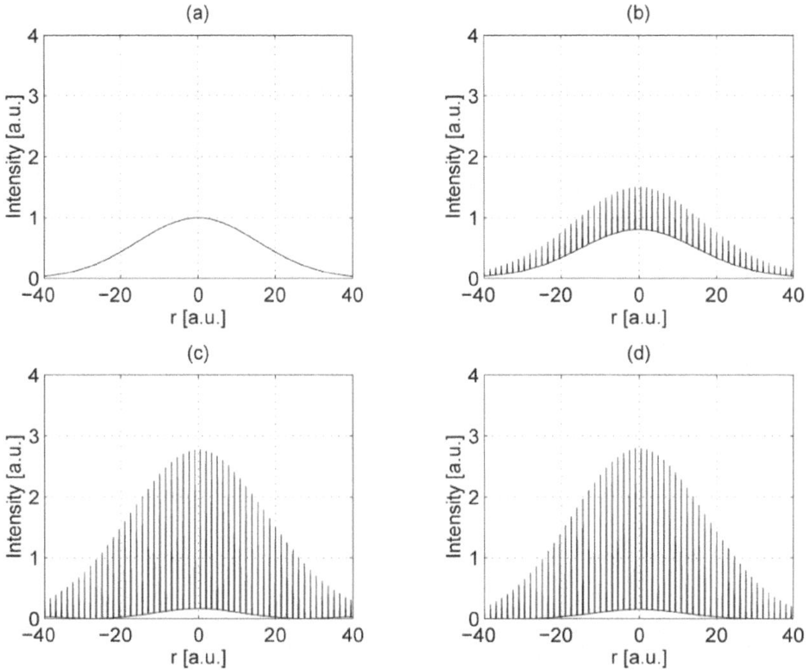

Fig. 9.2 Simulated radial output intensity linescan for a GPC-based array illuminator that uses a self-induced PCF on a Kerr medium. The simulations used an input phase parameter $\psi = \pi$ and an area ratio, $\Delta x_c \Delta y_c / \Delta x \Delta y = 1/10$. (a) $\theta_0 = 0$, (b) $\theta_0 = \pm\pi/2$, (c) $\theta_0 = \pm 5$ rad, (d) $\theta_0 = \pm 10$ rad

9.1.2 Kerr Medium with Saturable Nonlinearity

In physical systems, increasing illumination intensity levels will inevitably result in a saturation of the self-induced refractive index profile. At saturation, the smooth Gaussian refractive index profile will be truncated and transformed into a top-hat-like shape, which is highly similar to a standard GPC filter. The transfer function for a saturable nonlinear thin film is commonly described as:

$$T_0(f_x, f_y) = \exp\left(i \frac{2\pi L}{\lambda} n_2 \frac{\left|A_0(f_x, f_y)\right|^2}{1 + \left|A_0(f_x, f_y)\right|^2 / I_s} \right) \tag{9.10}$$

where I_s is a characteristic saturation intensity value of the medium in consideration. A normalized transverse profile of the induced phase shift at different intensity levels is shown in Fig 9.3. At intensity levels much smaller than I_s, Eq. (9.10) leads to the previous transfer function for a power-law Kerr-type behaviour. On the other hand, intensity levels much larger than I_s introduce the saturating top-hat refractive index profile,

$$T_0\left(f_x,f_y\right)\approx\exp\left(i\frac{2\pi L}{\lambda}n_2 I_S\right)=\exp\left(i\theta_S\right),\qquad(9.11)$$

where θ_S is the effective phase shift at saturation. This self-induced PCF is now equivalent to a static PCF in the standard GPC with Gaussian illumination (see Sects. 6.5.3 and 7.5), with the added benefit that this PCF is inherently both self-aligning and self-resizing. Thus, we can revert to a standard GPC formulation and describe the PCF as a saturated self-induced Fourier filter

$$H_s(f_x,f_y)=1+\left[\exp(\mathrm{i}\theta_s)-1\right]circ\left(\left(f_x^2+f_y^2\right)^{1/2}/\Delta f_C\right),\qquad(9.12)$$

where Δf_C is an upper cutoff for frequency components that are phase shifted together with the zero order. Using this filter yields an output

$$\left|o(x',y')\right|^2=\left|a(x',y')+\bar{\alpha}\left[\exp(\mathrm{i}\theta_S)-1\right]g(x',y')\right|^2,\qquad(9.13)$$

where $g(x',y')$ describes the SRW spatial profile.

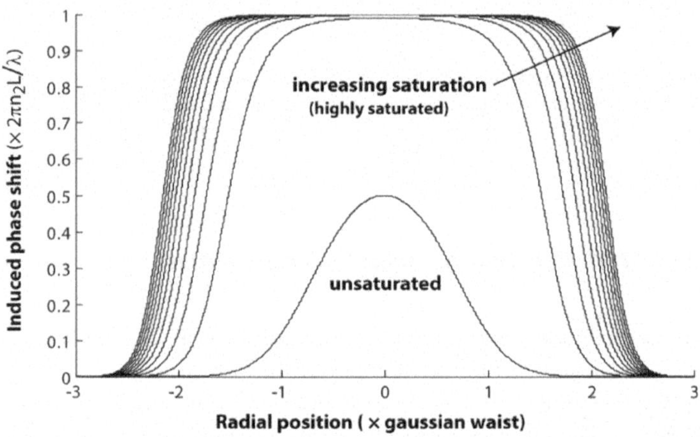

Fig. 9.3 Normalized transverse profile of the induced phase shift on a saturable thin nonlinear medium at different intensity levels.

For a highly saturated medium, the self-induced top-hat PCF transmits practically the entire Gaussian zero-order beam and the SRW spatial profile matches the image of the Gaussian illumination (see Sect. 6.5.3). The output is

$$\left|o(x',y')\right|^2=\frac{2P}{r_0^2}\exp\left(\frac{-2\pi r^2}{r_o^2}\right)\left|\exp\left[i\phi(x',y')\right]+\bar{\alpha}\left[\exp(\mathrm{i}\theta_S)-1\right]\right|^2,\qquad(9.14)$$

where $\phi(x',y')=\phi_0\sum_{mn}\mathrm{rect}\left(\frac{x'-m\Delta x}{\Delta x_c},\frac{y'-n\Delta y}{\Delta y_c}\right)$ is the image of the input phase modulation.

We can apply the optimization conditions developed in previous chapters for a self-induced PCF on a saturated nonlinear medium. For example, patterns with maximum contrast visibility are achieved using the input and filter conditions outlined by Eq. (6.36), which when adapted to the present case becomes

$$\begin{aligned}\bar{\alpha}_{real} &= \frac{1}{2}\\ \bar{\alpha}_{imag} &= \frac{1}{2}\cot\left(\frac{\theta_S}{2}\right)\end{aligned}, \tag{9.15}$$

A simple way to satisfy the condition for the real part of the normalized zero order is to use binary (0-π) encoding with an encoding fill factor F= ¼. This input phase encoding leads to $\bar{\alpha}_{imag}=0$, which is optimized by choosing the Kerr medium such that the saturation phase shift becomes $\theta_S=\pi$. The maximum output intensity reaches four times the maximum input intensity.

It is also possible to produce a high contrast pattern that is contrast-reversed with respect to the input modulation (i.e. π-encoded regions are dark while 0-encoded regions are bright). In this case, the optimal condition is revised into

$$\begin{aligned}\bar{\alpha}_{real} &= -\frac{1}{2}\\ \bar{\alpha}_{imag} &= \frac{1}{2}\cot\left(\frac{\theta_S}{2}\right)\end{aligned}, \tag{9.16}$$

which may be achieved using F = ¾ and $\theta_S=\pi$. As in the normal contrast mode, the maximum output intensity reaches four times the maximum input intensity.

Yet another interesting operation mode is to maximize the spot intensities while tolerating a non-zero background. This is achieved when using a saturated PCF phase shift of $\theta_S=\pi$ and an input with $\bar{\alpha}\rightarrow 1$, which occurs when $F\rightarrow 0$ for the π-encoded input region. In this case, Eq. (9.14) predicts a maximum output intensity that is 9 times larger than a background level that mimics the input intensity. A similar intensity ratio for a contrast reversed output is attained using $F\rightarrow 1$ to generate $\bar{\alpha}\rightarrow -1$ for 0-π binary encoding.

These cases are among the simple operating modes that can be realized when using a saturated Kerr medium at the PCF. It is also possible to use non-π input encoding, or even a greyscale phase input and achieve an optimum output visibility by satisfying the criteria in Eq. (9.15) for normal contrast, or Eq. (9.16) for contrast-reversed output. Some characteristic output intensity patterns are shown in Fig. 9.4.

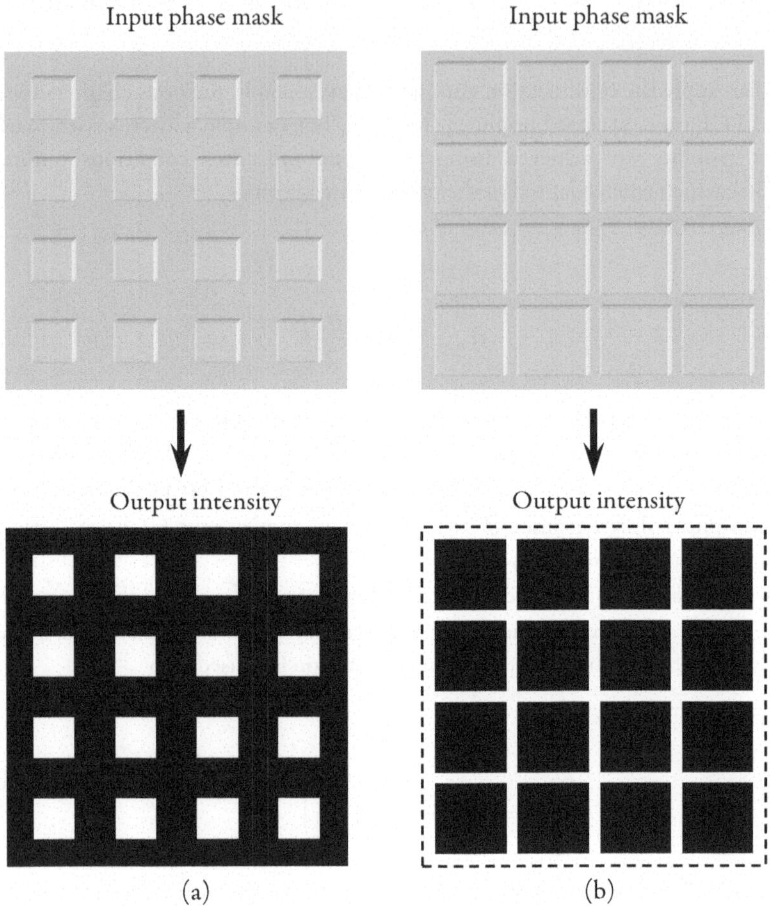

Fig. 9.4 Emerging intensity patterns for different encoding fill factors: (a) $0<F<1/2$ and (b) $1/2<F<1$.

9.1.3 Experimental Demonstration

A proof-of-principle demonstration of the proposed scheme was implemented with the GPC setup depicted in Fig. 9.1 using a HeNe laser illumination (power~ 30 mW, λ= 633 nm) and Fourier lenses with focal lengths f = 50 mm. A thin film Kerr medium replaced a prefabricated PCF. The nonlinear film was based on bacteriorhodopsin, which exhibits a large refractive index change at λ= 633 nm [2]. Moreover, a bacteriorhodopsin thin film acts as a negative Kerr medium at this wavelength, which makes it much easier to cope with than a positive Kerr medium. Bacteriorhodopsin is a highly nonlinear optical material, possessing extremely large third order nonlinearity. Optical Kerr coefficients in the range of $10^{-2}-10^{-4}\,\mathrm{cm}^2/\mathrm{W}$ have been reported [3, 4, 5]. The thin film used in the experiments was made from commercially available bacteriorhodopsin membrane suspensions. Bacteriorhodopsin membrane (12.5 mg) was sonicated and filtered before mixing with 2 ml of gelatine solution (80 mg gelatine dissolved in water). A thin film (~40 μm) of the mixture was cast on a $20\ \mathrm{mm}^2$ glass substrate.

The experiments used an *amplitude* mask at the input, instead of a phase mask. This is mathematically equivalent to a binary (0-π) phase mask that is superimposed with a uniform background and can, thus, illustrate the same features of the mathematical model above. Typical results are shown in Fig. 9.5. With the bacteriorhodopsin film placed far from the focal plane, there was a negligible nonlinear effect and the system worked as a standard *4f* imaging system. Thus, an image of the input mask, a periodic array of $\sim 100\ \mu\mathrm{m}\times 100\mu\mathrm{m}$ squares, was reproduced at the output as seen in Fig. 9.5a. A dramatic contrast reversal was observed as the thin film was moved towards the Fourier transform plane, with the dark squares becoming much brighter compared to the background as shown in Fig. 9.4b.

These results can be understood in terms of the GPC framework. Since the nonlinear medium is operated at saturation, it behaves as a static filter and the system exhibits its usual linear properties. Treating the amplitude mask as a superposition of a phase-modulated input with a background, each with amplitude a_0, we can designate the

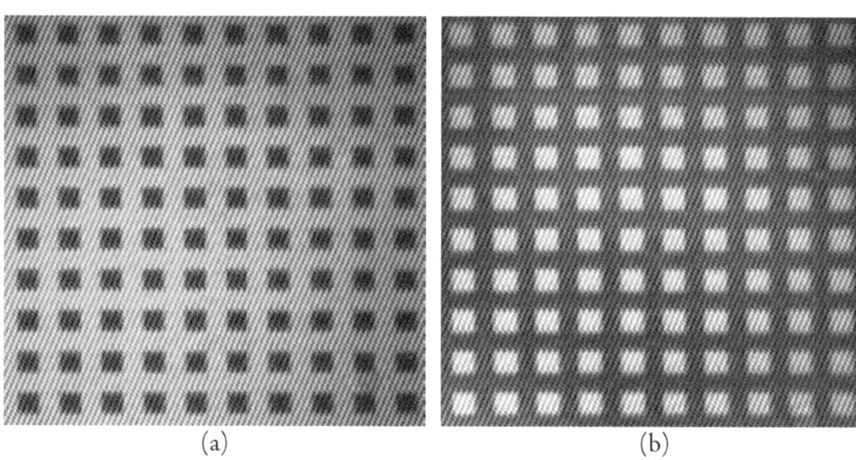

Fig. 9.5 Output patterns at different axial Kerr-medium filter locations: (a) An image of the input amplitude mask results when the Kerr-medium filter is axially shifted away from the focal plane (b) A contrast reversed image of the amplitude mask is generated when the Kerr-medium filter is placed at the focal plane. The squares have side length ~100μm.

bright regions by $2a_0$ amplitude or $4|a_0|^2$ intensity. The output is then modelled as a standard GPC with an additional SRW. Thus, the contrast reversed output bright regions will have $3a_0$ amplitude or $9|a_0|^2$ intensity while the background, coming from the extra SRW, will have a_0 amplitude or $|a_0|^2$ intensity. These features are illustrated in the results presented in Fig. 9.5, providing a proof-of-principle demonstration of the predicted phenomenon.

The foregoing theoretical analysis and experimental results illustrate that GPC can be implemented with a self-induced filter on a nonlinear Kerr-medium. A self-induced filter is inherently self-aligning and self-resizing, which provides for a robust and adaptive operation. Predictable operation that closely mimics standard GPC can be achieved by carefully selecting the self-induced filter parameters to operate in the saturated regime that ensures consistent optimal phase shift. However, care must be observed to ensure that the incident power is maintained at suitable levels. This adaptive scheme can provide the same advantages of high-efficiency projections using straightforward input phase encoding as in standard GPC. Electrically induced perturbations (via transparent electrodes) and/or external pumping beams in the filter plane can be further explored to synthesize even more interesting filtering schemes.

It is also worth noting that, when using amplitude masks, "GPC-with-extra-SRW" predicts output spot intensities exceeding twice the input intensity (i.e. $9|a_0|^2$ vs. $4|a_0|^2$). This points to yet another potential application of the GPC framework. Amplitude masks, although energy inefficient, continue to persist in various applications due to their simplicity. The downside is that energy efficiency of amplitude masks falls as the fractional size of the transmissive regions is reduced. By utilizing the contrast reversal demonstrated here, one can use a negative mask to transmit light within the wider, but otherwise absorbing regions, and achieve outputs with higher spot intensities even at low fill factors using amplitude mask. In the limit of low fill factors, the output would have spot intensities approaching $16|a_0|^2$ within a $4|a_0|^2$ background level. A positive mask at the output can be used to remove unwanted background if it cannot be tolerated by the application.

9.2 GPC Using a Variable Liquid-Crystal Filter

Whether using generalized phase contrast for sensing unknown wavefronts, or for synthesizing desired light projections, the filter phase shift that yields optimal output phase contrast depends on the input phase modulation. The dependence of the optimal filter phase shift on the input phase encoding was derived in Chapter 6 and is mathematically formulated as (c.f. Eq. (6.34))

$$\bar{\alpha}_{imag} = \frac{1}{2K}\cot\left(\frac{\theta_{optimal}}{2}\right), \tag{9.17}$$

where K may be set to 1, or another suitable value, by adjusting the filter size. Thus, a filter capable of imparting a user-defined phase shift makes a useful component for enabling adaptive operation.

Figure 9.6 shows one of several possible implementations of a phase contrast filter with variable phase shift, θ, on the frequencies close to the zero order with respect to the higher spatial frequencies. For practical purposes, the relative shift is achieved through a voltage-controlled phase shift of the higher orders relative to a fixed zero order without incurring spurious side effects.

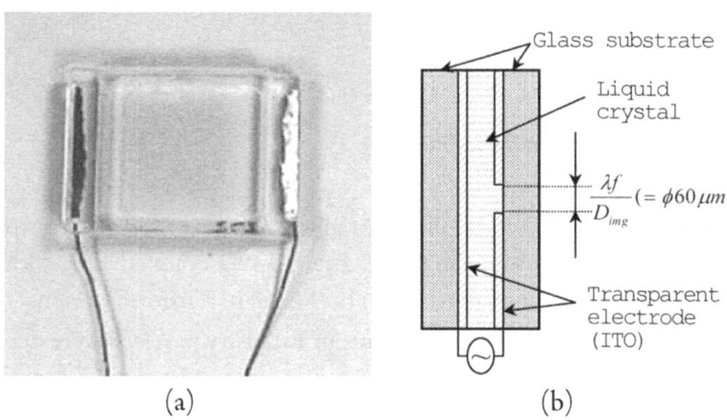

Fig. 9.6 Liquid crystal-based variable phase contrast filter: (a) Photograph (b) axial section

Figure 9.6(a) shows a photograph of a custom-fabricated 40 mm × 30 mm voltage-controlled liquid crystal phase contrast filter and Fig. 9.6(b) illustrates its structure. A 10 μm thick liquid crystal layer is sandwiched between indium tin oxide- (ITO) coated glass plates. The ITO coatings serve as transparent conductive electrodes for controlling the liquid crystal. To phase shift the zero order, a circular portion (60 μm diameter) of the ITO coating was removed from the middle part of one of the glass plates by laser ablation. Applying AC voltage on the ITO electrodes provides a relative phase shift between the zero-order and the higher orders. Figure 9.7 shows the measured voltage dependence of the phase modulation depth, θ. The results show that phase shifts within the range, $0 \le \theta \le 2\pi$, can be achieved using 0–4 V electrode voltage amplitudes. Applied voltages with 1 kHz and 10 kHz frequencies result in almost coincident transfer characteristics.

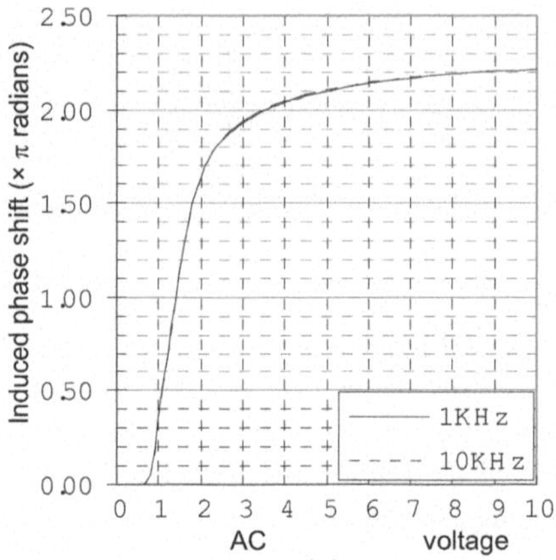

Fig. 9.7 Voltage dependence of the phase shift of the liquid crystal phase contrast filter.

9.2.1 Experimental Demonstration

The custom-fabricated, voltage-controlled, variable PCF was used in an experimental implementation of the GPC system in Fig. 9.1. An LCD-coupled PAL-SLM [6] was used to display phase-coded input patterns. The Fourier-transforming lenses in the $4f$ setup both had f=400 mm as focal lengths. Output intensity patterns were detected by a fibre-window CCD camera to eliminate interference noise effects. By using the integration function of an image processing unit (C3000/Hamamatsu Photonics) the detected intensity patterns were registered by 16 bits of analogue levels.

The system was evaluated using various binary inputs with striped and random dot patterns possessing different normalized zero-order strengths, $\bar{\alpha}$, which were modified according to the pattern duty cycles and phase depths. The filter was optimized according to the criterion in Eq. (9.17) to achieve output intensity patterns with optimum contrast and visibility. Examples of the experimental outcomes for different parameter settings are illustrated in Fig. 9.8.

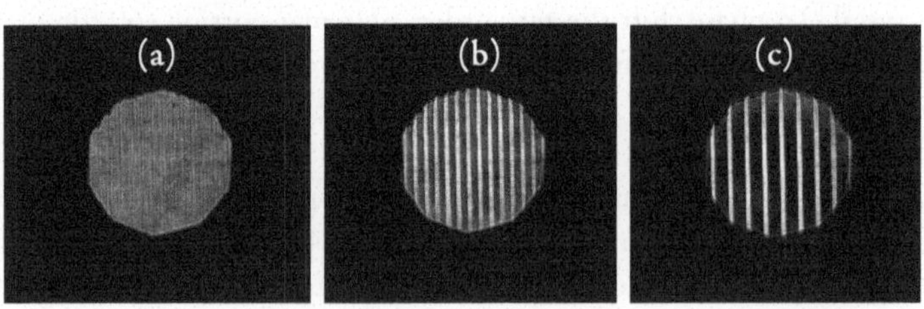

Fig. 9.8 Intensity patterns for different parameters: (a) PCF turned off, (b) binary phase pattern with $\phi = 1.1\pi$, duty cycle=40% and PCF phase $\theta = 0.6\pi$, (c) $\phi = \pi$, duty cycle=25% and $\theta = \pi$.

Figure 9.8(a) shows the output when the PCF was turned off: the system worked as a standard $4f$ imaging system and produced a nearly constant grey level output as expected. When the PCF was activated, but set to a phase depth not matching the optimal criterion, the output exhibited some phase contrast but with suboptimal visibility, as shown in Fig. 9.8(b). When the input phase modulation and the PCF phase shift finally satisfied the optimal criterion, a high contrast output pattern was obtained, as shown in Fig. 9.8(c).

9.3 Multibeam-Illuminated GPC With a Plurality of Phase Filtering Regions

In a typical generalized phase contrast setup with plane wave illumination (see Fig. 9.1), the light source may be considered as a point located at infinity and a phase contrast filter (PCF) imparts an optimal phase shift on the undiffracted light from this point source. The undiffracted spot is an image of the point source and is subject to diffractive broadening from intervening apertures. The GPC framework accounts for these diffractive effects and veers away from the point image model prevalent in numerous flawed phase contrast treatments. Thus, GPC provides better understanding of the output interferograms, which can lead to improved retrieval of unknown wavefronts. This also allows for finding appropriate experimental parameters such as the optimal PCF size and phase shift.

Most of the previous GPC analysis and applications considered on-axis point sources that generate axially propagating plane waves at the GPC input plane. However, it stands to reason that the same analysis applies to off-axis point sources. The same optimization criteria developed previously can also be applied when using off-axis point sources, with the obvious modification that the filter should likewise be shifted off-axis to properly align with the shifted image. The output intensity pattern is generally unaffected but would now possess a linear phase ramp since an off-axis point source generates an obliquely incident plane wave at the input plane of the GPC setup.

In this section we focus on a GPC implementation that is based on multiple point sources at infinity, as shown in Fig. 9.9 [7]. This superposition of point sources requires a matching constellation of phase shifting dots at the filter plane that are individually optimized using GPC-based criteria. Such sources may be realized using light sources such as fibre laser arrays, VCSEL-arrays, LED-arrays, and others. It is also possible to use a single laser source and generate such array by beam modulation techniques such as diffractive technologies. When these sources are placed at the front focal plane of a lens, they would generate multiple obliquely incident waves at the back focal plane to imitate point sources at infinity. This provides the multiple illuminations illustrated in Fig. 9.9.

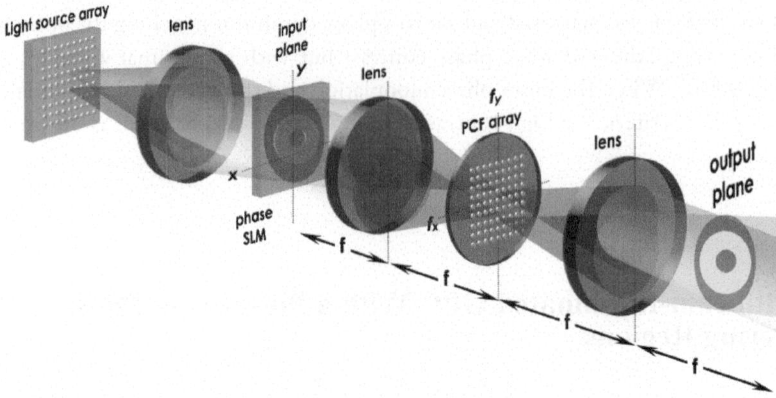

Fig. 9.9 GPC implemented with multiple beam illumination matched with multiple phase contrast filters.

The GPC output in this case is determined by the superposition of outputs from each source-PCF pair. Given N number of sources, we may write the output as (c.f. Eq. (6.9))

$$\begin{aligned} O(x,y)=\sum_{n=1}^{N} o_n(x,y)=\sum_{n=1}^{N} A_n\{a_n(x,y)\exp[i\phi_n(x,y)] \\ +\bar{\alpha}_n[\exp(i\theta_n)-1]g_n(x,y)\}\exp(ik_{x,n}x+ik_{y,n}y) \end{aligned}, \quad (9.18)$$

where $a_n(x,y)$, $\phi_n(x,y)$, $\alpha_n(x,y)$ and A_n are the effective input aperture, input phase modulation, normalized undiffracted order, and characteristic amplitude for each source; θ_n, and $g_n(x,y)$ are the corresponding PCF phase shift and resulting SRW profile; and $k_{x,n}$ and $k_{y,n}$ are wavevector components describing the tilt angle of each. Residual crosstalk due to the other PCFs at the Fourier plane was neglected, assuming moderate PCF densities. Each PCF parameter can be separately specified to optimize a given output metric. For example, optimal visibility for each source-PCF pair may be achieved by choosing parameters according to the criterion

$$\begin{aligned} \bar{\alpha}_{n,imag} &= \frac{1}{2g_n(0,0)}\cot\left(\frac{\theta_{optimal}}{2}\right) \\ \bar{\alpha}_{n,real} &= \frac{1}{2g_n(0,0)} \end{aligned}, \quad (9.19)$$

where $g_n(0,0)$ depends on the PCF size (see Fig. 3.3 in Chapter 3). The second criterion is particularly relevant in synthesis applications where the user can control and optimize the input phase.

For individually coherent yet mutually incoherent sources (e.g. array of independent fibre lasers), the output is taken as an incoherent superposition of output intensities

$$I(x,y)=\sum_{n=1}^{N} I_n(x,y)=\sum_{n=1}^{N} A_n^2\left|a_n(x,y)\exp\left[i\phi(x,y)\right]+\bar{\alpha}\left[\exp(i\theta_n)-1\right]g(x,y)\right|^2. \quad (9.20)$$

The capacity of GPC for broadband operation means the sources can have different wavelengths, if necessary, and can even be generated by pulsed sources. The output described in Eq. (9.20) indicates that the sources contribute identical output patterns and the superposition leads to higher output intensity. Increasing the output power levels in this way has several practical advantages over simply cranking up the power of a single input-illumination. For high-power applications, this method allows one to overcome maximum power limitations of commercially available lasers. This also allows one to use multiple lower-power sources to avoid the deteriorating beam quality of lasers with higher powers.

When using multiple oblique illuminations, the process of filtering the strong undiffracted order is distributed to multiple sites. This allows reaching output powers beyond the material breakdown level of a single PCF. Illumination from multiple angles also erases unwanted "shadows" from noise sources outside the input plane, which results in projections with minimal noise. Input illumination at oblique incidence also allows higher frequency components from the input plane to be transmitted to the output, which leads to sharper images. This is important in both sensing and synthesis applications of GPC. When projecting micropatterned light, such as for optical materials processing or optical trapping and manipulation, multibeam illumination allows the synthesis of patterns with finer feature sizes. In sensing applications, such as quantitative phase microscopy, the multibeam illumination enables imaging at higher imaging resolutions. Since GPC operates robustly over a wide wavelength range, multiple oblique illumination can also incorporate colour-coded incidence angles, which when matched with coloured detection, can lead to three-dimensional localization as described in ref. [8].

9.4 Miniaturized GPC Implementation via Planar Integrated Micro-Optics

The typical GPC implementation using discrete optical elements in a *4f* image processing geometry may be impractical for a certain class of applications that require robust portability and compactness. To address this class of applications, the GPC method can be implemented using planar integrated micro-optics and miniaturized in a single device. Such a miniaturized GPC device has been previously realized according to the schematic shown in Fig. 9.1 and was demonstrated for optical decryption applications, which will be discussed in detail in Chapter 10. All the essential micro-optical components of the miniature system were fabricated on a planar surface using lithographic techniques. A glass substrate is coated with a photosensitive polymer material called photoresist. The photoresist is exposed to patterned laser or electron beam illumination using lithographic masks or other pattern forming technologies. The illuminated areas of the photoresist layer are removed with the aid of developing chemicals. The patterned photoresist protects the underlying glass substrate during reactive ion etching (RIE), a combination of chemical reaction and physical impact of plasma ions, thus ending with a patterned substrate.

The final layout of the optical components on the planar GPC device is illustrated in Fig. 9.10. The miniaturized GPC system uses two diffractive microlenses (L1 and L2) in a folded *4f* geometry with a reflective phase contrast filter (PCF) at the common Fourier plane and coupling gratings at the object and image planes. The diffractive microlenses, quantized with four phase levels, were fabricated on the substrate using two lithographic masks. The microlenses are slightly elliptic for optimized imaging along a tilted optical axis. They are mildly astigmatic with slightly different focal lengths (f_x=25.58 mm and f_y=24.51 mm) along two perpendicular axes but with the f/#

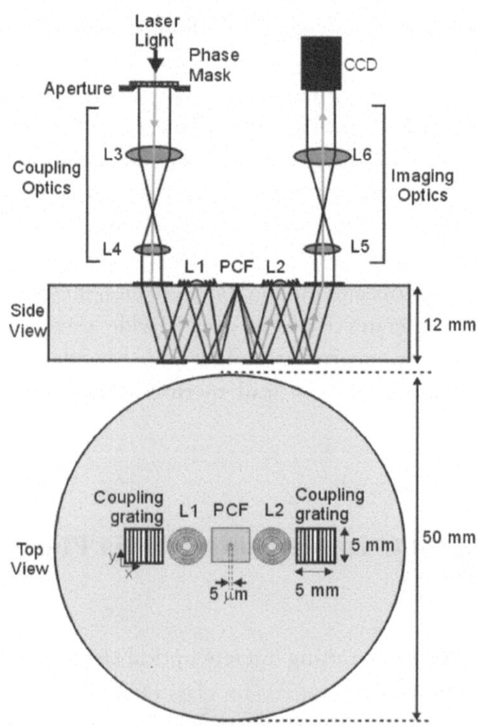

Fig. 9.10 Implementation of the GPC method in a 50-mm-diameter and 12-mm-thick planar-optical device. Two diffractive microlenses (L1 and L2), a 5-μm-diameter PCF and two input/output coupling gratings are arranged linearly on top of a glass substrate (top view). The beam path illustrates a folded version of the 4-f lens configuration (side view). External input and output (macro) optics are used to couple the input phase pattern and image the output intensity distribution, respectively.

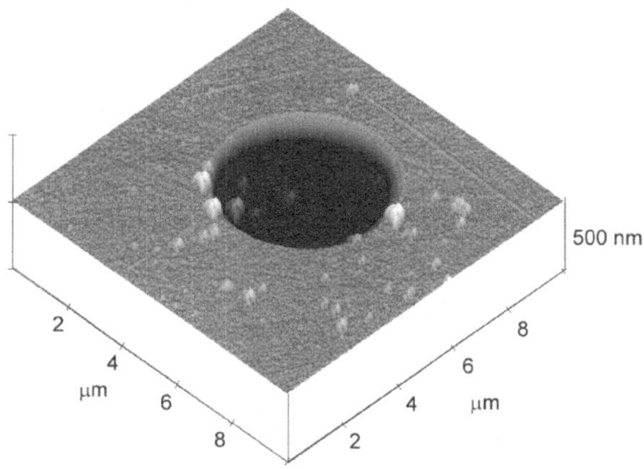

Fig. 9.11 Topographic image of the phase contrast filter (PCF) taken using an atomic force microscope. An anisotropic etching process is used to form the steep-edged 5-μm-diameter cylindrical hole with a depth designed to carry out a π-shift of the laser operating at wavelength $\lambda = 632.8$ nm.

maintained at ~5. The PCF is designed to operate at λ=0.633 μm and is etched as a hole with radius, R_1 = 2.5 μm, on the substrate. Light reflected through the hole acquires a π-phase shift with respect to directly reflected light. Figure 9.11 shows a topographic image of the PCF taken using an atomic force microscope. An anisotropic etching process is used to form a steep-edged cylindrical hole. The binary phase gratings are located at the object and image plane of the 4*f* system and couple light in and out of the system. The input grating, fabricated with a 2.13 μm period, deflects normally incident input light at 11.77° into the optical axis of the folded system. The output grating deflects obliquely incident output light so that it exits normal to the surface. During testing, remote phase modulation is imaged onto the input grating for coupling into the device and another set of imaging optics relays the intensity from the output grating onto a CCD camera.

The integrated planar device implementation of GPC operates in much the same way as a conventional GPC system. Light propagating along the optical axis from the input plane is focused by the microlens onto the phase-shifting hole. This is then collimated by the second microlens and creates a synthetic reference wave for the directly imaged input phase modulation. As a result, input phase information controls the contrast of the output intensity pattern. However, miniaturization can have unwelcome consequences on its performance. The planar layout, where the beam propagates obliquely through the folded optical system, can potentially lead to aberrations. The use of diffractive optical elements as coupling gratings and microlenses can suffer from the unwanted effects of phase quantization. This can be improved in future implementations by using fabrication techniques capable of producing continuous phase relief surfaces. This will increase the coupling efficiency of the gratings and avoid spurious orders and secondary foci in the microlenses.

9.4.1 Experimental Demonstration

A US Air force resolution target was used as input phase mask in a demonstration experiment, which π-shifts the target features with respect to the background. Figures 9.12.a-c show output images produced when the input beam is slightly tilted. The inputs are merely reproduced in the outputs, without any phase contrast, as the tilt causes the focused undiffracted beam to miss the PCF at the Fourier plane. The mask aperture diameter was varied from 4 mm to 6 mm corresponding to 1 mm to 1.5 mm radius at the surface of the planar optics. This aperture adjustment was done to illustrate the required size matching between the broadened zero-order and the phase-shifting PCF aperture when optimizing the output contrast. High contrast images, shown in Figs. 9.12(d)–(f), were obtained when the incidence angle was corrected to focus the undiffracted beam onto π-shifting PCF. Different sections of the phase mask were used for the different aperture settings to attain nearly identical spatial average values ($\overline{\alpha}$). The high contrast images (Fig. 9.12(d)–(f)) reveal the expected significance of matching the input and PCF apertures on the output contrast. Smaller input aperture (Fig. 9.12(d)), generates an output image with lower contrast with relatively no amplification in the maximum intensity as compared to the direct image in Fig. 9.12(a). The output contrast improves as the input aperture is increased, as shown in Fig. 9.12(f), having a maximum intensity that is almost twice of Fig. 9.12(c).

An optimized visualization of the input phase should result in an output distribution with a maximum intensity that is nearly four times the input intensity (corresponding to the intensity without the phase contrast filter in place; Figs 9.12(a)–(c)). The input aperture settings in the experiments effectively varied the PCF aperture relative of the broadened zero-order (mainlobe of the Airy pattern) and yielded η-values from 0.26 to 0.39. This range of η, however, is not the desired setting for generating output intensity distributions with optimum contrast [9]. Setting $\eta > 0.4$, which corresponds to $r > 3$ mm, would have yielded a higher contrast. However, the implementation with binary diffractive optical elements produced higher-order diffracted beams that interfered with and compromised the quality of the output image at bigger input apertures. This effect is evident in Fig. 9.12.f, which shows the image being corrupted by interference fringes from higher orders encroaching from the sides.

Aside from potentially disturbing the output, the higher orders due to the use of binary diffractive coupling elements also diverts energy away from the relevant order at the output. The measured first-order diffraction efficiency was only ~33% in the implementation used in the demonstrations. Together with the two four-phase-level microlenses and aluminium mirror coatings, having theoretical diffraction efficiencies of 81% and 85%, respectively, the overall transmission efficiency amounts to only ~4%. This low throughput is problematic for GPC applications requiring high intensities, e.g. optical lithography [10] or multiple-beam light-driven micromanipulation [11].

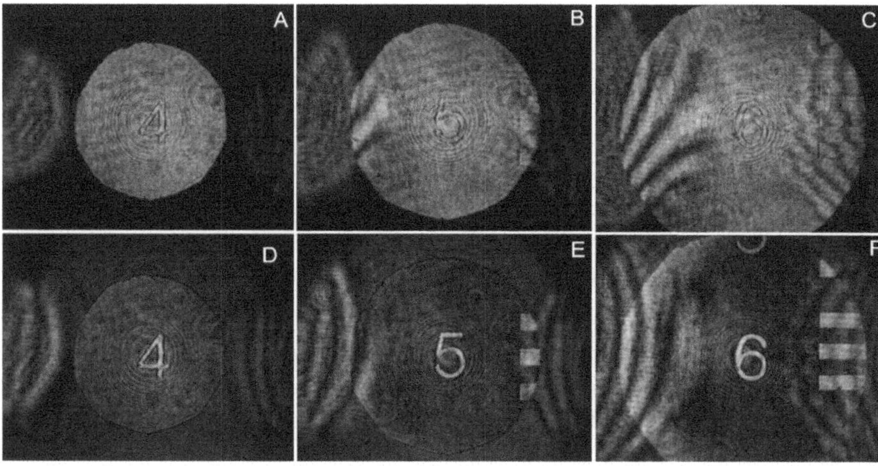

Fig. 9.12 Intensity distribution measured at the output for (a–c) without the spatial filtering effect of the PCF and (d–f) high-contrast images having η=0.26, 0.33 and 0.39, respectively. (f) Corrupted output image due to interference fringes that come from the higher-order diffracted beams of the binary input and output coupling-gratings.

Nevertheless, these experiments demonstrate functional imaging performance of the planar-optical system implemented with diffractive microlenses. No image deterioration due to aberrations was observed because of the moderate size of the image field (r ~3 mm) with relatively large features sizes. However, the effect of higher diffraction orders caused by the coupling gratings is a serious issue that has to be dealt with in future designs. System improvements, such as using more phase levels or continuous relief surfaces proposed in a previous work [12], can potentially elevate the overall transmission efficiency and minimize the unwanted effects of phase quantization. However, increasing the number of phase levels may not always be practical. For instance, using a four-phase-level input grating would require lithographic masks with a minimum feature size of less than 500 nm, which may be costly and impractical for applications requiring simple phase visualization. Thus, it is worth considering other possible optimizations within the GPC such as modifying the input aperture and phase contrast filter parameters according to criteria presented in the previous chapters. These optimizations remain relevant even with improved diffractive optical elements.

The preceding discussions and experimental demonstrations illustrate the feasibility of implementing the generalized phase contrast (GPC) method in planar integrated micro-optics. A miniaturized GPC system can be a vital element in applications that set a high premium on highly compact and robust implementations. When mainly interested in phase encoded information, such as visualising two-dimensional phase distributions, simply optimizing the PCF parameters can help turn a planar-optical implementation into a practical device. After all, GPC is a common-path interferometer and information fidelity takes precedence over energy efficiency in many interferometric systems. The application of miniaturized GPC in one such information-based application, optical encryption and decryption, will be explored in detail in Chapter 11. A miniaturized optical decryption system can have a significant impact in electro-optical data communications where security and authenticity verification are the principal motivations.

9.5 GPC in Combination with Matched Filtering

The generalized phase contrast (GPC) method processes a phase-modulated input using phase-only Fourier plane filtering to generate output patterns whose contrast and intensity levels are governed by the input phase information. In principle, the energy-efficient intensity patterns generated by GPC can be fed as input into another optical processing system. If the second optical processor also utilizes a Fourier filtering scheme, then linear systems theory asserts that the two filters – GPC and the second filtering operation – can be combined into a single hybrid-GPC filter. This hybrid-GPC filter performs the same function as the cascaded systems, but using less optics, resulting in a simpler and more robust and compact system. The second filter component of the hybrid-GPC filter can perform a variety of optical processing functions to enhance GPC applications in sensing and synthesis. When visualizing unknown phase objects, the second component can improve imaging through edge or contrast enhancement, and even pattern recognition. When synthesizing patterned light, the additional component can expand the repertoire of realizable patterns.

In this section, we explore the potential of hybrid-GPC filtering using the synthesis of reconfigurable light patterns as an illustrative example. The hybrid-GPC filter that we will consider incorporates the functions of a Van der Lugt optical correlator [13] as its second component. The resulting technique is dubbed as mGPC, which is short for "matched filtering GPC". Optical correlation is conventionally associated with pattern recognition applications such as for detecting and tracking objects with specific features within an input scene. In pattern recognition, the input scene is typically riddled with spurious information such that correlation is often faced with tradeoffs among various merit function such as efficiency, output noise, input distortion and background noise robustness, and input discrimination ability, among others.

Many optimization methods have been reported in optical correlation literature (see e.g. [14–18]) but the concept of a fully user-defined input is nearly alien since the emphasis is solely on pattern-recognition-based applications. Instead of conventional passive detection, mGPC combines GPC with optical correlation for active pattern synthesis. In our novel synthesis application, the user directly controls the input and, hence, not bothered by issues such as false positives and robustness to spurious noise and shape distortions, which are important in pattern recognition. Thus, instead of being constrained by a fixed detection target, the user can freely choose the detection target as a free optimization parameter. The rich literature on pattern recognition remains a fertile ground for reaping optimization solutions. As an illustrative example, we will explore the use of mGPC for generating reconfigurable array of light spots. Light arrays synthesized using mGPC can achieve much higher compression factors and peak intensities and still retain the same ease and convenience of phase encoding in conventional GPC.

9.5.1 The mGPC Method: Incorporating Optical Correlation into a GPC Filter

An mGPC filter combines the functions of a cascaded GPC and an optical correlator. The mGPC filter, $H(f_x,f_y)A^*(f_x,f_y)$, is determined by the product of the GPC filter, $H(f_x,f_y)$, and the correlation filter, $A^*(f_x,f_y)$. The process is equivalent to an expanded implementation, where a GPC output, $b(x,y)$, is optically correlated with another function, $a(x,y)$. Denoting the respective Fourier transforms of $a(x,y)$ and $b(x,y)$ as $A(f_x,f_y)$ and $B(f_x,f_y)$, it follows from Fourier transform properties that a cross-correlation $a(x,y) \star b(x,y)$ in the spatial domain corresponds to the product $A^*(f_x,f_y)B(f_x,f_y)$ in the Fourier domain. This is mathematically expressed as

$$\Im\{a(x,y)\star b(x,y)\} = A^*(f_x,f_y)B(f_x,f_y), \tag{9.21}$$

where $\Im$ represents the Fourier transform operation, $\star$ represents correlation and $A^*(f_x,f_y)$ is the complex conjugate of $A(f_x,f_y)$. This Fourier relation is the basis for the typical $4f$ optical correlation system illustrated in Fig. 9.13.

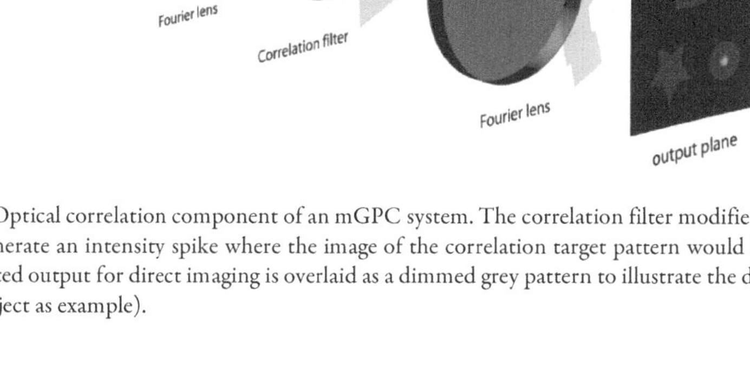

Fig. 9.13 Optical correlation component of an mGPC system. The correlation filter modifies the Fourier field to generate an intensity spike where the image of the correlation target pattern would have formed (the expected output for direct imaging is overlaid as a dimmed grey pattern to illustrate the detection of a circular object as example).

The first Fourier lens projects the Fourier transform of the input plane field, $b(x,y)$ at the back focal plane where it is multiplied by $A^*(f_x,f_y)$ using the filter. The second Fourier lens generates the correlation $a(x,y) \star b(x,y)$ at the output plane (tractable coordinate inversions were neglected for simplicity).

To illustrate the design principles in an mGPC-based pattern projection, let us consider the generation of reconfigurable multiple beams. We start with a user-defined correlation input, $b(x,y)$, which is generated using GPC. This consists of a superposition of spatially shifted copies of a characteristic pattern, $a(x,y)$:

$$b(x,y)=\sum_{m=1}^{M} k_m a(x-x_m, y-y_m). \tag{9.22}$$

When used with a correlation filter, $A^*(f_x,f_y)$, matched to the pattern, $a(x,y)$, the output of the optical correlation system is

$$\begin{aligned} a(r)\star b(r) &= \Im^{-1}\left\{\sum_{m=1}^{M} k_m A(f_x,f_y)\exp\left[-i2\pi\left(x_m f_x + y_m f_y\right)\right] c_0 A^*(f_x,f_y)\right\} \\ &= \Im^{-1}\left\{c_0 A^*(f_x,f_y) A(f_x,f_y)\right\} \otimes \sum_{m=1}^{M} k_m \delta(x-x_m, y-y_m) \\ &= c_0\left[a(r)\star a(r)\right] \otimes \sum_{m=1}^{M} k_m \delta(x-x_m, y-y_m), \end{aligned} \tag{9.23}$$

where $c_0 = \left|A^*(f_x,f_y)\right|_{\max}^{-1}$ is normalization constant that limits the maximum filter transmittance to 100%, denoting a filter without optical gain. The convolution at the output, denoted by $\otimes$, replicates the autocorrelation pattern at multiple positions, (x_m,y_m), which matches the user-defined configuration of the characteristic patterns at the input. The input strengths, k_m, is echoed in the corresponding peaks and can be tuned to desired greyscale levels.

The discussion above shows that an mGPC system exhibits linearity and space invariance. Space invariance implies that the output can be reconfigured through straightforward rearrangement of the patterns at the input. Thus, as in conventional GPC, dynamic reconfiguration in mGPC requires no computational overhead. This is an advantage in applications such as dynamic multiple beam steering and optical landscape projections [19, 20]. The mGPC can be further optimized according to relevant output metrics. For example, the filter normalization represents absorption losses in the filter, which may not be tolerable in some applications.

Owing to similar fundamental principles, optimization techniques developed for pattern recognition may be adapted for pattern projection. For example, one may use phase-only filters [14] to avoid absorption losses and optimize energy throughput. A phase-only filter, $\exp(-i\phi_{A(f_r)})$, eliminates absorption losses and generates an output

$$\begin{aligned} \tilde{a}(r)\star b(r) &= \Im^{-1}\left\{\sum_{m=1}^{M} k_m \left|A(f_x,f_y)\right| \exp\left[-i2\pi\left(x_m f_x + y_m f_y\right)\right]\right\} \\ &= \Im^{-1}\left\{\left|A(f_x,f_y)\right|\right\} \otimes \sum_{m=1}^{M} k_m \delta(x-x_m, y-y_m) \\ &= \left[\tilde{a}(r)\star a(r)\right] \otimes \sum_{m=1}^{M} k_m \delta(x-x_m, y-y_m), \end{aligned} \tag{9.24}$$

Neglecting the amplitude information in the filter alters the function that gets correlated with the input: $\tilde{a}(r)$ represents a modified pattern that matches the filter as a consequence of the alteration. However, since the phase provides a wealth of information about the original function, the essential features of the correlation are retained: peaks with controllable strengths are generated at designated positions.

Another advantage of using a phase-only correlation filter is seen by examining the second line of Eq. (9.24). This shows that the basic pattern replicated at the designated output plane positions is synthesized from the modulus of the target's Fourier components. A modulus operation results since the phase-only correlation filter cancels the phase of the incident Fourier transform field. For an on-axis input pattern, this creates an amplitude-modulated planar wavefront that is focused on-axis at the output plane. Characteristic input patterns shifted off-axis generate similar amplitude-modulated planar wavefronts upon filtering, but would contain linear phase ramps, and correspondingly focus at off-axis output points. The planar wavefronts generated at the Fourier plane result in correlation peaks with improved sharpness and higher peak intensities.

The integration of a phase-only correlation filter with a GPC filter creates a simple phase-only mGPC filter that additively combines the phase of both filters

$$H\left(f_x,f_y\right)A^*\left(f_x,f_y\right)=\exp[\mathrm{i}\phi_{H\left(f_x,f_y\right)}+\mathrm{i}\phi_{A^*\left(f_x,f_y\right)}]. \tag{9.25}$$

The use of a phase-only correlation filter, combined with phase-only input and filter in GPC, means the mGPC method can enable an energy-efficient system for generating dynamically reconfigurable light spikes. However, the correlation component of an mGPC filter is generally a continuously varying function. Phase quantization imposed by practical constraints can introduce undesirable artefacts. A more practical system that requires only binary phase elements at the input and filter planes can be created by exploiting the symmetry properties of Fourier transforms.

9.5.2 Optimizing the mGPC Method

When optimizing the mGPC method, we can exploit the fact that the optical correlation is no longer constrained to predefined characteristic patterns. Thus, choosing a characteristic input pattern becomes an essential part of the optimization process – an option generally not available in conventional pattern recognition. For example, we can choose a characteristic pattern whose matching filter is inherently binary and does not require further quantization. We can exploit the symmetry properties of the Fourier transforms to design inputs that inherently require binary phase elements. Patterns that are described by real and even functions have a purely real Fourier transform characterized by a binary phase (π for negative regions and 0 otherwise). Patterns described by real and odd functions similarly have purely imaginary Fourier transforms and inherently require binary filters.

A phase-only correlation filter creates amplitude modulated plane waves at the Fourier plane. To generate outputs with sharper peaks and higher intensities, the filtered field must be distributed over a wider frequency range. Inverse filters can generate a uniform distribution over a wide frequency range and ideally produce very sharp peaks. However, without optical gain, an inverse filter merely dampens Fourier amplitudes down to the lowest non-zero Fourier component. Thus, the projections are highly energy inefficient and the sharp peaks have low intensities. To generate sharp peaks with high intensities, one must choose a characteristic input pattern that possesses a broadband spatial spectrum and design a matching phase-only correlation filter for it.

Finding an input pattern with a broadband spatial spectrum is reminiscent of a converse design process in computer-generated holography. When designing computer-generated phase holograms, the Fourier plane phase is treated as a free parameter that is tweaked to satisfy the illumination constraints (usually uniform). We can adapt these optimization concepts to find characteristic input patterns that can uniformly distribute energy in the Fourier plane. In addition, we can incorporate more constraints to address other practical issues. For example, we can impose symmetry constraints that will only require binary phase elements during implementation. A sensible constraint is to impose rotational symmetry since, aside from exhibiting even-type symmetry, its matching filter is also rotationally symmetric, which is more convenient for optical alignment.

The picture of a GPC-to-correlator cascade is only one among several ways of interpreting the mGPC operation. It is equally valid to treat it as a correlator-to-GPC cascade, if necessary. Still another way is to interpret the mGPC filter as consisting of two complementary components acting on different aspects of the input pattern: (1) the correlator filter acts on higher frequency components arising from the structural features of the basic characteristic pattern, and (2) the GPC filter acts on low frequencies around the zero-order, which is determined by the global properties of the input field. The third interpretation allows us to ignore global effects and background levels when designing an optimal characteristic pattern. Thus we can use phase-modulated inputs when designing the correlator component of the mGPC filter (e.g. subtracting the average level from a binary amplitude input results in a binary phase modulated field).

As GPC optimization has been treated extensively in previous chapters, we will now mainly focus on the correlation aspect of mGPC optimization. Figure 9.14 presents results from numerically simulated optical correlation. It starts with a circular input phase disc and illustrates how input modifications affect the Fourier amplitude distribution and the output correlation. The simulations considered binary characteristic patterns with circular symmetry that are encoded as 0- and π- phase-shifts onto a circular amplitude input and then used with matched phase-only filters. The patterns in Fig. 9.14(*b–e*) are phase-encoded onto a circular amplitude profile having the same size as the disc in Fig. 9.14(*a*), which also matches the size of the outermost phase ring. The inner radius of the pattern in Fig. 9.14(*b*) is determined to minimize the zero-order. The pattern in Fig. 9.14(*c*) is generated by adjusting the size of the dark central disc to maximize the output peak.

Introducing randomness along the radial coordinate disperses energy at the Fourier plane. The pattern of alternating circular rings in Fig. 9.14(*d*) yielded the highest output peak among several tested configurations of alternating rings whose radii are obtained from computer-generated random numbers. Input ring configurations with higher output peaks may be obtained by searching from a larger set of random configurations but this time-consuming process is effectively a crude direct search optimization. In this case, it will be better to apply direct search optimization or iterative Fourier transform algorithms as a more systematic alternative.

The phase pattern in Fig. 9.14(*e*) was determined using a standard iterative Fourier Transform optimization algorithm based on the Gerchberg-Saxton method [21–23]. To impose circular symmetry, the iterations utilized quasi-discrete Hankel transforms [24],

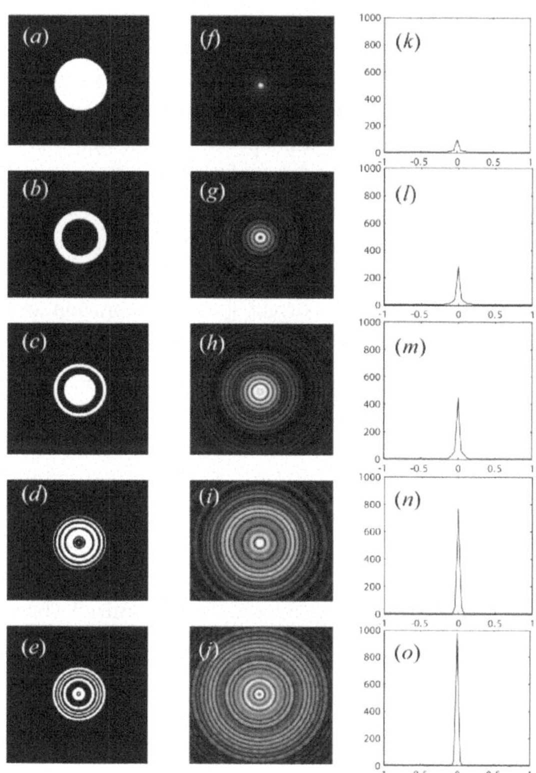

Fig. 9.14 Optimizing output peak intensity and sharpness through input phase modulation that broadens the Fourier domain energy distribution. (a–e) Binary phase input patterns (white: π, black: 0); (f–j) respective Fourier amplitude distributions; and (k–o) Intensity linescans through the output correlation peaks over the expected disc image diameter, expressed relative to the input intensity. The input phase patterns are all encoded onto a circular amplitude pattern with the same dimension as the pattern in (a).

instead of two-dimensional Fourier transforms. Phase binarization was applied as a final step. In theory, a very sharp output spike requires an infinite plane wave. In practice, such a distribution will incur severe losses from the finite optical apertures. It is therefore more suitable to constrain the Fourier plane distribution within a realistic region based on the expected numerical aperture (NA) during the iterative design. This was anticipated and included during optimization and, consequently, the dominant Fourier components of an iterated input are bounded within a circular region, as shown in Fig. 9.14(*j*). To generate multiple sharp peaks, one can exploit the linearity and shift invariance of the correlation process and mimic the desired output configuration when arranging copies of the optimized characteristic pattern at the input.

The main strength of the proposed pattern projection system lies in the ease with which a user is able to modify the output characteristics while maintaining highly efficient energy throughputs. The strength of each output peak is directly controlled by local modifications to the encoded input. The sharpness of each output peak is also independently adjustable through localized changes at the input. The versatility of the technique makes it highly suitable for rapidly reconfigurable projections. A glimpse of this versatility is provided by the image sequence presented in Fig. 9.15, which was taken from a simulated movie of dynamic mGPC operation. The figure shows snapshots from a dynamic input stream alongside their corresponding output snapshots. The simulations used the correlation target pattern illustrated in Fig. 9.14(e) with its matching phase-only mGPC filter. Even though a continuously updated diffractive element can also generate patterns with similar dynamics, it will require input optimization to generate each updated frame with the correct spot configuration and uniformity. This frame-by-frame optimization for dynamic diffractive elements can be a serious bottleneck for high speed pattern reconfiguration with high fidelity. In contrast, no further optimization is required for an mGPC system once the optimal characteristic pattern has been obtained.

A prototype demonstration illustrating the correlation component of the mGPC method was implemented using a standard 4*f* optical processing system. The setup was similar to a standard GPC demonstration, but with a correlation filter taking the place of the GPC filter. A programmable phase-only spatial light modulator (Hamamatsu PAL-SLM) was used to phase-encode an array of optimized characteristic patterns, similar to Fig. 9.14(e), onto an incident collimated laser beam (λ=532nm). Another phase-only SLM positioned at the Fourier plane of the first SLM encodes a matching phase-only correlation filter. Typical results are illustrated by screen captures of the graphical user interface (GUI) during the experiments, as shown in Fig. 9.16. The figure shows a captured image of the output and includes a plot of patched intensity linescans along four lines overlaid on the image. Figure 9.16(a) shows direct imaging of the phase input through the 4*f* system when the correlation filter is deactivated in the second SLM. A residual intensity modulation arises from the spatial frequency cutoff from the finite optics apertures, which weakly visualizes the multiple input phase rings. Figure 9.16(b) shows the output when the mGPC filter is activated, illustrating a sharp spike with a definitive intensity gain relative to the directly imaged pattern. Although two SLMs were used for this prototype demonstration, only one single SLM is actually required for dynamic operation since the filter can be fabricated as a static element to match the chosen correlation target.

The relative widths of the correlation spike and the weakly imaged disc observed in Fig. 9.16 suggest that the distance between spikes adjacent spikes must be much greater the spike width to prevent patterns from overlapping at the input. This means that array illumination with very high compression factors can be achieved. Now, what if we want the output peaks to be closer to each other? Numerical simulations show that an mGPC-based system can tolerate overlapping input patterns. An intuitive way to understand this is to treat two overlapping discs as a superposition of a "full moon" signal with an adjacent "new moon" noise, where the signal and noise assignment is interchangeable based on symmetry. Thus, the system can detect two overlapping "full moons", without modifying the spatial filter. Distinct correlation peaks appear at their expected positions during overlap with minimal changes in peak sharpness. However, input signal overlap leads to a lower total energy, causing a proportionate drop in the generated output peaks. For applications that cannot tolerate intensity variation, the predictable effects of overlapping may be compensated using suitable amplitude or phase encoding of the input.

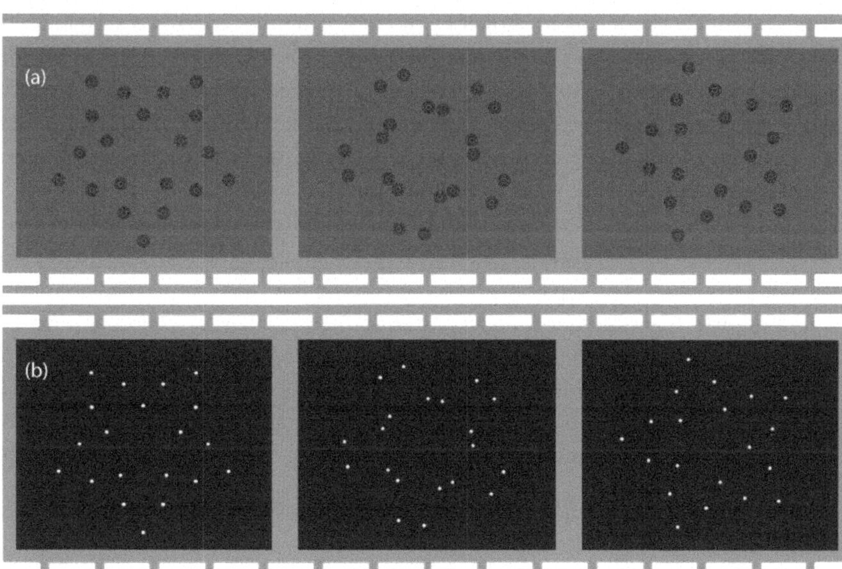

Fig. 9.15 Snapshots from a simulated movie of dynamic output patterns generated using the mGPC-based projection system. (a) Encoded input phase (grey: zero; black: π); (b) Projected output intensity peaks (the actual spots are more sharply peaked, but were broadened to anticipate printing resolution issues).

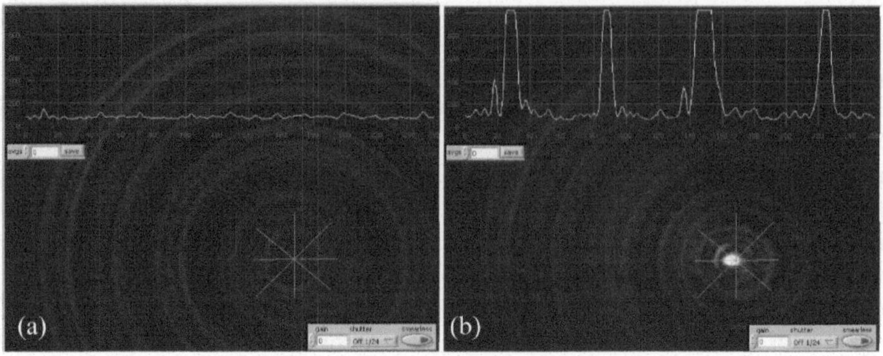

Fig. 9.16 Experimental proof-of-principle of mGPC pattern projection. Screen captures of the user interface are shown, zooming in on a single spike generated by the $4f$ system. The overlaid traces show patched intensity linescans along the paths indicated by the four intersecting lines. (a) $4f$ imaging of a single phase-encoded correlation target; (b) Intensity spike generated when the spatial filtering is activated.

In summary, this section illustrates the potential of incorporating generalized filtering functions into a GPC filter. The case of mGPC, which combines GPC and matched filtering, was highly promising as shown by numerical simulations and a simple proof-of-principle experiment. These results may be exploited in optimized implementations using contemporary spatial modulation technologies to generate rapidly reconfigurable pattern projections for various applications in optics and photonics. The linearity and space-invariance of the mGPC approach leads to straightforward output reconfiguration. It is highly encouraging to note that optical correlation remains a vibrant field, which can potentially benefit mGPC. Relevant contemporary developments in optical pattern recognition and optical correlation may be explored to further enhance the functionality of mGPC-based pattern projection and tailor them for particular applications. The additional design freedoms in pattern projection may be explored to investigate the optimization limits. Integrating other filtering functions into GPC can potentially offer other functionalities. An implementation using a dynamic filter that can shift between different operating modes (pure GPC and hybrid-GPC) can potentially address an even wider range of applications requiring efficiently generated multiple light spots (see e.g. [25–28]).

9.6 Summary and Links

In this chapter we explored alternative ways of experimentally implementing GPC, which can offer practical advantages. Using a voltage-controlled liquid crystal filter, we illustrated the idea of an adjustable phase contrast filter (PCF), which can provide adaptive performance and ensure that the PCF phase shift matches an arbitrary phase input to maintain optimal outputs. We also showed that an adaptive and auto-aligning PCF may be achieved using a light-induced PCF (e.g. on a Kerr medium). If needed, it is also possible to implement GPC using multiple input illuminations at different incidence angles, separately filtering the focused undiffracted components from each illumination to create multiple overlapping outputs. This can reduce noise from out-of-focus light and help scale the operating power beyond the filter breakdown for a single focused spot. We also discussed another method for achieving higher output peak intensities by integrating a matched filtering function into the GPC filtering. Integrating other image processing functions can be considered in the future. Furthermore, we discussed the possibility of miniaturization by implementing GPC as a single device based on planar integrated microoptics. We will demonstrate the applications of this miniaturized device for optical cryptography in Chapter 11. The alternative schemes considered here may also be considered when the common path interferometer is optimized for the reverse phase contrast method, which is discussed in the next chapter.

References

1. D. Weaire, B. S. Wherrett, D. A. B. Miller, and S. D. Smith, "Effect of low-power nonlinear refraction on laser-beam propagation in InSb," Opt. Lett. **4**, 331-333 (1979).
2. D. Zeisel and N. Hampp, "Spectral relationship of light-induced refractive index and absorption changes in bacteriorhodopsin films containing wildtype BR_{WT} and the variant BR_{D96N}," J. Phys. Chem. **96**, 7788–7792 (1992).
3. C. Brauchle, N. Hampp, and D. Oesterhelt, "Optical applications of bacteriorhodopsin and its mutated variants," Adv. Mater. 3, 420–428 (1991).
4. O. Werner, B. Fischer, and A. Lewis, "Strong self-defocusing effect and four-wave mixing in bacteriorhodopsin films," Opt. Lett. **17**, 241-243 (1992).
5. P. S. Ramanujam and L. R. Lindvold, "Dark spatial solitons in bacteriorhodopsin thin films," Appl. Opt. **32**, 6656-6658 (1993).
6. N. Mukohzaka, N. Yoshida, H. Toyoda, Y. Kobayashi, and T. Hara, "Diffraction efficiency analysis of a parallel-aligned nematic-liquid-crystal spatial light modulator," Appl. Opt. **33**, 2804–2811 (1994).
7. J. Glückstad, "Generation of a desired wavefront with a plurality of phase contrast filters," International Patent Application No. PCT/DK2004/000452 (priority date 26 June 2003).
8. J. S. Dam, I. R. Perch-Nielsen, D. Palima, and J. Glückstad, "Three-dimensional imaging in three-dimensional optical multi-beam micromanipulation," Opt. Express **16**, 7244-7250 (2008).
9. J. Glückstad and P.C. Mogensen, "Optimal phase contrast in common-path interferometry," *Appl Opt* **40**, 268-282 (2001).
10. C. Mack, "Phase contrast lithography," *Proc SPIE* **1927** (ed. Cuthbert J D Optical/Laser Microlithography VI) 512-520 (1993).

11. R.L. Eriksen, V.R. Daria and J. Glückstad, "Fully dynamic multiple-beam optical tweezers," *Opt Express* **10,** 590-602 (2002).
12. V.R. Daria, J. Glückstad, P.C. Mogensen, R.L. Eriksen and S. Sinzinger, "Implementing the generalized phase-contrast method in a planar-integrated micro-optics platform," *Opt Lett* **27** 945-947 (2002).
13. A. Vander Lugt, "Signal Detection by Complex Spatial Filtering," *IEEE Trans. Inf.* Theory **IT-10,** 139–145 (1964).
14. J.L. Horner and P.D. Gianino, "Phase-only matched filtering," *Appl. Opt.* **23**, 812–816, (1984).
15. BVKV Kumar and L. Hassebrook, "Performance measures for correlation filters," *Appl. Opt.* **29**, 2997–3006 (1990).
16. P. Refregier, "Filter design for optical pattern recognition: multicriteria optimization approach," *Opt. Lett.* **15**, 854–856 (1990).
17. G. Ravichandran and D.P. Casasent, "Minimum noise and correlation energy optical correlation filter," *Appl. Opt.* **31**, 1823–1833 (1992).
18. R.R. Kallman and D.H. Goldstein, "Phase-encoding input images for optical pattern recognition," *Opt. Eng.* **33**, 1806–1812 (1994).
19. M. MacDonald, G. Spalding and K. Dholakia, "Microfluidic sorting in an optical lattice," *Nature* **426**, 421–424 (2003).
20. J. Glückstad, "Microfluidics: Sorting particles with light," *Nature Materials* **3,** 9–10, (2004).
21. R.W. Gerchberg and W.O. Saxton, "A practical algorithm for the determination of the phase from image and diffraction plane pictures," *Optik* **35**, 237–246, (1972).
22. F. Wyrowski and O. Bryngdahl, "Iterative Fourier-transform algorithm applied to computer holography, *J. Opt. Soc. Am. A* **5**, 1058–1065 (1988).
23. O. Ripoll, V. Kettunen and H.P. Herzig, "Review of iterative Fourier-transform algorithms for beam shaping applications," *Opt. Eng.* 43, 2549-2556 (2004).
24. M. Guizar-Sicairos and J.C. Gutiérrez-Vega, "Computation of quasi-discrete Hankel transforms of integer order for propagating optical wave fields," *J. Opt. Soc. Am. A* **21**, 53–58, (2004).
25. A . Egner, V. Andresen and S.W. Hell, "Comparison of the axial resolution of practical Nipkow-disk confocal fluorescence microscopy with that of multifocal multiphoton microscopy: theory and experiment," *J. Microsc.* **206**, 24–32 (2002).
26. K.L. Tan, S.T. Warr, I.G. Manolis, T.D. Wilkinson, M.M. Redmond, W.A. Crossland, R.J. Mears and B. Robertson, "Dynamic holography for optical interconnections. II. Routing holograms with predictable location and intensity of each diffraction order," *J. Opt. Soc. Am. A* **18**, 205–215 (2001).
27. J. Kato, N. Takeyasu, Y. Adachi, H.B. Sun and S. Kawata, "Multiple-spot parallel processing for laser micronanofabrication," *Appl. Phys. Lett.* **86** 044102 (2005).
28. P.J. Rodrigo, V.R. Daria and J. Glückstad, "Real-time three-dimensional optical micromanipulation of multiple particles and living cells," *Opt. Lett.* **29**, 2270–2272 (2004).

Chapter 10
Reversal of the GPC Method

Our main interest in the book, up till now, has been about the theoretical formulation and practical applications for transforming a spatially phase-only encoded wavefront into output intensity variations through the generalized phase contrast. We shall now shift gears in this chapter by asking the question: is it possible to start with amplitude-only modulation at the input and use spatial filtering to generate an output whose phase-only modulation is governed by the input amplitudes? The answer is yes, and the fundamental idea is based on the generalized formulation already developed for the GPC-method. As the function is a reversal of the conventional phase contrast, we shall refer to this approach as the reverse phase contrast (RPC) method. Although reversed in terms of its function, the same derivations used for the GPC method remains highly relevant and will be adopted in the formulation and optimization of the RPC method.

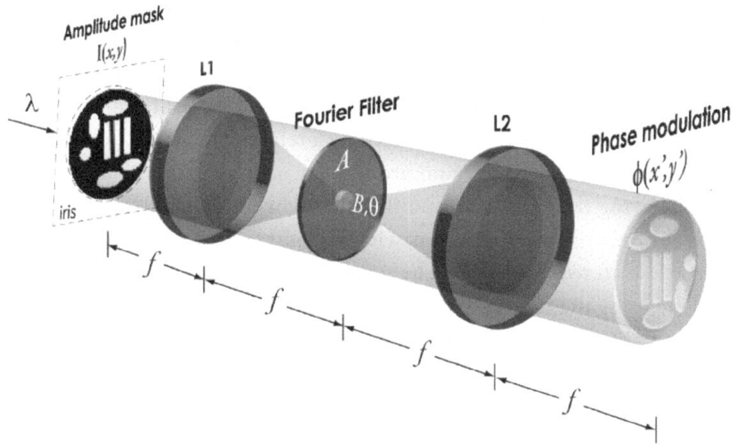

Fig. 10.1 Generic RPC setup based around a $4f$ optical system (lenses L1 and L2). The input amplitude modulation is shown as an aperture truncated binary amplitude mask, which generates a binary phase modulation, at the observation plane by a filtering operation at the Fourier plane. The values of the filter parameters (A, B, θ) determine the type of filtering operation.

Why develop a reversed phase contrast? The main motivation for the development of the RPC method is to obtain dynamic high-performance spatial phase modulation using simple and widely available amplitude based spatial modulators. Reconfigurable spatial phase modulation of a light field is required in a number of areas in optics, including phase modulation for holographic multiplexing, storage and encoding [1], phase-only encryption [2] and the testing of focus in optical apparatus [3]. In addition, the RPC technique can be used with a binary amplitude mask acting as the input information to create interchangeable but static phase distributions.

This chapter is devoted to a presentation of the theoretical basis and supporting experimental demonstrations of the RPC method. Theoretical foundations provide a basis for understanding the constraints and operating requirements in practical optical systems, which allows us to work out design recipes for developing an optimized RPC system. These prescriptions have been adopted and applied in the experimental demonstrations that are likewise presented in the chapter.

10.1 Amplitude Modulated Input in a Common-Path Interferometer

The set-up in which we will consider the implementation of the RPC method is shown schematically in Fig. 10.1. This setup exhibits essentially the same geometry as that used in the standard GPC method except that an amplitude mask at the RPC input replaces the input phase mask used in GPC.

Although intended to perform a reverse function, the optical system in Fig. 10.1 may still be considered as a common-path interferometer (CPI) with an amplitude input. Hence, the system can be described by the same generalized CPI formulation previously developed in Chapter 3 for GPC. The first lens (L1) performs a spatial Fourier transform of the amplitude-modulated input and projects the spatial frequency components onto an RPC filter. The filter is characterized by a relative phase shift, θ, and amplitude transmittance, B, within a small region around the optical axis, and another amplitude transmittance, A, for the rest. The undiffracted light components from the input are focused onto the on-axis filtering region, whereas light generated by the spatially varying components are scattered elsewhere. The second lens (L2) synthesizes the output at the back focal plane using the filtered Fourier components.

The output light field in the filtering system illustrated in Fig. 10.1 is, in general, complex-valued and exhibits both amplitude and phase modulation. By tuning the size, relative phase shift and amplitude transmission between the two regions of the filter, we can optimise RPC to get an output with uniform intensity and phase-only modulation. The resulting output phase modulation can then be set to mimic the encoded input amplitude pattern.

To find optimal parameters for achieving phase-only modulation using the RPC method, we must first derive how the output light field depends on the input modulation and the filter parameters. We will assume that a uniform, axially propagating monochromatic plane wave of wavelength, λ, illuminates the input amplitude mask through a circular input aperture with radius, Δr. The incident light amplitude, $a(x,y)$, at the entrance plane of the RPC optical system is then described by

$$a(x,y)=\mathrm{circ}(r/\Delta r)\alpha(x,y), \tag{10.1}$$

where $\alpha(x,y)$ denotes the amplitude modulation due to the input mask and the aperture truncation is represented by circ-function, defined to be unity within the region, $r=\sqrt{x^2+y^2}\le\Delta r$, and zero elsewhere. Similarly, we assume a circular spatial filter centred on-axis, described by a transfer function (c.f. Eq. (3.2))

$$H(f_x,f_y)=A\left[1+\left(BA^{-1}\exp(j\theta)-1\right)\mathrm{circ}(f_r/\Delta f_r)\right], \tag{10.2}$$

which applies an amplitude transmittance, B, and phase shift, θ, to frequency components within the range $-\Delta f_r\le f_r\le\Delta f_r$ and amplitude transmittance, A, elsewhere. The filter action is again understood as a combination of (1) transmission of all frequency components from the input, subject to amplitude-scaling by a factor A, and (2) synthesis of a reference wave from the low frequency components of the input. The output is then formed by the superposition of an amplitude-scaled image of the input and a synthetic reference wave. Thus, we obtain an expression for the light distribution, $o(x',y')$, at the observation plane within the boundaries, $\mathrm{circ}(r'/\Delta r)$ (c.f. Eq. (3.14)):

$$o(x',y')=A\left[\alpha(x',y')+K\bar{\alpha}\left(BA^{-1}\exp(j\theta)-1\right)\right], \tag{10.3}$$

where $\alpha(x',y')$ is the image of the input amplitude mask, $\bar{\alpha}$ is a zero-order strength normalized relative to the zero-order without input modulation, and K denotes an SRW scaling factor determined by the size of the low-pass filtering region, which may be taken as the on-axis value of the low-pass filtered input:

$$g(x,y)=\Im^{-1}\left\{\mathrm{circ}(f_r/\Delta f_r)\Im\{a(x,y)\}\right\}. \tag{10.4}$$

As it proved sufficient for many GPC applications, we can again approximate the low-pass filtered input by

$$g(r')=2\pi\Delta r\int_0^{\Delta f_r}J_1(2\pi\Delta rf_r)J_0(2\pi r'f_r)\mathrm{d}f_r, \tag{10.5}$$

which again leads to a simplified expression $K=g(0)=\left[1-J_0(2\pi\Delta r\Delta f_r)\right]$.

Working with an amplitude-modulated input leads to a simpler calculation of the normalized zero-order strength, $\bar{\alpha}$, which is now guaranteed to be nonzero and real-valued. For an input illumination having unit strength, $\bar{\alpha}$ is simply the spatial average of the real-valued input:

$$\bar{\alpha}=\left(\pi(\Delta r)^2\right)^{-1}\iint_{\sqrt{x^2+y^2}\leq\Delta r}\alpha(x,y)\mathrm{d}x\mathrm{d}y \tag{10.6}$$

Having seen its convenience for the GPC method, we will similarly define a combined filter parameter, $C=BA^{-1}\exp(j\theta)-1$, and reformulate Eq. (10.3) with the simplified SRW term:

$$o(x',y')=A\left[\alpha(x',y')+K\bar{\alpha}|C|\exp(j\psi_C)\right]. \tag{10.7}$$

10.2 CPI Optimization for the Reverse Phase Contrast Method

Having developed a working formulation for describing a CPI with an amplitude-modulated input, we now turn to its optimization so as to create an RPC system that can generate a purely phase-modulated output using an amplitude-modulated input. We will consider an arbitrary input amplitude modulation within a circular input aperture with area of $\pi(\Delta r)^2$. We will consider binary amplitude modulation at the input, where a user-defined region $\Re_{\min}$ is encoded with a lower amplitude level, $\alpha_{\min}\in[0;\,1]$, while the remaining region $\Re_{\max}$ is assigned the upper amplitude level, $\alpha_{\max}=1$, corresponding to full transmission. We can write the modulation function as

$$\alpha(x,y)=\begin{cases}\alpha_{\min} & \forall\ (x,y)\in\Re_{\min}\\ \alpha_{\max}=1 & \forall\ (x,y)\in\Re_{\max}\end{cases} \tag{10.8}$$

Adopting the same notation to denote the respective areas of the regions for simplicity, the spatial average of this binary amplitude modulation over the input aperture is found to be

$$\bar{\alpha}=\left[\pi(\Delta r)^2\right]^{-1}(\Re_{\min}\alpha_{\min}+\Re_{\max})=1+F(\alpha_{\min}-1), \tag{10.9}$$

where the fill factor, $F=\left[\pi(\Delta r)^2\right]^{-1}\Re_{\min}$, indicates the fractional area occupied by the $\Re_{\min}$ region relative to the total input aperture area.

The principal goal of the RPC method is use an input amplitude pattern to dictate the spatial features of an output phase modulation that is embedded onto a uniform background intensity. Having formulated the CPI output as a superposition, it is highly useful to visualize this superposition using complex phasors on an Argand diagram, as shown in Fig. 10.2. A purely phase-modulated output requires that the $\Re_{\min}$ and $\Re_{\max}$ regions map to image regions with the same intensity. We can formally write this constant intensity criterion as

$$\left|\alpha_{\min}+K\bar{\alpha}|C|\exp(j\psi_C)\right|=\left|1+K\bar{\alpha}|C|\exp(j\psi_C)\right|. \tag{10.10}$$

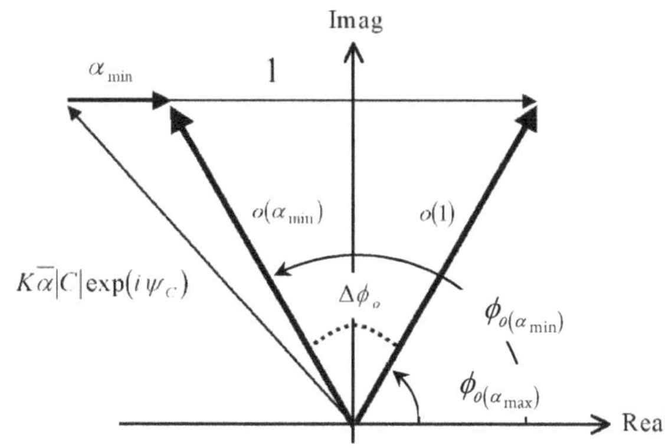

Fig. 10.2 Argand diagram showing the solution vectors, $o(\alpha_{\min})$ and $o(\alpha_{\max}=1)$, for Eq. (10.10) under the constant background intensity condition upon which the desired phase modulation, $\Delta\phi_o$, is present.

Since the superpositions on both sides of Eq. (10.10) have the same imaginary component, a common magnitude results when their real parts have the same magnitude. We can thus rewrite Eq. (10.10) as

$$\begin{aligned} \left|\alpha_{\min}+K\bar{\alpha}|C|\cos(\psi_C)\right| &= \left|1+K\bar{\alpha}|C|\cos(\psi_C)\right| \\ \alpha_{\min}+K\bar{\alpha}|C|\cos(\psi_C) &= \pm\left[1+K\bar{\alpha}|C|\cos(\psi_C)\right]. \end{aligned} \tag{10.11}$$

The (+) sign on the right-hand side yields the solution, $\alpha_{\min}=1$, which corresponds to a trivial amplitude-to-phase-conversion where a uniform input amplitude generates an output with a uniform intensity and uniform phase. The (–) sign yields the nontrivial case and allows us to write the criterion for achieving phase-only output modulation as

$$K|C|\cos\psi_C = -\frac{1+\alpha_{\min}}{2\bar{\alpha}} = -\frac{1+\alpha_{\min}}{2\left[1+F(\alpha_{\min}-1)\right]}, \tag{10.12}$$

where we have rearranged the variables to group the filter parameters on the left-hand side and the input parameters on the other side. This expression requires that the phase of the combined filter parameter satisfy the constraint, $\psi_C > \pi/2$, since the right hand side of Eq. (10.12) is negative for any physical choice of input amplitude modulation parameters.

When aiming to develop an RPC system designed to function as an amplitude-to-phase converter, it is useful to find an expression explicitly showing such effect. Using the vector geometry depicted in Fig. 10.2, we can write such an expression as

$$\phi(x,'y') = \operatorname{arccot}\left[\frac{\alpha(x,'y')+K\bar{\alpha}|C|\cos\psi_C}{K\bar{\alpha}|C|\sin\psi_C}\right], \tag{10.13}$$

where $\phi(x',y')$ denotes the output phase and $\alpha(x,'y')$ is a replica of the input amplitude modulation. This equation also describes how the input and filter parameters affect the conversion.

In most practical applications, only the phase modulation depth is relevant and not the absolute phase values. Referring to the Argand diagram in Fig. 10.2, the two solution vectors, $o(\alpha_{\min})$ and $o(1)$, have identical lengths when the constant intensity criterion is satisfied. The angle, $\Delta\phi_o$, between these vectors corresponds to the output phase modulation depth. Applying trigonometric considerations allows us to write an expression describing the phase modulation depth as

$$\cot\left(\Delta\phi_o/2\right)=\frac{2K\bar{\alpha}|C|\sin\psi_C}{1-\alpha_{\min}}. \tag{10.14}$$

By reconciling this modulation depth expression from the phasor geometry in Fig. 10.2 with the input function as described in Eq. (10.8), we can write the output spatial phase modulation in terms of the input amplitude modulation and filter parameters as

$$\phi(x,' y')=2\arctan\left[\frac{1-\alpha(x,' y')}{2K\bar{\alpha}BA^{-1}\sin\theta}\right]+\phi_{\alpha(\max)}, \tag{10.15}$$

where $\phi_{\alpha(\max)}=\arctan\left[\mathrm{Im}(C)/(1+\mathrm{Re}(C))\right]$ is the phase angle in the $\Re_{\max}$ image region.

An alternate expression for the phase modulation depth can be derived from the phasor diagram in Fig. 10.2. We can see that the angle, $\Delta\phi_o$, satisfies the relationship $\exp\left(i\varphi_{o(1)}\right)=\exp\left(i\varphi_{o(\alpha_{\min})}-i\Delta\phi_0\right)$. Assuming that the uniform intensity criterion is satisfied, we can apply basic properties of complex exponentials and rewrite this phasor relationship as $o(1)=o(\alpha_{\min})\exp(-i\Delta\phi_0)$. This can be expanded as

$$\exp\left(j\Delta\phi_o\right)=\frac{o(\alpha_{\min})}{o(1)}=\frac{\alpha_{\min}+K\bar{\alpha}|C|\exp\left(j\psi_C\right)}{1+K\bar{\alpha}|C|\exp\left(j\psi_C\right)}, \tag{10.16}$$

where the output phase modulation depth, $\Delta\phi_o$, is again expressed in terms of the experimental parameters.

Designing an optimized RPC system requires an explicit relationship that allows us to determine the combined filter parameter, *C*, according to the input and output parameters. We have previously obtained that the real component of *C* is determined according to the constant intensity criterion (see Eq. (10.12)), while the imaginary component of *C* is related to the phase modulation depth (see Eq. (10.14)). Combining these components, we can write the combined filter parameter as

$$\begin{aligned} C&=\frac{-1-\alpha_{\min}+j(1-\alpha_{\min})\cot\left(\Delta\phi_o/2\right)}{2K\bar{\alpha}} \\ &=\frac{-1-\alpha_{\min}+j(1-\alpha_{\min})\cot\left(\Delta\phi_o/2\right)}{2K\left[1+F(\alpha_{\min}-1)\right]}. \end{aligned} \tag{10.17}$$

Knowing the combined filter parameter, we can then use it to determine the filter absorption parameters, A and B, and the phase shift, θ, using (c.f. Eq. (3.17)):

$$\begin{cases} BA^{-1} = |C+1| \\ \theta = \arg(C+1) \end{cases} \tag{10.18}$$

Having found the full expression for C, it is now possible to calculate the expected uniform intensity of the phase-modulated output. First, we express the synthetic reference wave in terms of the input parameters and the phase modulation depth:

$$SRW = K\bar{\alpha}C = -\frac{1}{2}(1+\alpha_{\min}) + j\frac{1}{2}(1-\alpha_{\min})\cot(\Delta\phi_o/2). \tag{10.19}$$

This allows us to write the two solution vectors as

$$\begin{aligned} o_{(\alpha\min)} &= SRW + \alpha_{\min} = \frac{1}{2}(1-\alpha_{\min})\left[-1 + j\cot(\Delta\phi_o/2)\right] \\ o_{(\alpha\max)} &= SRW + \alpha_{\max} = \frac{1}{2}(1-\alpha_{\min})\left[1 + j\cot(\Delta\phi_o/2)\right] \end{aligned} \tag{10.20}$$

These expressions verify that the two vectors indeed have a common intensity, in accordance with the uniform intensity criterion, which is given by

$$A^2\left|o_{(\alpha\min)}\right|^2 = A^2\left|o_{(\alpha\max)}\right|^2 = \frac{A^2}{4}(1-\alpha_{\min})^2\left[1+\cot^2(\Delta\phi_o/2)\right]. \tag{10.21}$$

This uniform intensity is specified purely in terms of the input amplitude modulation depth, $(1-\alpha_{\min})$, and the output phase modulation depth, $\Delta\phi_o$. At first glance, it seems as if the output intensity is maximized for $\alpha_{\min}=0$, i.e. when the minimum input amplitude level is set to zero. However, this is not always true since the transmission parameter, A, cannot be chosen arbitrarily but is determined by the other parameters (see Eq. (3.17). This is true when $\cot(\Delta\phi_o/2)=0$, which corresponds to a vital phase modulation depth setting, $\Delta\phi_o=\pi$. However, $\cot(\Delta\phi_o/2)$ assume high values for very low modulation depths, which might require a higher $\alpha_{\min}$, as shown in Eq. (10.14)), if one is to maintain a maximal filter transmittance A.

Having thoroughly analyzed the various aspects of RPC optimization, we can now formulate a design procedure for choosing the input amplitude modulation and filter parameters to generate a given binary spatial phase modulation of phase, $\Delta\phi_o \in [0;\ \pi]$. This RPC design procedure can be summarised in the following manner:

- The first step in an output-oriented design is, naturally, to specify the desired output binary phase modulation, $\phi(x,'y')$. The spatial phase distribution should specify a characteristic phase modulation depth, $\Delta\phi_o$, chosen within the range $[0;\ \pi]$, and determines the fill factor, F, to be used in later calculations.

- The next step is to choose a value for $K=\left[1-J_0\left(2\pi\Delta r\Delta f_r\right)\right]\leq 1$, which is a ubiquitous parameter in the optimization analysis. The *K*-value depends on the input aperture and filter diameter as discussed in Chapter 3. It is worth considering the range $K\leq ½$, since these *K*-values yield the most uniform output intensity profiles.

- The third step is to choose a value for $\alpha_{\min}$ and apply the other previously chosen parameters to determine the filter parameter *C* using Eq. (10.17)). There is flexibility at this stage since multiple $\alpha_{\min}$ values can have matching filter parameters for generating the required phase modulation depth. If possible, one should use the lowest possible $\alpha_{\min}$ to minimize absorption and optimize the energy throughput.

- Finally, the value of *C* is then used to determine the filter parameters (A, B, θ).

We now give some sample calculations for designing RPC systems. We will use $K=½$ to generate outputs with minimal intensity inhomogeneity.

In the first example, we wish to generate a binary phase-modulated output pattern with a 50% duty cycle characterized by 0-π phase shift within a uniform intensity background. We will modulate the input amplitude using a high-contrast binary amplitude mask designed to fully block light at designated regions. These conditions correspond to the following set of parameters: $\Delta\phi_o=\pi$, $F=½$, $\alpha_{\min}=0$. Using Eq. (10.17) to determine the combined filter parameter and then applying Eq. (3.17) afterwards, we can obtain the filter parameters required for this task. In this case, $C=2 \Rightarrow A=B=1,\ \theta=\pi$ which corresponds to a fully transmissive phase-only filter in the Fourier plane of the setup in Fig. 10.1.

In the second example, we again generate a binary phase-modulated output with 50% duty cycle pattern from a high contrast input amplitude mask. However in this case we would like the binary output phase to exhibit a $\pi/2$ modulation depth. Substituting $\Delta\phi_o=\pi/2$, $F=½$, $\alpha_{\min}=0$ and $K=½$ in Eq. (10.17), we obtain the following specifications for the filter parameters: $C=2\sqrt{2}\exp\left(j3\pi/4\right) \Rightarrow A=1/\sqrt{5},\ B=1,$ $\theta=\pi-\sin^{-1}\left(2/\sqrt{5}\right)$.

A quick glance at these examples seems to suggest that RPC is irreversible. This is seen in the first example, where a binary 50% duty cycle on-off input intensity pattern resulted in a spatially identical binary 0-π modulated phase-only pattern using a lossless π-phase RPC filter. To illustrate reversibility, we should be able perform a reverse operation where applying a similar 50% duty cycle binary 0-π phase pattern reproduces the high-contrast amplitude pattern. However, we know from GPC that such phase modulation will not produce any zero order and, since there will be no generated SRW, the output will simply reproduce the same invisible phase modulations instead of a high-contrast intensity pattern. However, the contradiction in the foregoing analysis arose because it simply considered the phase modulation created within the image region of the input aperture and ignored the tail of the SRW that extend beyond this image region. When the SRW tail is included in the analysis, a zero-order appears correctly and the high-contrast input can be reproduced as a superposition of the phase-modulated image and the resulting SRW. In fact, a similar effect has been analyzed and experimentally demonstrated (see Sect. 7.4), where a phase-modulated region that would otherwise not generate any zero-order successfully created a high-contrast output after being enclosed within a bigger, non-modulated region whose size is chosen to get the correct zero-order properties.

A final thing to point out is that all derivations and examples of the RPC method are based on the assumption that a perfect binary amplitude modulator is applied at the input. In a practical implementation, however, we may have to cope with aberrations due to small-scale phase perturbations introduced by a non-perfect input modulator. Fortunately, these can in most practical cases be modelled as slowly varying in comparison with the pixellation of the applied spatial light modulator. This implies that neighbouring pixels will essentially be subjected to the same impact of a given phase perturbation. Hence, the desired RPC output phase modulation is only slightly perturbed as can be quickly verified by use of the Argand diagram in Fig. 10.2.

10.3 Experimental Demonstration of Reverse Phase Contrast

The previous theoretical analysis allowed us to specify design procedures for optimising an RPC system. We will now illustrate this optimization in an actual RPC system for demonstrating the predicted amplitude-to-phase conversion. These experiments demonstrate the viability of RPC as a technique for producing wavefronts that are phase-modulated according to user-defined spatial patterns. Before proceeding, it is worth considering why one would wish to use amplitude modulation to generate a desired phase modulation rather than attempting to modulate the phase directly. First, we must realize that using amplitude to encode a modulated phase is neither rare nor new. They are routinely encountered in conventional holograms that are, essentially, "photographic slides" of interference patterns [4–6]. Reconstructing phase from amplitude holograms, however, suffers from artefacts, such as the undiffracted readout light and so-called twin image. In Gabor's pioneering holograms [4], the extraneous artefacts combined with the desired reconstruction. On the other hand, while Leith and Upatnieks' off-axis holograms are able to isolate desired reconstruction from the spurious components [5], it requires storing a carrier frequency that, subsequently, trades off resolution and information capacity [7]. The proposed RPC technique is not haunted by these issues.

The RPC technique can directly encode a desired phase modulation on the propagating beam and thus minimize energy loss to spurious orders. Any relatively high contrast amplitude modulation, or intensity distribution can serve as the input template for producing the desired phase modulation pattern. In the case of a fixed phase distribution, a major advantage of the use of amplitude masks to define the required phase pattern is the relative simplicity with which they can be manufactured when compared to phase-only elements. The use of standard chrome-on-glass mask technology would make it possible to achieve high-resolution phase patterns, the phase shift of which would be controlled by the filtering system. In fact, it is possible to tune the output phase shift via the contrast ratio of the mask and the filter parameters. If a dynamic phase modulator is required, then an amplitude modulator, in the form of a commercially available liquid crystal display (LCD) projector element, or possibly a MEMS (Micro Electronic Mechanical System) type device can be used. The use of one such dynamic device, a DMD (digital micromirror-array device) will be demonstrated in the next section. In the remainder of this section, we describe the basis of the RPC technique, including a brief theoretical treatment. We describe the experiments that have been undertaken using both fixed and dynamic modulation and discuss important criteria for the experimental implementation of the RPC technique. The results are interpreted with reference to the theoretical background and the RPC technique is compared qualitatively with alternative methods for the generation of phase-only modulation.

We will now describe experiments undertaken to characterise the performance of the RPC technique using first, a fixed amplitude mask, and then an SLM for encoding the input amplitude modulation. Using chrome-on-glass as fixed mask offers the advantage of achieving high contrast, nonpixellated input amplitude patterns as compared with a liquid crystal SLM (meaning that the term $\alpha_{\min}=0$ can be obtained). On the other hand, an SLM can generate dynamically reconfigurable phase modulation patterns, which is vital for certain applications. Moreover, an SLM can also exploit the fact that $\alpha_{\min}=0$ is not always the optimal input setting.

10.3.1 Experimental Setup

In the first case we consider the use of a fixed amplitude mask as the input modulator. A schematic diagram of the experimental setup is shown in Fig. 10.3. The RPC system is implemented on one arm of an interferometer to enable the measurement of the generated output phase. In Fig. 10.3, the beam splitters (BS1 and BS2) and mirrors (M1 and M2) form the reference arm for a Mach-Zender interferometer in which the output fringes are recorded on a CCD camera. Light of wavelength λ=635nm from a laser diode (LD) was spatially filtered, expanded and collimated using a beam expander (BE) to approximate a plane wave illumination. A beam splitter (BS1) diverted part of the collimated beam into the reference arm while the transmitted beam illuminated an amplitude mask through an iris (IR1). The mask was positioned at the input plane of a RPC system formed by lenses L1 and L2 (f=200mm) and a phase-only filter located at their common focal plane. The output of the RPC system interfered at an angle with the reference beam at the output plane where a CCD camera detects interference fringes. The reference beam size can be controlled using a second iris (IR2) and can be blocked altogether to check whether the phase modulation is embedded within a uniform intensity background.

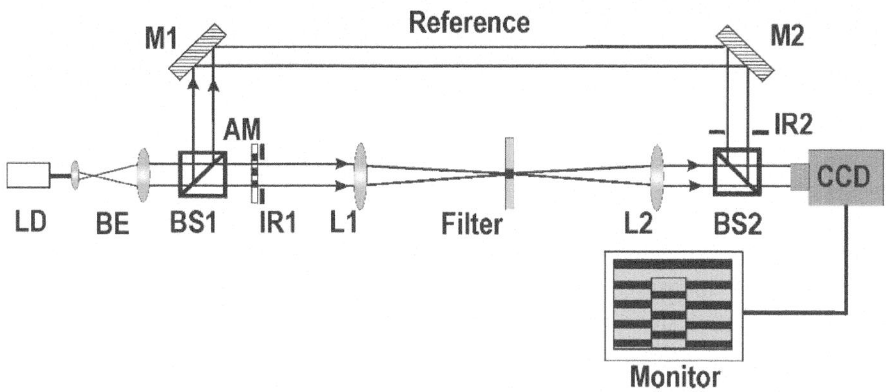

Fig. 10.3 Experimental set-up for the implementation and characterization of the RPC method. A plane wavefront produced by a laser diode (LD) and beam expander (BE) is incident on spatial filtering $4f$ system (lenses L1 and L2) that uses a phase-only filter to generate a phase modulation from an amplitude mask (AM) placed in the same plane as an iris IR1. The output phase distribution is visualised by an interferometer, the reference arm of which is formed by the mirrors (M1 and M2) and the beamsplitters (BS1 and BS2) and the diameter of which is controlled by an iris (IR2). The resulting fringe pattern is recorded with a CCD camera.

10.3.2 Matching the Filter Size to the Input Aperture

The RPC filter used for the experiments was a circularly symmetric phase-only filter with no amplitude damping ($A=B=1$). The central region of the filter contained a 60μm diameter phase shifting region, the thickness of which was calibrated to yield a phase shift of π at 633nm. These filter parameters yielded a combined filter parameter, $C=-2$. As in the generalised phase contrast method, obtaining optimal performance from an RPC system requires matching the size of the input aperture with the phase-shifting region of the filter is critical. The theoretical aspects of correct filter and input aperture matching have been described previously [8]. We will now focus on important practical considerations to illustrate how these theoretical considerations are adopted into practice.

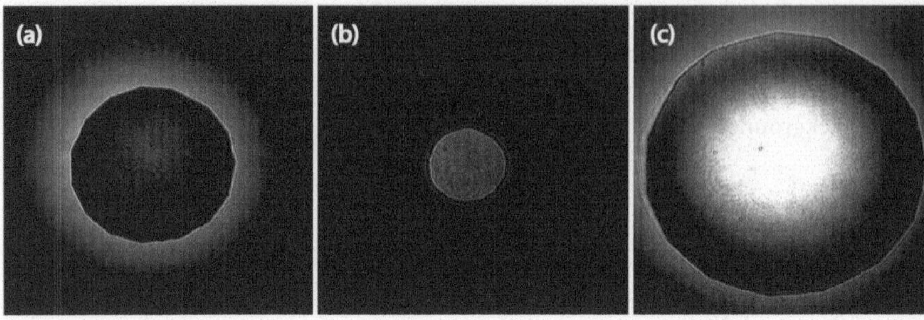

Fig. 10.4 Experimental results for matching of the input aperture (IR1 in Fig. 10.3) to the phase filter. These examples show (a) a correctly matched aperture and (b, c) incorrectly sized apertures that are either too small (b) or too large (c) for the filter size being used.

Implementing an optimized RPC system requires setting a proper *K*-value. The input amplitude mask was first removed and the reference beam blocked as they were not needed and can even disturb the initial stages of the optimization process. Figure 10.4 shows the output generated using different input aperture sizes, defined by the iris IR1 in Fig. 10.3, which is akin to changing the value of *K*. In Fig. 10.4(a), the RPC filter was correctly positioned on-axis within the system and the input iris was adjusted until the image region of the input iris appeared uniformly dark. The intensity within this region may be treated by the interference between the directly imaged input aperture and the synthetic reference wave as described in Eq. (10.7)). As can be directly verified by applying the parameters $\alpha(x',y')=1$ and C=−2 into Eq. (10.7), a dark output, $o(x',y')=0$, is obtained under the condition $K=1/2$. Using a smaller aperture, as shown in Fig. 10.4(b), means operating at a value of $K<1/2$ and, conversely, a larger aperture as in Fig. 10.4(c) corresponds to $K>1$ operation characterized by a very non-uniform output intensity distribution. Changing the ratio of the iris and the filter diameter can have a dramatic influence on the flatness and strength of the SRW and, consequently, on the uniformity and contrast of the output. Although it is also possible to produce a phase-modulated output at different *K*-settings, it is in the $K=1/2$ regime that one should ideally operate the RPC system to achieve a good trade off between SRW flatness and light throughput. The impact of *K* and the SRW in common-path interferometry, discussed in the context of a phase-modulated input in Chapter 3, maintain its relevance in RPC.

10.3.3 RPC-Based Phase Modulation Using a Fixed Amplitude Mask

After having correctly aligned the RPC filter and adjusted the input aperture size to operate sufficiently close to the $K=1/2$ point, the amplitude mask was then positioned as close as possible to the input plane of the RPC system defined by IR1 in Fig. 10.3.

The chrome-on-glass mask used in the demonstration showed a standard USAF target, where the section transmitted through the input aperture was a group of three equally spaced metal bars the approximate size of which are 2.5 × 0.5 mm.

Figure 10.5 shows a set of output images experimentally obtained from the RPC system using different input aperture sizes similar to those in Fig. 10.4. If this was a simple imaging operation, the three vertical bars of the target would appear dark on a light background. To meet the uniform intensity criterion, the aperture in Fig. 10.5(a) was adjusted to minimize the intensity difference between the imaged regions of the transparent and opaque portions of the input mask. Aside from modifying the K-value, changing the input iris diameter also affects the $\bar{\alpha}$ through the associated change in the fill factor F in Eq. (10.9). Reducing the aperture size violated the uniform background criterion and the system resembled a simple imaging system, as shown by the output

Fig. 10.5 Experimentally obtained output plane images for different iris sizes using an amplitude input mask (a section of USAF resolution target). (a) A correctly matched iris size gives a constant amplitude background; (b) a smaller iris generates an image of the mask; (c) a larger iris generates a contrast-reversed image of the mask.

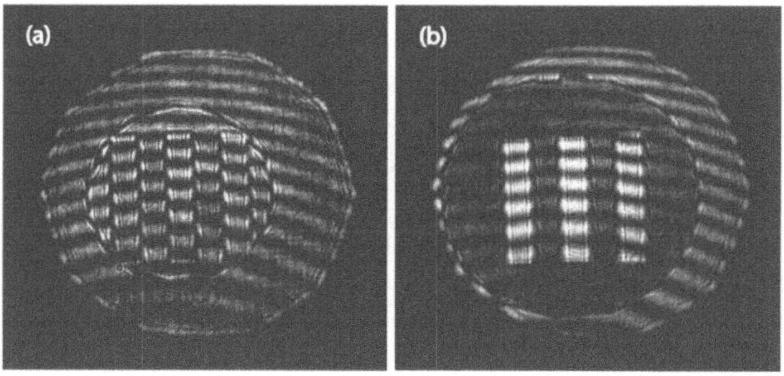

Fig. 10.6 Interferometric measurement of spatial phase modulation generated by the RPC method. (a) Fringe pattern shows π phase shift within the designated regions on a constant amplitude background when using matched aperture and amplitude object (c.f. Fig. 10.5(a)); (b) No phase shift occurs when the aperture is no longer correctly matched.

in Fig. 10.5(b). Likewise, increasing the aperture size also violated the uniform background criterion and generated an output that shows contrast reversal of the input mask, shown in Fig. 10.5(c). We will explain these results later, together with the phase results, using the framework developed earlier.

Reverting to the correct input aperture setting, the reference arm of the interferometer is then activated to measure the output phase perturbations encoded by the RPC technique. The interferogram shown in Fig. 10.6(a) corresponds to the intensity distribution shown in Fig. 10.5(a) and visualizes the embedded phase perturbations as sheared fringes between image regions corresponding to different intensity regions of the input mask. The extent of the reference beam, defined by the aperture IR2, is seen as the larger of the two circular apertures apparent on the interferogram. The SRW extended beyond the image region of the input aperture (IR1) and interfered with the reference beam to form the surrounding fringes. Increasing the input aperture IR1 up to the contrast reversal regime in Fig. 10.5(c) resulted in a notable loss of the phase shift in the output image, as shown in Fig. 10.6(b). The fringe visibility is poorer beyond the bars, but we can clearly see that the fringes are continuous within the extent of the aperture defined by IR1, which means that there is no longer a phase shift associated with the structure of the input image.

The experimental results shown in Fig. 10.5 and Fig. 10.6, including the contrast reversal and the disappearance of the phase shift, can be analysed and understood using the Argand diagram representation we developed earlier for the modelled output in Eq. (10. 7).

In Fig. 10.7, we analyse the experimental results using the Argand diagram introduced in Fig. 10.2(b). In this case we are working with a further simplified system where we use a high contrast mask so that $\alpha_{\min} = 0$ and a lossless $\theta = \pi$ filter which leads to C=−2. In Fig. 10.7(a) we consider the case of a system where the input conditions match the filter parameter such that the constant background intensity criterion of Eq. (10.10) is fulfilled. This describes the case shown in Fig. 10.5(a) and Fig. 10.6(a) in which the different regions of the output are characterized by a balanced intensity, $|o(1)| = |o(0)|$, and relative phase shift of π between them. When the input aperture diameter is too small, an imbalance results between the output vectors such that $|o(1)| > |o(0)|$. This case, which is represented in Fig. 10.7(b), effectively resembles an imaging operation where the transmissive regions of the mask appear brighter than the non-transmissive regions. In Fig. 10.7(c) we depict the case where the use of an oversized aperture resulted in mismatched output vectors such that $|o(1)| < |o(0)|$. Although this results in contrast reversal, similar to the output shown in Fig. 10.5 (c), there is still a phase shift of π between the different regions of the output image regardless of the intensity levels, similar to the cases depicted in Fig. 10.7(a) and (b). Altogether, these three cases illustrate that the generation of phase-only modulation superimposed on a uniform intensity background in an optimized RPC system is simply a special case of an otherwise generally complex output. Figure 10.7(d) depicts the case when the input aperture is much larger than the K=1/2 setting so that $|o(0)| >> |o(1)|$. This generates a contrast-reversed output without an accompanying phase shift between the regions, which corresponds to the result shown in Fig. 10.6(b).

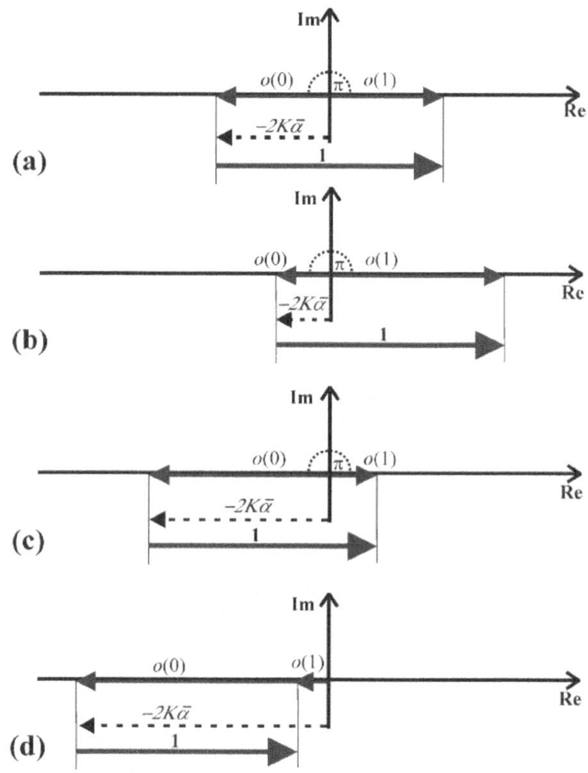

Fig. 10.7 Simplified Argand diagrams for the experimental system where $C = -2$ and $\alpha_{\min} = 0$ showing the conditions for: (a) a correctly matched aperture fulfilling the constant intensity background condition, (b) an aperture that is too small giving the condition $|o(1)| > |o(0)|$ corresponding to a standard imaging operation of the input amplitude mask, (c) shows the result of using an oversized aperture which produces a contrast reversal between the input and output of the system such that $|o(1)| < |o(0)|$ where dark sections of the input mask appear brightest and vice versa. In (d) we show what can happen when the input aperture is very much larger than it should be, so that $|o(0)| >> |o(1)|$ and there is no phase shift between the two resultant vectors $o(0)$ and $o(1)$. Referring to the interferometric measurement of the output image shown in Fig. 10.6(b), we see that we have a condition of contrast reversal without phase shift.

10.3.4 RPC-Based Phase Modulation Using an SLM as Dynamic Amplitude Mask

The optical set-up shown in Fig. 10.3 was modified to replace the fixed mask with a reflection-type spatial light modulator, which was operated as a dynamic amplitude mask. A Hamamatsu parallel-aligned liquid crystal modulator was inserted between a polarizer and analyzer to operate in amplitude modulation mode. In general, such an SLM will have a lower contrast than a fixed mask and the resolution of the resulting phase distribution will be limited to that of the modulator. The experiment used a HeNe laser as the light source and the same Fourier plane filter as was used in the earlier experiment with a fixed amplitude mask.

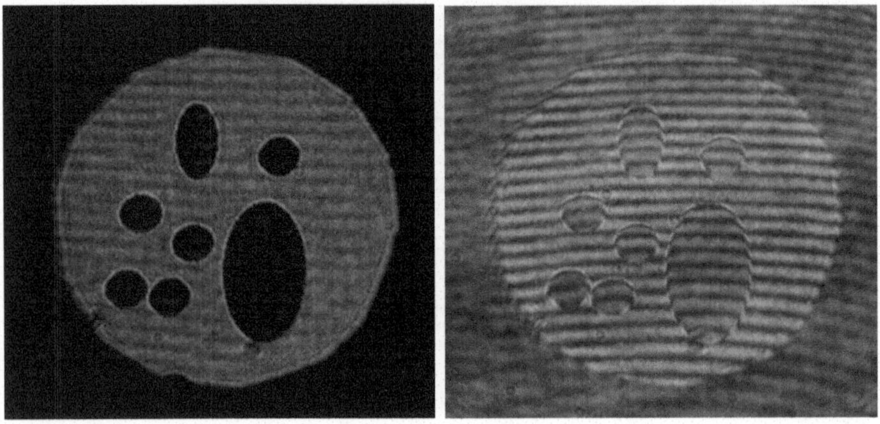

Fig. 10.8 Experimental results for the generation of phase modulation using an SLM operating as the input amplitude modulator. These show (a) an image of the input amplitude distribution without the filter in place and (b) interference fringe measurement of the output phase modulation. Examination of the fringes reveals that we achieve a phase modulation of π.

In Fig. 10.8(a), we show the output generated when the RPC filter is removed and the reference arm deactivated, such that the system simply reproduces an image of the input amplitude modulation encoded by the SLM. The image consists of a number of circular and ellipsoidal dark regions on a light background. The iris setting has been optimized according to a similar procedure as for the fixed mask. The iris was slightly out of focus due to a small axial displacement between the SLM and iris and some slight interference fringes are visible due to stray light scattered off the beam-splitter placed in front of the SLM. Although it is also possible to use the SLM to encode an iris, a physical iris was used to avoid the side-effects of the imperfect contrast that can disturb the zero-order and to block spurious scattering from the other regions of the device. After aligning the filter into place, the reference arm is activated and the interferometric result is shown in Fig. 10.8(b). Looking at the fringes, it can be seen that the generated output exhibits uniform intensity with an embedded binary phase modulation that echoes the spatial features of the input amplitude pattern. The fringe spacing indicates that we have a phase shift of approximately π at the output modulation and have thus successfully converted an input amplitude distribution into a spatially identical phase distribution. As before, we observe fringes in the region outside the aperture arising from the SRW tail scattered beyond the image region of the iris.

These results demonstrated that the reverse phase contrast technique is a viable method for the generation of a binary phase distribution by the conversion of a spatially varying amplitude distribution using a Fourier plane filtering technique. The results also illustrate that the theoretical framework developed earlier can sufficiently explain the form of the output of a common-path interferometer with an amplitude modulated output, which generates optimization principles for an RPC technique that can compete in terms of robustness and accuracy with either fixed or dynamic phase elements.

10.4 Reverse Phase Contrast Implemented on a High-Speed DMD

The viability of RPC was experimentally established in the previous section both with the use of a static amplitude mask and a dynamic amplitude mask implemented with a tandem of an LC-based SLM and a polarizer [9]. However, the use of an LC-SLM-polarizer tandem as input amplitude modulator is, at best, a proof-of-principle demonstration of the RPC potential for dynamic phase modulation. After all, the LC-SLM can be used in phase mode to directly encode the desired phase information onto an incident wavefront. However, regardless of whether using an LC SLM as an amplitude modulator to modulate the output phase in an RPC-based scheme or using the LC SLM to directly encode the phase, the dynamics of the resulting phase modulation is hampered by the relatively slow response time of the LC device, thus preventing high-speed phase modulation.

In this section, we demonstrate a fast and robust phase modulation scheme using an RPC system based on a state-of-the-art digital micromirror-array device (DMD™) [10] as input amplitude modulator. The DMD (Texas Instruments) consists of an array of light-deflecting aluminium micromirrors (1024×768 square pixels; 13.68 μm pixel pitch; 88% active area fill), each of which can achieve two positional states, ON or OFF, corresponding to electromechanically induced diagonal mirror tilt angles, γ, of $+12°$ or $-12°$. The individually addressable micromirrors can be switched between bistable states in ~15 μs. In addition, the DMD also has significantly higher illumination power tolerances and supports a very wide spectral region from 350–2000 nm and, furthermore, sets much less stringent requirements on the polarization of the incident field as compared to an LC-based SLM.

10.4.1 Setup

Figure 10.9 shows a picture of a compact setup that converts DMD-encoded amplitude-only patterns into geometrically identical phase patterns using the RPC method. It employs a collimated monochromatic laser beam (λ = 1065.7 nm, Ytterbium fibre laser, IPG Laser GmbH) to read out the amplitude-only object generated by the computer controlled DMD. The diagonal tilt angles of the micromirrors dictates the rather unorthodox orientation of the DMD chip, seen in Fig. 10.9(a), which ensures that the micromirror tilt axes are perpendicular to the optical table. In this geometry, the incident relevant propagating beam and the chip normal is along the optical axis of the $4f$ setup. Alternately, when an expanded, collimated beam is incident at an angle of ~24° with the chip normal (Fig. 10.9(b)), the strongest Fraunhofer diffraction order when all micromirrors are ON propagates parallel to the optical axis into the $4f$ system and is the only order collected by the remaining optics. All output intensity data are normalized to the intensity of this diffraction order.

Let us now briefly go through a mathematical description of a DMD-based RPC setup to identify possible nuances in the implementation. A given input object encoded on the DMD is described by a real-valued amplitude transmittance $e(x,y)$,

$$e(x,y) = \text{circ}(r/\Delta r)\alpha(x,y), \tag{10.22}$$

where $\alpha(x,y)$ is an amplitude-modulated signal truncated by a circular aperture denoted by a circ function that is unity at radial positions, $r = \sqrt{x^2 + y^2}$, within a circular region of radius Δr, and zero otherwise. Unlike the previously described RPC implementation, the input iris is now generated using the input-encoding device itself (i.e. micromirrors outside a defined circular iris region on the DMD are set to OFF-state). Therefore, the input iris and the amplitude input signal, $\alpha(x,y)$, are both imaged in focus and are dynamically adjustable via computer-controlled electronic addressing.

The RPC experiments employed a non-absorbing spatial Fourier filter located at the midplane between the two $4f$ setup lenses (Fig. 10.9(b)). The RPC filter imparts a π-phase shift onto spatial frequencies $f_r = \sqrt{f_x^2 + f_y^2}$ within a circular region of radius, Δf_r, centred on-axis. The extent of the phase shifting region, defined by Δf_r, corresponds to a physical spatial radius, R, which depends on the focal length and illumination wavelength according to the relation $\Delta f_r = (\lambda f)^{-1} R$.

The intensity $I(x',y')$ at the output plane of the RPC $4f$ setup is derived as [9, 11]

$$I(x',y') = \left[\text{circ}(r'/\Delta r)\alpha(x',y') - 2\bar{\alpha} g(r')\right]^2, \tag{10.23}$$

which is considered as an interference between the directly imaged input pattern (the first term) and a synthetic reference wave, (SRW, the second term). The amplitude of the SRW depends on the spatial average of the input amplitude pattern, $\bar{\alpha}$, given by

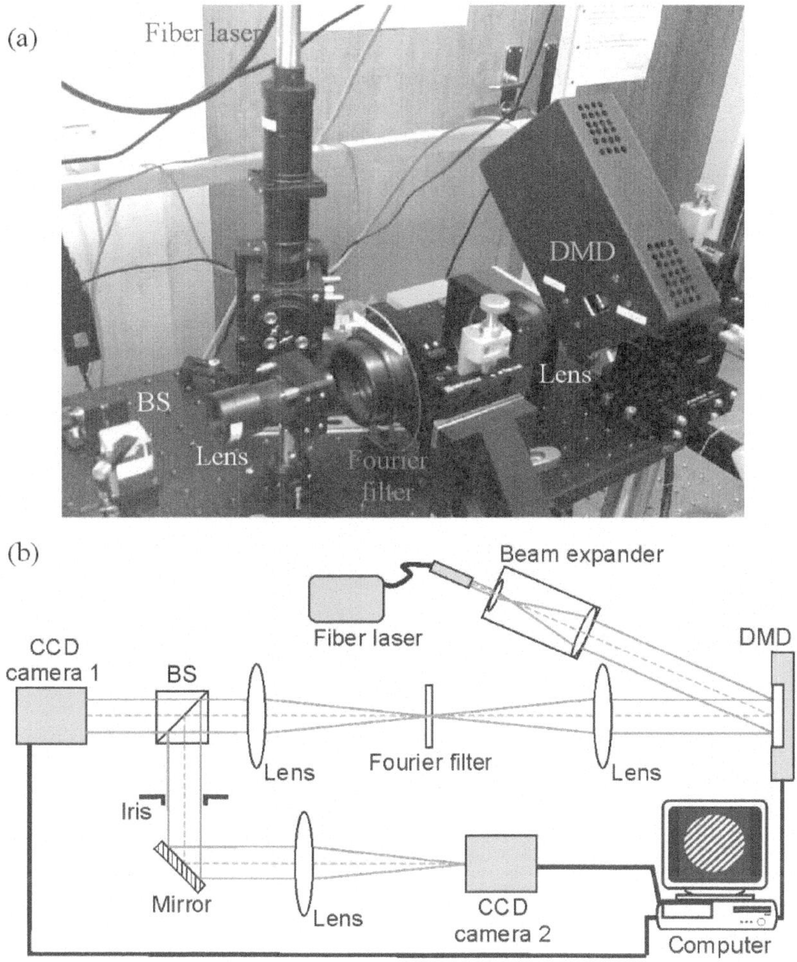

Fig. 10.9 (a) Photograph of the reverse phase contrast (RPC) $4f$ setup for converting an amplitude-only pattern displayed on a digital micromirror-array device (DMD) into a spatially similar phase pattern at the output plane. (b) Schematic diagram of the whole setup. The expanded and collimated laser beam is made incident to the DMD chip at an angle of ~ 24°, twice the micromirror tilt angle $|\gamma|$, such that the beam coming out normal to the chip (at ON-state) is the strongest Fraunhofer diffraction order and the only order that passes through the optical train. CCD camera 1 detects the intensity at the output plane. CCD camera 2 captures the optical Fourier transform of the iris-truncated output pattern. Identical lenses with 100-mm focal length are used. A phase-only filter (made from an optical flat with a tiny circular pit) is used to create a phase shift of π over an on-axis circular region (diameter, $2R$ ~ 39 μm) in the common Fourier plane of the two $4f$ setup lenses. BS indicates a beam splitter.

$$\bar{\alpha} = \left(\pi(\Delta r)^2\right)^{-1} \iint_{\sqrt{x^2+y^2}\leq\Delta r} \alpha(x,y)\mathrm{d}x\mathrm{d}y\,. \tag{10.24}$$

The SRW exhibits a characteristic spatial profile,

$$g(r') = 2\pi\Delta r \int_0^{\Delta f_r} J_1\left(2\pi\Delta r f_r\right) J_0\left(2\pi r' f_r\right)\mathrm{d}f_r\,, \tag{10.25}$$

10.4.2 Results and Discussion

Optimizing the performance of the RPC system requires correctly matching the size of the input aperture with the size of the π phase-shifting region of the Fourier filter. As illustrated in the earlier section, a good starting point is to satisfy the so-called dark-background condition. Using an optimal input aperture radius, Δr_{op}, that satisfies this condition nulls the output intensity within the boundary of the aperture image when an

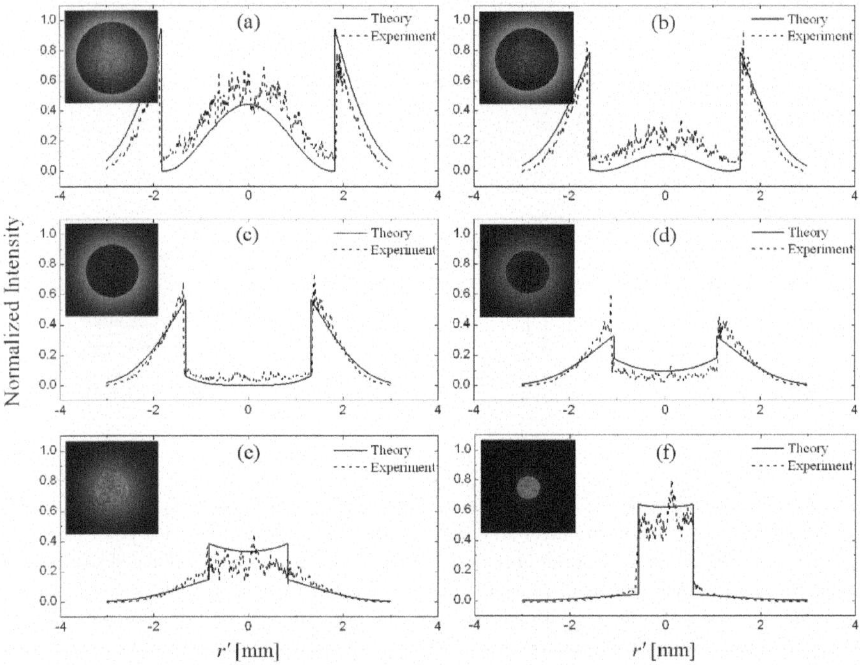

Fig. 10.10 Comparison of the theoretical and experimental intensity profiles at the output plane of the $4f$ setup that images a circular iris of diameter, $2\Delta r$ = (a) 3.65 mm, (b) 3.15 mm, (c) 2.65 mm, (d) 2.15 mm, (e) 1.65 mm, and (f) 1.15 mm with the phase-only filter centred at the Fourier plane common to the two lenses. Each experimentally obtained intensity profile is a diagonal line-scan through the center of the CCD-captured image (inset).

unmodulated, uniform input signal $\alpha(x,y)=1$ is used. When using a physical aperture, determining Δr_{op} for the setup in Fig. 10.9 is done by monitoring the output intensity at CCD camera 1 for different input iris radii Δr. For a DMD-based aperture, we simply encoded $e(x,y)=\text{circ}(r/\Delta r)$ on the DMD where, in this case, Δr is electronically (rather than mechanically) tuned and is easily measurable based on the addressed DMD pixels. Figure 10.10 shows the results for different input aperture sizes. Intensity linescans through the experimentally obtained output images, shown in the insets, are plotted together with the corresponding numerically calculated curves using $I(x',y')=\left[\text{circ}(r'/\Delta r)-2g(r')\right]^2$ for comparison.

The results clearly show a good agreement between the experiment and the model, which both give the same optimal aperture radius, $\Delta r_{op}\sim$ 1.33 mm, as shown in Fig. 10.10(c). The observed uniformity of the achieved dark background indicates that the SRW exhibits a reasonably flat spatial profile, $g(r')$, which justifies a constant-value approximation based on its value at the centre: $K=g(r'\to 0)=1/2$.

Setting the input iris at the optimal radius, $\Delta r_{op}\sim$ 1.33, we tested the RPC effect by encoding a 50% duty-cycle binary amplitude grating as input pattern into our DMD-based RPC system. A high-contrast image of the test object, obtained using the same *4f* setup but without the RPC Fourier filter, is shown in the upper inset of Fig. 10.11. Upon aligning the on-axis phase-only Fourier filter into its correct position, the CCD camera 1 detects a low-contrast output intensity shown in the lower inset of Fig. 10.11. This output can be explained by our RPC model defined in Eq. (10.23) considering that, for the particular 50% duty cycle grating pattern where $\bar{\alpha}\sim 1/2$ (see Eq. (10.24)), the output intensity is given by

$$I(x',y')=\begin{cases}\left[0-g(r')\right]^2 & \text{for } (x',y')\in\Re_{\text{OFF}}\\ \left[1-g(r')\right]^2 & \text{for } (x',y')\in\Re_{\text{ON}}\end{cases}, \tag{10.26}$$

where $\Re_{\text{ON}}$ and $\Re_{\text{OFF}}$ denote regions where the ON and OFF micromirrors are imaged. The spatial profile, $g(r')$, within the interior of the circular iris' image, may be approximated by a constant, K = 1/2. Thus, we expect a four-fold intensity reduction and equalization between $\Re_{\text{ON}}$ and $\Re_{\text{OFF}}$ regions, which is more accurate at the center and is indeed what we obtain experimentally in Fig. 10.11. It is also worth to note that the DMD-defined circular input aperture achieved a much better contrast compared to the LC-SLM based RPC system requiring an extra polarizer, which usually has a limited extinction ratio.

Along with the intensity equalization, it is also apparent from Eq. (10.26) that we have transformed our amplitude-only binary grating into a periodic distribution of (approximately) +1/2 and −1/2 amplitude values, which corresponds to a 0-π binary phase grating that mimics the input pattern: $-\frac{1}{2}\exp[-i\,\pi e(x',y')]$. We have previously established, using interferometric detection, that the RPC output indeed exhibits a binary phase pattern. For the DMD-based RPC system, we used an alternative verification scheme where we experimentally demonstrated that the output indeed behaves similar to conventional binary phase grating. We took an optical Fourier transform of the RPC output, taking care to exclude the residual halo in the RPC pattern using a truncating iris of radius Δr_{op}, to determine and record its far-field diffraction pattern. A conventional 50% fill 0-π binary phase grating is characterized by a far-field diffraction pattern with highly suppressed zero-order component and where the dominant first orders each have ~0.41 of the total power flux. The experimental result, seen in Fig. 10.11, shows that the results obtained for the RPC output is, indeed, very close to the theoretically expected far-field diffraction pattern. Therefore, the binary amplitude grating encoded by the DMD successfully generated a binary 0-π phase grating, thus illustrating the amplitude-to-phase conversion in RPC. The theoretical model developed in the previous section show that it is also possible to generate binary phase patterns characterized by other modulations depths, $\Delta\phi < \pi$. The optimization procedure previously described can ensure that, with the inherent loss due to the use of an amplitude input, synthesis of a desired phase pattern can optimize the light throughput in the system.

These results confirm that the reverse phase contrast (RPC) method can be implemented using a digital micromirror-array device (DMD). The DMD enables a robust and possibly the fastest amplitude-only 2D spatial modulation that the RPC system converts into a spatially identical phase modulation. We have described and illustrated the advantages gained from a DMD-based RPC system, particularly the enhanced optimization of electronically tuning the DMD-defined input aperture that matches the RPC Fourier filter. Arbitrary binary amplitude patterns may be encoded by a DMD at promisingly high-speed refresh rates and then converted to corresponding phase patterns by RPC [7, 8] to address various applications [1, 12–16].

10.5 Summary and Links

In this chapter we revisited the basic framework for CPI analysis, developed in Chapter 3, and determined how it can be optimized when using amplitude-modulated inputs to operate in a reversed phase contrast mode. We have coined this approach reverse phase contrast (RPC) since it effectively works as a reverse version of the GPC method. The RPC method enables a given spatial binary intensity distribution to be converted into a binary phase distribution with a spatially uniform intensity profile, the phase step of which is determined by a Fourier plane filtering operation. We were able to obtain a design recipe for an RPC system by building upon the basic CPI framework, which has proved useful in the optimization of various GPC systems. Unlike fringe-based amplitude encoding of phase information that tends to have lower spatial resolution than the modulating device, the spatial resolution of an RPC system follows that of the modulating device. High quality amplitude masks can be employed for static phase modulation

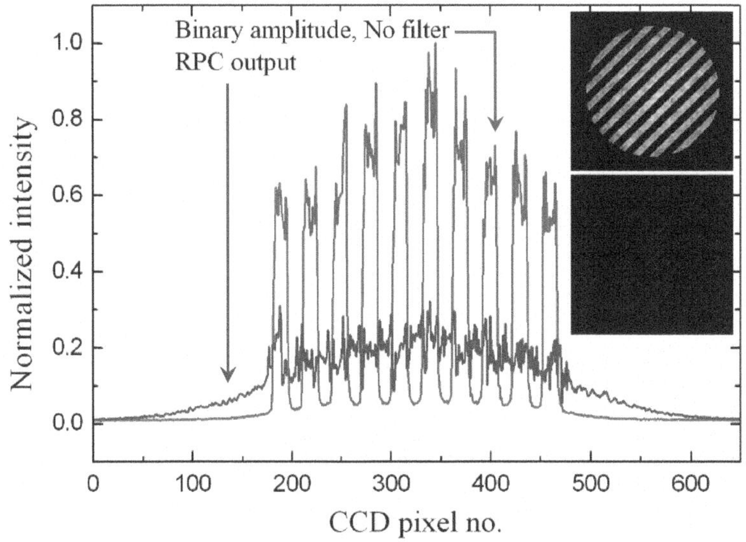

Fig. 10.11 Intensity profiles measured along a diagonal (perpendicular to grating bars) for the DMD-encoded binary amplitude grating and for the corresponding phase pattern produced via RPC when the filter is centred at the Fourier plane. The CCD-captured 2D images for the binary amplitude grating and the RPC output are shown in the upper and bottom insets, respectively.

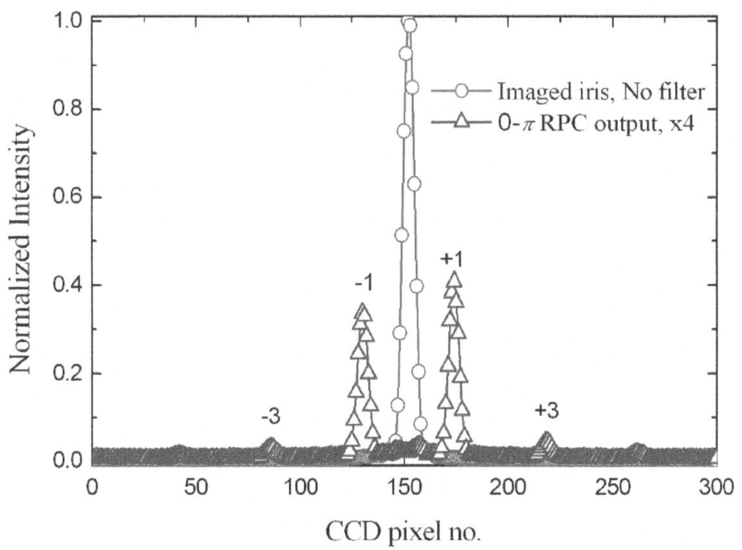

Fig. 10.11 Measured far-field diffraction profiles of the circular iris' image at the output of the $4f$ setup without the phase-only filter (circles) and the binary phase pattern produced via RPC (with binary amplitude-only input from the DMD) when the filter is centred at the Fourier plane (triangles). The latter shows the suppressed 0th and even diffraction orders, and the dominant +1 and −1 orders each with strength (four times the actual) approximately equal to the theoretical value of ~0.41 for a 50% duty-cycle, 0-π binary phase pattern.

while high-speed amplitude modulators can be used for dynamic phase modulation, as in the digital micro-mirror device (DMD) based demonstration discussed in the chapter. The RPC method offers the possibility of achieving high performance phase-only spatial light modulation without the need for a phase-only spatial light modulator. In the next chapter, we will integrate an RPC with a GPC system into a single CPI system intended for applications in optical cryptography.

References

1. C. Denz, G. Pauliat, G. Roosen, T. Tschudi, "Volume hologram multiplexing using a deterministic phase encoding method", Opt. Comm. **85**, 171-176 (1991).
2. N. Towghi, B. Javidi, Z. Luo, "Fully phase encrypted image processor", J. Opt. Soc. Am. A, 16, 1915-1927 (1999).
3. S. Jutamulia, "Phase-only Fourier transform of an optical transparancy", Appl. Opt. **33**, 280-282 (1994).
4. D. Gabor, "A new microscopic principle," Nature (London) **161**, 777–778 (1948).
5. E.N. Leith and J. Upatnieks, "Reconstructed Wavefronts and Communication Theory," J. Opt. Soc. Am. **52**, 1123-1128 (1962).
6. Adolf W. Lohmann, "A Pre-History of Computer-Generated Holography," Optics & Photonics News **19**, 36-47 (2008).
7. A. Macovski, "Hologram Information Capacity," J. Opt. Soc. Am. **60**, 21-27 (1970).
8. J. Glückstad and P. C. Mogensen, "Optimal phase contrast in common-path interferometry", Appl. Opt. **40**, 268-282 (2001).
9. P. C. Mogensen and J. Glückstad, "Reverse phase contrast: an experimental demonstration," Appl. Opt. **41**, 2103-2110 (2002).
10. L. Yoder, W. Duncan, E.M. Koontz, J. So, T. Bartlett, B. Lee, B. Sawyers, D.A. Powell, and P. Rancuret, "DLP™ Technology: Applications in Optical Networking," Proc. SPIE, **4457**, 54-61 (2001).
11. J. Glückstad and P. C. Mogensen, "Reverse phase contrast for the generation of phase-only spatial light modulation," Opt. Commun. **197**, 261-266 (2001).
12. P. C. Mogensen and J. Glückstad, "Phase-only optical encryption," Opt. Lett. **25**, 566-568 (2000).
13. P. C. Mogensen, R. L. Eriksen, and J. Glückstad, "High capacity optical encryption system using ferro-electric spatial light modulators," J. Opt. A: Pure Appl. Opt. **3**, 10-15 (2001).
14. P. C. Mogensen and J. Glückstad, "Phase-only optical decryption of a fixed mask" Appl. Opt. **40**, 1226-1235 (2001).
15. R. John, J. Joseph, and K. Singh, "Holographic digital data storage using phase modulated pixels," Opt. Lasers Eng. **43**, 183-194 (2005).

16. T. D. Wilkinson, W. A. Crossland, and V. Kapsalis, "Binary phase-only 1/f joint transform correlator using a ferroelectric liquid-crystal spatial light modulator," Opt. Eng. **38**, 357-360 (1999).
17. J. Glückstad, "A method and an apparatus for generating a phase modulated wavefront" US patent application 60/257,093 (priority date 22 Dec. 2000).

Chapter 11
Optical Encryption and Decryption

Cryptography entails recording or transmission of concealed information where only the application of a correct key enables the comprehension of the original information. The art of cryptography dates back to ancient times where secret information has been transmitted in terms of symbols and sketches. Through the years, cryptography has evolved and the medium by which it has been implemented changes depending on the state of science of a particular era. The current state of technology rests on specialized electronic data processing machines and computers. Light is also exploited for information transport and storage. Using light to encode digital information has proven as a highly efficient technology that has radically revolutionized modern-day data communications. Hence, it is a natural course to incorporate optical cryptographic techniques into contemporary optical data communications and storage. Moreover, optical cryptographic methods may be future solutions to problems related to intellectual property protection, product authentication, falsified bankcards and identification cards, and other similar predicaments.

Optical cryptographic techniques exploit the coherent nature of a laser beam. These techniques have proven to yield efficient ciphered information in addition to extremely fast decryption via parallel optical processing [1]. Javidi, *et al* have proposed a number of optical cryptographic schemes involving the use of phase masks for: (1) encrypting amplitude information based on the double-phase encoding scheme [2, 3]; (2) encryption of phase-encoded information [4]; and (3) holographic storage of encrypted information [5]. These schemes require recording of encrypted masks containing both amplitude and phase information. Optical cryptography can also be achieved by operating on a single lossless parameter that allows for full optical reversibility: the phase [6] or polarization [7–9] of a coherent light carrier.

Phase-only cryptography is based on the direct superposition of a phase mask containing the original data and an encrypting phase key and vice-versa [10–14]. This encryption process also implies that all operating light fields in general have at least a full 2π-phase cycle of modulation. Since optical phase is undetectable by the eye or by standard light-capturing devices, an encrypted phase array is invisible in addition to its incomprehensible format. Upon decryption, visualizing an invisibly decrypted field can be achieved by an efficient conversion of the field into a high-contrast intensity image. The generalized phase contrast (GPC) method plays a vital role in phase-only optical cryptography as it is used to visualize the decrypted but invisible field. The phase contrast technique proposed by Nobel Laureate Fritz Zernike [15] can only view phase images correctly having less than $\pi/3$ phase modulation while decrypted phase can have a much larger phase stroke. The GPC method [16–18] resolves the limitations of Zernike's phase contrast method by setting a more elaborate analytic model of the process. Thus, the GPC method can provide optimized visualization of the decrypted phase information.

This chapter describes the fundamentals of phase-only optical cryptography and the visualization of decrypted information using the GPC method. It also exploits the planar-integrated micro-optics platform implementation of the GPC method in a miniaturized device [13, 19, 20] to demonstrate the feasibility of a highly compact and robust GPC scheme for optical cryptography. The miniaturized GPC method is a particularly robust implementation that is not prone to position tolerances and alignment problems, which are major issues when using discrete optical components [21]. Real-world applications require robust and easily producible systems, which can be achieved by pre-designing the optical system in an integrated device in addition to making it compact. A similar rationale came about when integrated circuits revolutionized the electronic industry. Hence, integrating and miniaturizing an optical cryptographic system provides for a much more realistic set of applications and even enables a direct interface to micro-opto-electronic-based devices.

11.1 Phase-Only Optical Cryptography

In phase-only optical cryptography, one encodes amplitude image information, $o(x,y)$, as a two-dimensional phase distribution, $O(x,y)=\exp\left[j2\pi o(x,y)\right]$, which is subsequently encrypted by scrambling with a random phase pattern, $R(x,y)=\exp\left[j2\pi r(x,y)\right]$, to yield an encrypted field, $E(x,y)$, given by:

$$E(x,y)=O(x,y)R(x,y)=\exp\left[j\left(2\pi\left(o(x,y)+r(x,y)\right)\right)\right] \tag{11.1}$$

where $o(x,y)$ and $r(x,y)$, are two-dimensional matrices containing element values normalized within the interval [0;1].

To perform decryption, the encrypted field, $E(x,y)=\exp\left[j2\pi e(x,y)\right]$, is applied with a decrypting key generated using the complex conjugate of the encrypting phase, $R^*(x,y)$, thus retrieving the phase-encoded signal:

$$O(x,y)=E(x,y)R^*(x,y)=\exp\left[j\left(2\pi\left(e(x,y)-r(x,y)\right)\right)\right] \tag{11.2}$$

Figure 11.1 shows a graphical illustration of the encryption and decryption procedures. The phase encoded signals, $O(x,y)$, $R(x,y)$, $E(x,y)$ and $R^*(x,y)$ are, technically, invisible to the naked eye or image acquisition devices such as cameras. However, these phase-encoded signals are depicted by their respective phase patterns for ease of visualization. The grey and white pixels indicate the relative phase-shifts and are not representative of the amplitude of the operating light fields. In principle the signal, $O(x,y)$ represents a 2D matrix of coded information and, hence, will be more suitably depicted by a pattern that, like $R(x,y)$, is visually incomprehensible. However, for clarity of presentation and as a convenient visual aid in the discussion, we have chosen to depict $O(x,y)$ as a conceptually visible image pattern depicting the word "RiSØ", as seen in Fig. 11.1.

The geometry of the system for implementing phase-only optical encryption is shown in Fig. 11.1(b). A plane-polarized wavefront is incident on the original phase image to set the field $O(x,y)$. Aligning the phase-modulated field, $O(x,y)$, with a phase mask, $R(x,y)$, generates the encrypted field $E(x,y)$. The encrypted information can, in principle, be directly transmitted or first stored optically onto appropriate recording devices and then retrieved for decryption. Phase-only optical decryption is achieved by aligning the encrypted, phase-modulated field, $E(x,y)$, with the decrypting phase key, $R^*(x,y)$, to retrieve the unencrypted field, $O(x,y)$. The decrypted signal information remains indiscernible to the naked eye or any image acquisition device as it is maintained as a phase-only modulation of a light field.

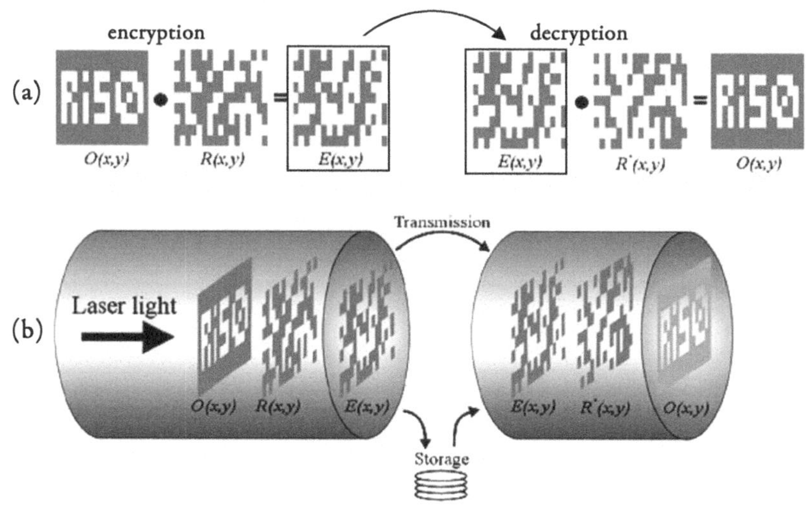

Fig. 11.1 Phase-only optical cryptography.

11.2 Miniaturization of the GPC Method via Planar Integrated Micro-Optics

A highly compact system can be advantageous for some applications of phase-only cryptography. To cater to these applications, the GPC method can be miniaturized into a single device using planar integrated micro-optics, as previously discussed in Sect. 9.4. Let us briefly describe the miniaturized GPC system used in cryptographic demonstrations discussed in the following section. Using lithography, the optical components were fabricated as micro-elements on a planar surface as shown schematically in Fig. 11.1 and by the picture on Fig. 11.2 (right). To form structured coatings or surface profiles, a glass substrate is first coated with a photoresist – a photosensitive polymer material. Then, a laser or electron beam, intensity modulated either using lithographic masks or other methods, illuminates a desired pattern on the photoresist layer, which is subsequently removed with the aid of developing chemicals. The formed photoresist pattern is transferred to the glass substrate by reactive ion etching (RIE), where a combination of chemical reaction and physical impact of plasma ions removes the substrate material from regions where it is not protected by the photoresist coating.

The microlenses (L1 and L2) were fabricated by two lithographic masks to form four-phase-level diffractive lenses on the topside of the substrate. The microlenses have been optimized for imaging along a tilted optical axis and thus are slightly elliptic and have slightly different focal lengths (f_x=25.58 mm and f_y=24.51 mm) along the two perpendicular lateral directions while the f/# is maintained at f/#~5.

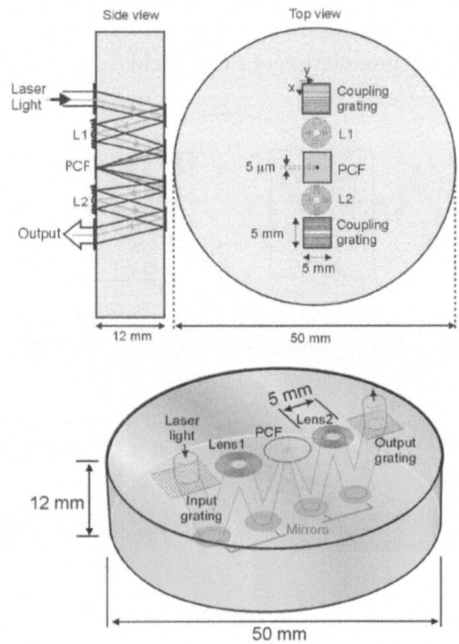

Fig. 11.2 The miniaturized GPC system.

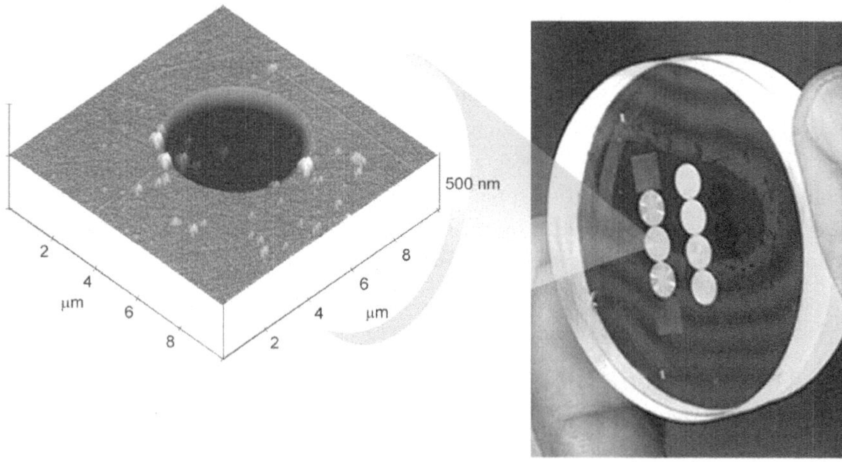

Fig. 11.3 The phase contrast filter (left) fabricated on the integrated planar-optical device (right).

The two diffractive microlenses are arranged linearly on top of a glass substrate as shown in Fig. 11.2 (top view). The beam path through the folded 4*f* system is depicted on the side and perspective views of the miniaturized GPC system in Fig. 11.2. The object plane of the 4*f* system is located at the surface of the input grating. Light incident normal to the planar-optical device is coupled into the substrate through a binary phase grating. The binary coupling gratings were fabricated with a 2.13 μm period and deflects an incident beam by 11.77° onto a steering mirror that then redirects it to lens L1, travelling a total path length equivalent to the focal length from the coupling grating to lens L1. The converging beam from lens L1 reflects off another mirror and focuses at the substrate where a reflection coated PCF has been fabricated. The focus is located at the Fourier plane and the PCF performs a half-wave phase shift of the on-axis, zero-order region of the focused light. The filter is designed for operation at λ=0.633 μm and is etched as a hole with radius, R_l = 2.5 μm, on the substrate. Figure 11.3 shows a topographic image of the PCF taken using an atomic force microscope. An anisotropic etching process is used to form a steep-edged cylindrical hole. After the PCF, the reverse Fourier transform is performed in the second half of the symmetric system.

Miniaturizing the GPC method to an integrated diffractive micro-optical system can have unwelcome consequences (see Sect. 9.4). The quality of finite aperture diffractive optical elements (DOE) affects the resolution as well as the size of the image field in an imaging system. Although favored for practical reasons, such as compactness and compatibility with standard photolithography techniques, the use of DOEs influences the imaging behavior and spatial resolution of the optical system because of discrete phase quantization. The finite number of quantization steps results in the distribution of the input light to undesired higher-order diffraction beams, thus reducing light efficiency. On the contrary, miniature refractive optical elements can be more difficult to fabricate due to the need for thick deposition of phase structures in order to achieve the necessary optical function. Another issue is the aberration caused by the oblique orientation of the optical axis. This comes as a result of the arrangement of the planar integrated components, where the beam propagates at a certain angle in the optical system that is folded into a two-dimensional layout.

However, the other unwelcome factors, such as low light throughput, have diminished relevance for phase-only cryptography. Phase-only cryptography basically requires efficient visualization of the decrypted spatial phase modulation. Hence, a virtually lossless light propagation provided by the GPC method compensates for the inherently low throughput of planar integrated micro-optical devices.

11.3 Miniaturized GPC Method for Phase-Only Optical Decryption

The performance of the miniaturized GPC method for phase-only optical decryption is examined using the experimental system shown in Fig. 11.4. The external macro-optical system is composed of three 4*f* lens setups. The first 4*f* setup, using L1 and L2, is for imaging the decrypting key from the SLM to the phase mask with the encrypted information. The second 4*f* lens setup, using L3 and L4, is for coupling the decrypted field to the input grating of the device. The third setup using L5 and L6, placed after the GPC, is for scaling the intensity distribution from the GPC output plane to the CCD camera. The decrypting key information is encoded using a phase-only spatial light modulator (SLM), which is illuminated by an expanded HeNe laser beam (λ=0.633 μm). The SLM is a parallel-aligned nematic liquid crystal (Hamamatsu Photonics), which can modulate phase of at least 2π at λ=0.633 μm. The SLM is optically addressed by an XGA-resolution (768×768 pixels) liquid crystal projector that is controlled from the video output of a computer. To facilitate phase-only optical decryption, lenses L1 and L2 project the phase image of the decrypting key to the encrypted phase pattern. The decrypted phase pattern is scaled and directed to the miniaturized GPC-system via lenses L3 and L4. The truncating circular aperture, placed just after the decrypting key, governs the central spot size of the beam at the filtering region at the Fourier plane [18]. The contrast-enhanced intensity distribution generated by the GPC planar optics device is scaled and relayed by lenses L5 and L6 to the CCD camera.

To fabricate the encrypted phase mask, the optical flat was initially coated with antireflection film for operation at 633 nm. A layer for creating π-phase shifting pixel elements was added by spin-coating a 490 nm thick photoresist layer. The thickness of the photoresist layer was chosen to achieve an optical path difference that is half of the operating wavelength when compared to free-space propagation. The phase-shifting pixels, each approximately 176 × 333 μm in size, were added by etching selected portions of the photoresist layer with the desired decrypting phase pattern using a direct laser writing method. The resulting encrypted phase mask on the optical flat contained a 17×9 array of binary phase pixels, which were either 0 or π.

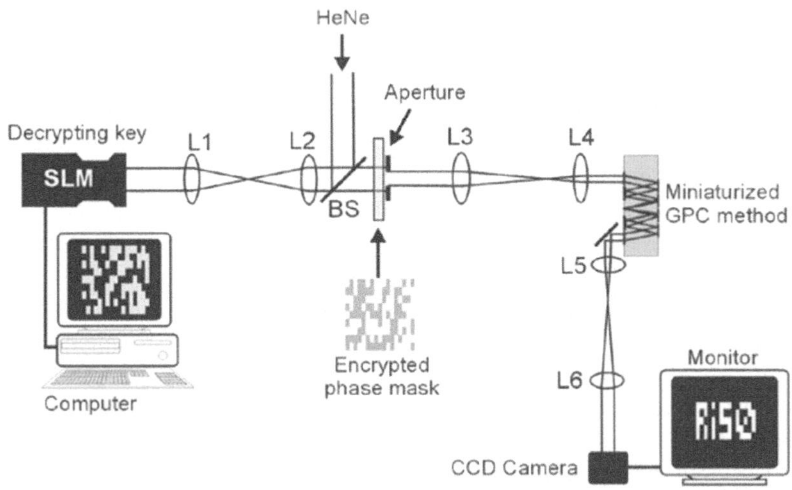

Fig. 11.4 System for testing phase-only optical decryption using the miniaturized GPC method.

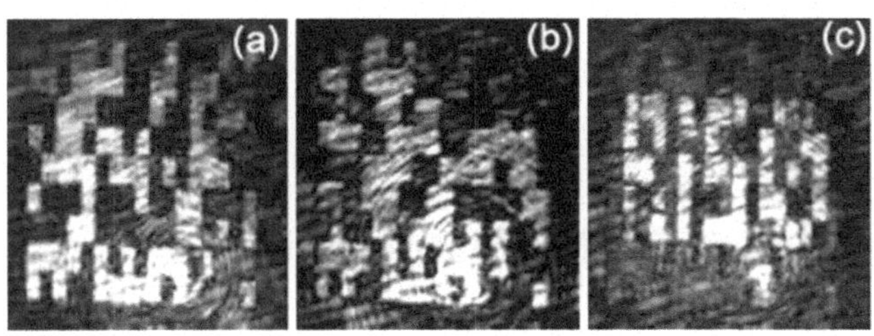

Figure 11.5 Performance of the phase-only optical decryption using the miniaturized GPC method. (a) Visualization of the correct decrypting phase key (b) Visualization of the encrypted phase mask using a wrong key exhibiting a uniform phase (c) Output showing successful optical decryption.

Figure 11.5 shows the successful decryption of a 17×9-pixel phase pattern using the setup shown in Fig. 11.4. The embedded information consists of four 5×3-pixel letters depicting the word "RiSØ". The decrypting key encoded at the SLM is imaged through the macro-optical setup (lenses L1 to L4) and projected onto the encrypted phase mask. When the phase mask is removed, the decrypting phase key can be visualized by phase contrast imaging using the miniaturized GPC method as, shown by the high contrast intensity pattern depicted in Fig 11.5(a). Inserting the encrypted phase mask after the first 4*f* lens setup (L1 and L2) and deactivating the SLM to uniformly illuminate the phase mask visualizes the encrypted information as a contrasted intensity pattern at the output as shown in Fig. 11.5(b). This output can also be interpreted as an unsuccessful decryption due to a wrong key (i.e. a uniform phase key is incorrect). Successful phase-only decryption is achieved when the SLM is activated to encode the correct decrypting key, leading to a visualization of the original unencrypted information as a high-contrast image, as shown in Fig. 11.5(c). It is important to note that the high diffraction orders caused by the binary coupling gratings [19] do not affect the intensity pattern in the field of view of the output. The use of a larger PCF on the planar integrated micro-optics, which requires slightly smaller input aperture diameters, results in better contrast images. The details of this optimization have been discussed in a previous work [20]. The low quality of visualization is due to tilt and alignment errors for both the encrypted and the key patterns. A slight tilt of the decrypting phase mask will result in uneven phase visualization as shown in Fig. 11.5(b). Such error propagates through the decryption process and contributes to the poor visualization of some of the pixels in Fig. 11.5(c).

It should be noted that the pixels of the phase mask, each approximately 176 × 333 μm, are relatively large and cannot be truly used as basis for demonstrating the imaging limitations of the system. The imaging performance of the planar-optical device can resolve feature sizes smaller than 10 μm [22, 23]. Theoretically, the resolution of the miniaturized GPC can be estimated depending on the operating numerical aperture (NA ~0.28) to resolve features as small as 2.3 μm. Considering aberrations, it is safe to assume that the miniaturized GPC-system can handle decryption and visualization of a phase encrypted information with 300×300 pixels having pixel sizes of 5 μm. This assumption, however, only covers the imaging performance of the miniaturized GPC system. A further limitation on the number of pixels can be attributed to the current state of technology of SLMs.

11.4 Phase Decryption in a Macro-Optical GPC

A compact, robust and portable implementation can be desirable for some applications of optical phase encryption. However, as we noted above, the demonstrated performance of the miniaturized system is susceptible to fabrication artifacts which can be separately treated and optimized. For comparison, we have carried out the entire phase decryption using a standard macro-optical GPC [12] and the results are displayed in Fig. 11.6.

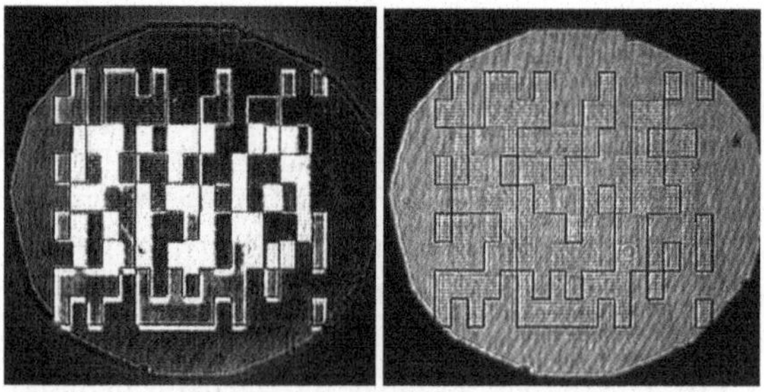

Fig. 11.6 Decryption of the 17×9 pixel fixed mask with a 17×9 pixel phase key using a standard macro-optical GPC system. The left image shows the successful decryption which reveals the text RiSØ. If the PCF is misplaced then the decrypted phase information is not visualized as displayed to the right.

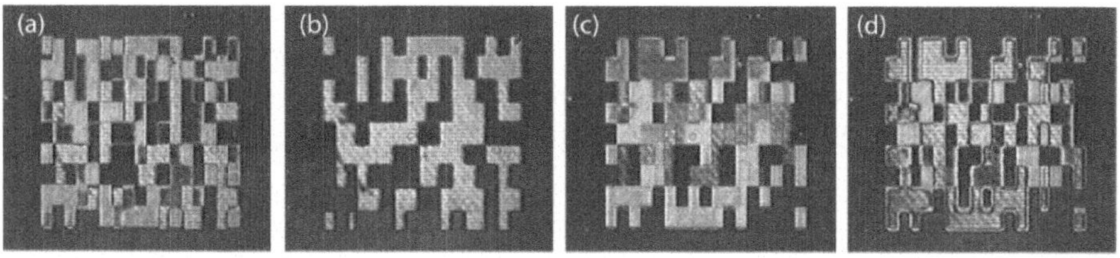

Fig. 11.7 Unsuccessful decryption of the fixed mask occurs (a) when the incorrect key (b) is applied to the fixed mask. In this case, the incorrect key is simply the correct key rotated through an angle of 180°. An incorrect phase shift in the key will also fail to yield a complete decryption of the fixed mask. Examples of the application of geometrically correct keys but with approximate phase shifts of $\pi/2$ and 2π are shown in (c) and (d) respectively.

If the incorrect key is applied to decrypt the encrypted phase mask then, as expected, the encrypted information will not be correctly recovered. Examples of the application of a number of different incorrect keys are shown in Fig. 11.7.

In Fig. 11.7(a), we have applied a rotated form of the correct key. It can be seen that there is not a discernible decryption of any of the characters in the encrypted information. It is also interesting to note that the visibility is slightly poorer when compared with the successful decryption result shown in Fig. 11.6. Even though we are using a rotated version of the correct key, when the mask and key are combined they do not give the correct overall phase shift for which the aperture is matched and so we observe a moderate reduction in contrast. In Fig. 11.7(c) and Fig. 11.7(d), we are using aligned, spatially correct, keys that, in this case, have the incorrect binary phase shifts of $\pi/2$ and 2π respectively by changing the phase shift addressed on the SLM. Thus, although the alignment of the optical system is perfect, the information is not decrypted and visualized because the binary phase shift of the key and the fixed mask do not match. This emphasises the fact that, with a dynamic phase element such as an SLM acting as the phase key, it is possible to tune the phase shift of the key to match an imperfect phase shift in the fixed element. This further adds to the flexibility of this approach for a practical system. Since the decryption operation requires pixel-to-pixel mapping, any relative lateral displacement between the mask and the key, even by as little as one pixel, will result in an incorrect decryption. This system is thus robust against the application of an incorrect key. Furthermore, it is also clear from Fig. 11.6 that the PCF is a crucial element in the decryption process and use of an incorrect PCF – one that has the wrong size, phase shift or is laterally misplaced – will result in an unsuccessful decryption.

11.5 Envisioning a Fully Integrated Miniaturized System

The previously described "proof-of-principle" experimental setups consisted of an external macro-optical setup necessary to scale both the encrypted and the key patterns to the appropriate sizes for imaging using the miniaturized GPC system. To include the entire optical setup into a fully integrated micro-optical system, both the encrypted and the key patterns have to be fabricated at the appropriate scale. Implementing the system into a fully integrated planar-optical device will ease up alignment and tilt problems, which are common difficulties encountered when using discrete optical components.

Figure 11.8 shows the intended implementation of the whole opto-electronic setup in planar integrated optics using a two-stage *4f* lens setup. An image of a phase-encrypted pattern is projected on the decrypting phase key pattern using the first *4f* lens setup. An encrypted phase-only pattern, recorded on a bankcard, a passport, a currency note, or other items requiring security, can be instantly verified for authenticity by subjecting it to this planar integrated setup. The phase-only key can be dynamically encoded on a compact, electronically controlled Liquid Crystal on Silicon (LCoS) SLM. Compared with the system illustrated in Fig. 11.4, the positions of the encrypted pattern and key patterns have been interchanged in Fig. 11.8. Successful decryption of the encrypted data only requires that the encrypted pattern and the correct key pattern are properly aligned. Thus, their positions can be interchanged, and the actual positioning can be adapted to the practical demands of particular implementations. Moreover, the functionalities of the encrypted pattern and the key pattern can also be interchanged such that the phase key can be written on a static phase mask to decrypt a dynamic stream of opto-electronic data. An initial calibration cycle by a non-mechanical alignment of the two phase-only patterns can be achieved by an automated electronic scrolling of the pattern encoded on the LCOS-SLM. The decrypted phase data is then converted into an intensity pattern using the GPC method via the second *4f* setup with a PCF at the Fourier plane. The intensity pattern at the output can subsequently be acquired and recorded using a detector array, which can be interfaced to electronic components of the security system.

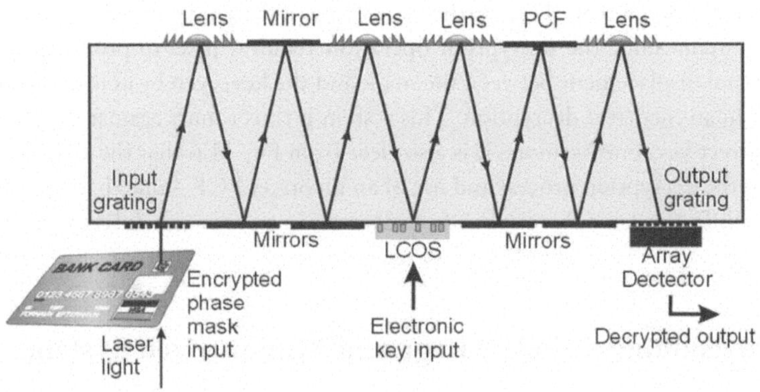

Fig. 11.8 Fully integrated phase-only optical decrypting system.

11.6 Decrypting Binary Phase Patterns by Amplitude

This last section brings into use all the "nuts and bolts" developed out of GPC, RPC, phase encryption and the miniaturization ideas and assembles and integrates them into a new method for decrypting a binary phase-only mask using an amplitude-only spatial light modulator key. The approach is based on a double-passage through a GPC setup. On the first passage, the setup operates in RPC mode and converts a binary amplitude key to phase for decrypting a binary phase-only encrypted pattern. On the second passage, the setup operates in GPC mode to convert the decrypted phase pattern to an intensity pattern that can be suitably detected by an image acquisition device. A compact dual-path system, applicable in reflection geometry, is suggested for this decryption operation.

Most systems for optical encryption and decryption are inherently complex and rely on highly sophisticated opto-electronic devices that could limit their practical application outside the laboratories. Ideally, an optical encryption system should possess all the powerful features of *phase-only* parallel processing while only applying widely available, cheap and robust opto-electronic components. The aim of realizing optical cryptography through delicate, reconfigurable and wave-retarding opto-electronics seems to be in contradiction with the equally important goal of using low-cost and robust devices applicable for a widespread real-world implementation. This very fact has motivated us to introduce a novel approach for decrypting phase-only encrypted information using a simple and widely available display device: the amplitude-only spatial light modulator (SLM).

The fundamental idea in this section exploits the transformation of a spatial amplitude modulation into a spatially similar phase-only modulation using a Fourier filtering process based on RPC as described in Chapter 10. The dynamic range of the phase modulation can be adjusted arbitrarily by using a liquid-crystal based device as phase contrast filter. The combination of an amplitude SLM with a tuneable phase filter results in a high performance phase-only decryption in which the spatial light modulation and the optical phase shift are effectively decoupled. Moreover, this combination makes it possible to tune the decrypting phase modulation either by adjusting the contrast ratio of the SLM or by tuning the filter parameters. The decrypted phase pattern is subsequently converted to an intensity pattern using an additional, yet similar, spatial filtering operation according to the standard implementation of the GPC method. A dual-path system is hereby possible to apply in a reflection geometry using a single spatial filtering setup for first converting the amplitude modulation to a phase decryption pattern in the forward path and, subsequently, for converting the decrypted phase into a visible intensity pattern in the return path.

The remainder of this chapter presents the theoretical basis for this new method and discusses the overall constraints and operating requirements for its implementation. We also present a recipe for designing a system that will be supplemented with an illustrative example based on a compact dual-path spatial filtering realization of the system operated in reflection-type geometry. The complete setup is then analysed numerically using a highly accurate one-dimensional FFT-based modelling that includes all system aperture truncations. Finally, we conclude by exploring an advanced implementation scheme employing a miniaturized version carried out as a fully integrated planar micro-optics device.

11.6.1 Principles and Experimental Considerations

The conceptual light transformations and operations needed in this system are illustrated in Fig. 11.9. Each optical processing step illustrated in Fig. 11.9 has been previously demonstrated by the authors, both theoretically and experimentally [10–14, 18–20, 24–1]. Accordingly, our present focus is to analytically devise and design a way of combining these experimentally viable steps into a combined and fully functional module, aimed at taking optical cryptography technology out of the laboratories and into real-world implementations.

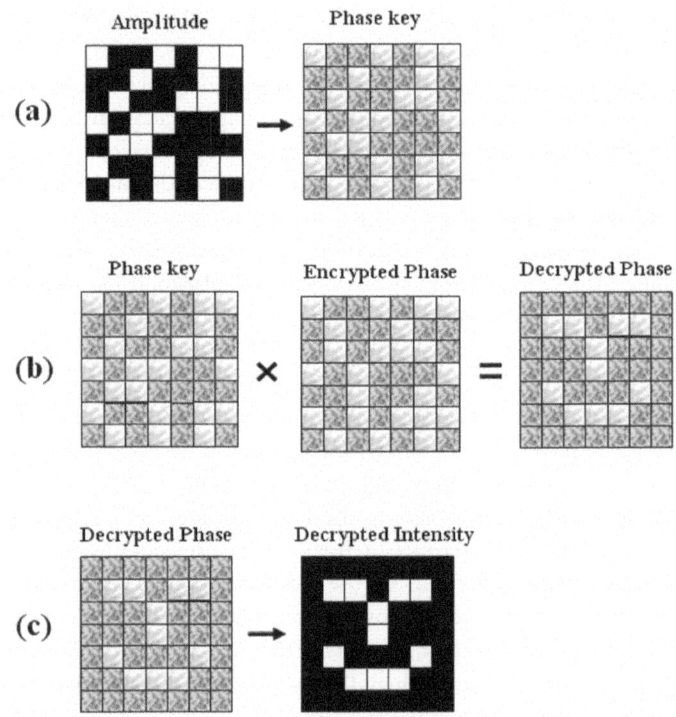

Fig. 11.9 Schematic outline of the processing steps for decrypting a phase-only mask with an amplitude-only spatial light modulator. (a) A reconfigurable binary amplitude pattern is first converted to a binary phase decryption key using the reverse phase contrast (RPC) method. (b): The generated phase key decrypts a pre-encrypted phase-only mask. (c): The decrypted phase information is finally converted into an intensity pattern using GPC.

Figure 11.10 shows a schematic diagram of a system suitable for implementing the optical processing steps outlined in Fig. 11.9. It can be seen that the optical system is based on two standard 4*f* systems with similar, though not necessarily identical, spatial filters in each of the two Fourier planes. The input is an aperture-truncated amplitude-only distribution, which is generated by a plane wave incident through an iris (Ir1) on an amplitude-modulating device such as an SLM. Ideally, the output of the first 4*f* system based on lenses L1 and L2 is a two-dimensional modulation spatially distributed according to the input amplitude-SLM pattern, and with phase values determined by the parameters of the first Fourier filter (i.e. the transmittance parameters A, B and phase shift θ) and the relative input amplitude levels. If the input amplitude distribution is a binary high contrast modulation, we can obtain a binary phase distribution at the output of the first 4*f* system.

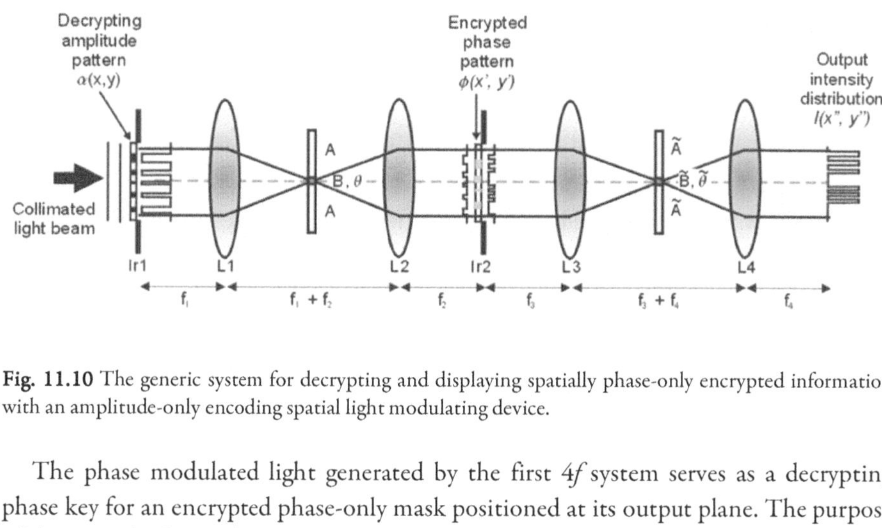

Fig. 11.10 The generic system for decrypting and displaying spatially phase-only encrypted information with an amplitude-only encoding spatial light modulating device.

The phase modulated light generated by the first 4*f* system serves as a decrypting phase key for an encrypted phase-only mask positioned at its output plane. The purpose of the second 4*f* setup (L3 and L4) is to render this decrypted phase pattern visible by an additional GPC-based filtering operation using a filter characterized by transmittance parameters $\left(\tilde{A},\tilde{B}\right)$ and phase shift, $\tilde{\theta}$. This second 4*f* setup has its own aperture truncation superposed with the decrypted phase indicated by the second iris, Ir2.

We will now get hold of expressions for the optical field as it propagates through the combined setup in Fig. 11.10 and formally describe its operation for amplitude-only based decryption of an encrypted phase-only mask. To optimize this system, we must derive a relationship between the input amplitude values and the resulting light field, which is in general complex valued. We can see that the first 4f system essentially implements the RPC method, which we have described in Chapter 10. To achieve continuity and to develop a description that is tailored to the present context, we will begin by first describing the light propagation through this RPC subsystem.

In Fig. 11.10 a monochromatic field of wavelength, λ, illuminates an input amplitude-SLM through a truncating circular iris with radius, Δr. We can describe the incident light amplitude, $a(x,y)$, at the entrance plane of the optical system as,

$$a(x,y)=\mathrm{circ}(r/\Delta r)\alpha(x,y), \tag{11.3}$$

where $\alpha(x,y)$ is a binary input amplitude modulation and the circ-function is defined as unity within the region, $r=\sqrt{x^2+y^2}\leq\Delta r$, and zero elsewhere. The circular input aperture is matched by a circular, on-axis centered spatial filter which we can write as

$$H(f_x,f_y)=A\left(1+\left(BA^{-1}\exp(i\theta)-1\right)\mathrm{circ}\left(f_r/\Delta f_r\right)\right) \tag{11.4}$$

where $B\in[0;1]$ is the filter transmittance for the focused light, $\theta\in[0;2\pi]$ is the phase shift applied to the focused light and $A\in[0;1]$ is a filter parameter describing the field transmittance for the off-axis scattered light as indicated in Fig. 11.10. The circ function specifies the radius of the phase-shifting region in the spatial frequency domain as Δf_r. The physical dimensions of this filter can be determined from the equations, $(f_x,f_y)=(\lambda f)^{-1}(x_f,y_f)$ and $f_r=\sqrt{f_x^2+f_y^2}$, which relate the spatial frequency coordinates to physical spatial coordinates.

The optical Fourier transform of the input field from Eq. (11.3) is multiplied with the filter parameters in Eq. (11.4) and the filtered field is then subjected to a second optical Fourier transform, corresponding to an inverse Fourier transform with inverted coordinates. These processes generate a complex amplitude light distribution, $o(x',y')$, at the intermediate coordinate plane, (x',y'). Within the boundaries, $\mathrm{circ}(r'/\Delta r)$, describing the image of the input aperture, the light distribution may be written as

$$o(x',y')=A\left[\alpha(x',y')+K\bar{\alpha}|C|\exp(i\psi_C)\right] \tag{11.5}$$

where the complex term $|C|\exp(i\psi_C)$ describes a combined filter parameter:

$$C=|C|\exp(i\psi_C)=BA^{-1}\exp(i\theta)-1 \tag{11.6}$$

In Eq. (11.5), $\alpha(x',y')$, describes the image formed by the encoded binary input amplitude modulation, $\alpha(x,y)$, at the intermediate coordinate plane (x',y'). Dividing the input aperture into two regions, $\Re_{\max}$ and $\Re_{\min}$, which may be discontinuous, the binary spatial amplitude modulation may be specified as

$$\alpha(x,y)=\begin{cases}\alpha_{\min} & \text{for } (x,y)\in\Re_{\min}\\ 1 & \text{for } (x,y)\in\Re_{\max}\end{cases} \tag{11.7}$$

where the upper input amplitude modulation level in the region $\Re_{\max}$ is described by a transmission coefficient equal to unity for optimum energy throughput.

The term K in Eq. (11.5), incorporates the properties of the truncating apertures in the input and phase-shifting regions of the first 4*f* system. The relationship between K and the respective radii, Δr and Δr_f, of the input aperture and the first filter aperture have been previously established in Chapter 3. For an illumination with wavelength λ and a Fourier transforming lens L1 with focal length, f_1, we can adopt these previous results and write

$$K = 1 - J_0\left(2\pi \Delta r \Delta r_f \lambda^{-1} f_1^{-1}\right) \tag{11.8}$$

Finally, the term $\bar{\alpha}$ in Eq. (11.5) can be thought of as the spatial average value of the input amplitude modulated wavefront. In the case of binary amplitude input modulation, $\bar{\alpha}$ is a function of the fraction of the input aperture associated with the regions $\Re_{\max}$ and $\Re_{\min}$. This can be expressed in terms of a fractional area F of the aperture associated with the transmission coefficient $\alpha_{\min}$ as shown in Eq. (11.9):

$$\bar{\alpha} = \frac{\left(\Re_{\min}\alpha_{\min} + \Re_{\max}\right)}{\pi\left(\Delta r\right)^2} = 1 + \left(\alpha_{\min} - 1\right)F \tag{11.9}$$

The terms K and $\bar{\alpha}$ are inextricably linked in a practical system and it is often more useful to think in terms of the $K\bar{\alpha}$ product when analysing the operation of the first 4*f* filtering system.

Since the first 4*f* system will serve to illuminate the encrypted phase mask with the phase key, an important requirement on this system is to achieve a flat output intensity distribution upon which the spatial phase modulation is present. The 4*f* output must therefore satisfy the uniform intensity condition,

$$\left|\alpha_{\min} + K\bar{\alpha}\left|C\right|\exp\left(i\psi_C\right)\right| = \left|1 + K\bar{\alpha}\left|C\right|\exp\left(i\psi_C\right)\right|, \tag{11.10}$$

which is obtained from equating the modulus of Eq. (11.5) for the two input modulation amplitudes represented by $\alpha_{\min}$ and 1 from Eq. (11.7). The phase modulation $\Delta\phi_0$ obtained at the intermediate coordinate plane, (x', y'), is given by:

$$\exp\left(i\Delta\phi_o\right) = \frac{\alpha_{\min} + K\bar{\alpha}\left|C\right|\exp\left(i\psi_C\right)}{1 + K\bar{\alpha}\left|C\right|\exp\left(i\psi_C\right)} \tag{11.11}$$

A handy way to solve this expression is to use an Argand diagram, where the complex vectors on either side of Eq. (11.10) are plotted.

Figure 11.11 illustrates a plot fulfilling the requirement for constant intensity, where the solution-vectors, $o(\alpha_{\min})$ and $o(1)$, have the same length and the angle between them denotes the output phase modulation depth, $\Delta\phi_0$, as given by Eq. (11.11). Binary phase encryption/decryption systems typically require a phase modulation $\Delta\phi_0 = \pi$ as we saw in the previous section. In this somewhat simplified situation, the imaginary components of the complex vectors in Eq. (11.10) disappear and we are left with real-valued solution vectors with equal lengths and phase shift π. For this case we have $\psi_c = \Delta\phi_o = \pi$ so that the output described by Eq. (11.5) can be simplified to

$$o(x',y')=\alpha(x',y')-|C|K\bar{\alpha}=\left|\alpha(x',y')-|C|K\bar{\alpha}\right|\exp\left(i\pi\left(\frac{\alpha(x',y')-\alpha_{\min}}{1-\alpha_{\min}}\right)\right) \quad (11.12)$$

Referring to the case described by Eq. (11.12), we see that the product, $K\bar{\alpha}$, together with the modulus of the combined filter parameter, $|C|$, and SLM normalized minimum amplitude level, $\alpha_{\min}$, are the critical parameters which determine the relationship between the input and output wavefronts of the first 4*f* filtering system (L_1 and L_2).

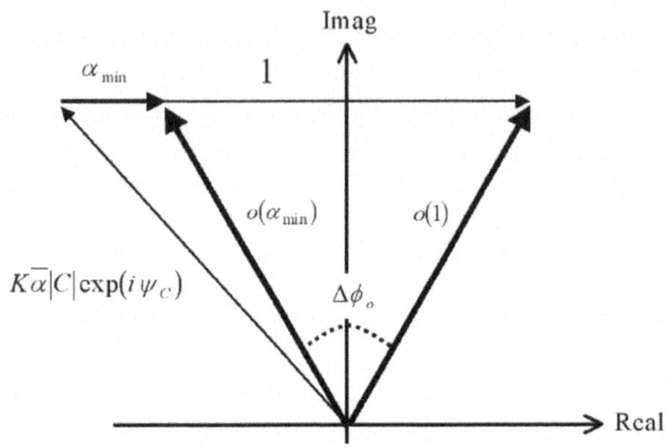

Fig. 11.11 Argand diagram showing the complex solution vectors $o(\alpha_{\min})$ and $o(1)$ plotted from Eq. (11.10) for the constant intensity background condition with an achieved phase modulation of $\Delta\phi_0$.

The derived binary phase modulation described by the last term of Eq. (11.12) can now serve as the decrypting phase of the encrypted binary phase mask with phase, $\phi(x',y')=\pi\beta(x',y')$ where $\beta(x',y')$ is a binary function with values 0/1, positioned at the primed coordinate plane in Fig. 11.10. An additional iris, Ir2, truncates the superposed phase patterns (ideally equal to the decrypted phase pattern) and this wavefront enters the second 4f filtering setup composed of lenses L_3 and L_4. Following the exact same derivation procedure as for the first 4f setup we can write the combined filter parameter for the second filter as:

$$\tilde{C}=|\tilde{C}|\exp(i\psi_{\tilde{C}})=\tilde{B}\tilde{A}^{-1}\exp(i\tilde{\theta})-1 \quad (11.13)$$

and similarly express the aperture related term.

$$\tilde{K}=1-J_0\left(2\pi\Delta\tilde{r}\Delta\tilde{r}_f\lambda^{-1}f_3^{-1}\right), \quad (11.14)$$

where $\tilde{K}$ deals with the properties of the truncating apertures in the second 4*f* system where $\Delta\tilde{r}$ and $\Delta\tilde{r}_f$ are the radii of the aperture governed by iris, Ir2, and the second filter aperture, respectively, and f_3 is the focal length of the Fourier transforming lens L_3.

Combining Eq. (11.10) with the phase shift introduced by the encrypted phase mask and feeding this into the GPC-filtering setup of the second 4*f* system, we obtain the following expression for the intensity at the output detector array shown in Fig. 11.10:

$$I(x'',y'') \propto \Bigg| |\alpha(x'',y'') - |C|K\bar{\alpha}| \exp\left(i\pi\left(\beta(x'',y'') + \left(\frac{\alpha(x'',y'')-\alpha_{\min}}{1-\alpha_{\min}}\right)\right)\right) - |\tilde{C}|\tilde{K}\left(\overline{|\alpha(x'',y'') - |C|K\bar{\alpha}| \exp\left(i\pi\left(\beta(x'',y'') + \left(\frac{\alpha(x'',y'')-\alpha_{\min}}{1-\alpha_{\min}}\right)\right)\right)}\right)\Bigg|^2 \tag{11.15}$$

where the large bar indicates the operation of a spatial averaging.

Using the constraints imposed by Eq. (11.10), we can simplify Eq. (11.15) into:

$$I(x'',y'') \approx |1-|C|K\bar{\alpha}|^2 \Bigg| \exp\left(i\pi\left(\beta(x'',y'') + \left(\frac{\alpha(x'',y'')-\alpha_{\min}}{1-\alpha_{\min}}\right)\right)\right) - |\tilde{C}|\tilde{K}\left(\overline{\exp\left(i\pi\left(\beta(x'',y'') + \left(\frac{\alpha(x'',y'')-\alpha_{\min}}{1-\alpha_{\min}}\right)\right)\right)}\right)\Bigg|^2 \tag{11.16}$$

which is, apart from a constant, recognized to be exactly the expression for GPC-filtering a decrypted $0/\pi$ binary phase pattern as we saw in the earlier chapters with the decrypted phase function of the form:

$$\phi_{decrypt}(x'',y'') = \pi\left(\beta(x'',y'') + \left(\frac{\alpha(x'',y'')-\alpha_{\min}}{1-\alpha_{\min}}\right)\right) \tag{11.17}$$

By choosing the lower state amplitude modulation of the input SLM to be $\alpha_{\min} = 0$, Eq. (11.16) can be further simplified using the imposed constraint that $|C|K\bar{\alpha} = ½$ from Eq. (11.10) to finally obtain:

$$I(x'',y'') \approx \frac{1}{4}\Big|\exp\left(i\pi\left(\beta(x'',y'')+\alpha(x'',y'')\right)\right) - |\tilde{C}|\tilde{K}\left(\overline{\exp\left(i\pi\left(\beta(x'',y'')+\alpha(x'',y'')\right)\right)}\right)\Big|^2 \tag{11.18}$$

which we recognize as an intensity scaled-down version (the factor ¼ due to the use of amplitude at the input and not phase) of the decryption operation as demonstrated in refs. [7–9].

For a configuration with $C = \tilde{C}$ we can fold the system shown in Fig. 11.10 to operate the encrypted phase mask in reflection geometry and use the same 4f lenses for illuminating the encrypted mask with the phase key and, upon illumination, for visualizing the resulting phase modulation. This is illustrated in Fig. 11.12, where the combination of a polarizing beamsplitter and a quarter waveplate serves the function of isolating the return path from the forward path to enable proper detection and avoid unwanted feedback into the laser.

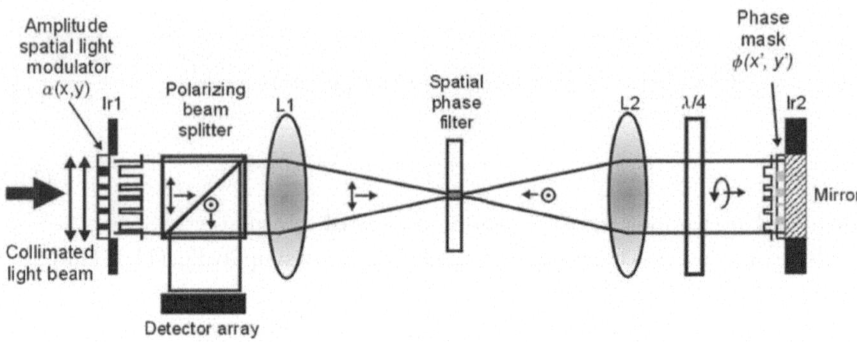

Fig. 11.12 Dual-path system working in reflection geometry using a quarter waveplate and a polarizing beamsplitter to isolate the polarization of the forward and reversed propagating beams.

An alternative scheme that implements the unfolded optical system in a very compact and robust way is demonstrated in Fig. 11.13, which utilizes a planar-integrated micro-optics platform to implement the whole system. This miniaturized approach, discussed in Sect. 11.5, was originally described in ref. [13] and its application to phase-only decryption experimentally demonstrated in ref. [14].

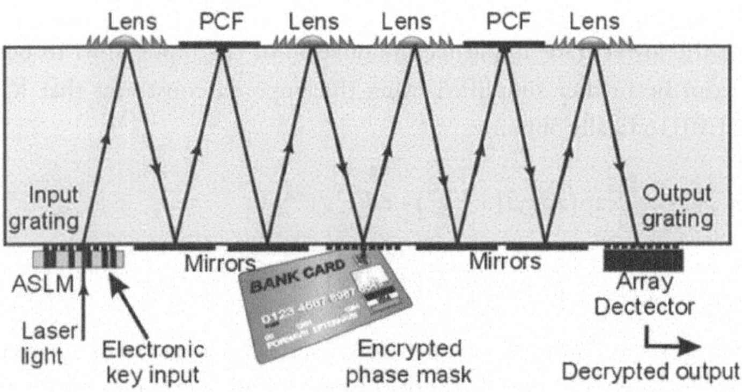

Fig. 11.13 Envisaged planar-integrated micro-optics implementation of the generic setup in Fig. 11.10. An amplitude-only spatial light modulator (ASLM) encodes the decrypting key and propagates through cascaded phase-only spatial filtering systems comprised of two lossless phase contrast filters (PCF).

11.6.2 Numerical simulations

In order to demonstrate the robustness of the optical decryption approach analytically derived by Eq. (11.12) we have performed a high-resolution FFT-based simulation. To obtain high numerical accuracy the simulations have been carried out for densely sampled one-dimensional signals and filters. Numerical results corresponding to the various processing steps for decrypting a phase-only mask with an amplitude-only spatial light modulator will be presented and discussed.

Figure 11.14(a) shows a segment of the input binary amplitude pattern with a uniform random distribution providing for an average value of a half. Following the processing step schematized in Fig. 11.9(a) this pattern is converted by the first 4f system into a binary phase key on top of an approximately flat amplitude value equal to the average value of the input as shown in Fig. 11.14(b). The simulation parameters of the phase filter in the first 4f system has been chosen according to the constraint, $|C|K\bar{\alpha} = ½$, leading to the use of a π-phase shifting filter with a size providing for $K = ½$.

Results corresponding to the two left-most schematics in the processing step illustrated in Fig. 11.9(b) are presented in Fig. 11.15, which show (a) the generated phase key and (b) the encrypted phase pattern. The decrypted phase, encoded on top of an approximately flat amplitude value equal to the average value of the input, is shown in Fig. 11.16(a). This is subjected to a final visualization step by the second 4f system, which converts the phase information into intensity variations, providing for the decrypted periodic pulse array shown in Fig. 11.16(b). The simulation parameters of the phase filter in the second 4f system has been chosen according to optimal light efficiency and contrast leading to the use of a π-phase shifting filter with a size providing for $K = 1$.

The reliability of the proposed optical decryption process can be ascertained by examining the fidelity of the decrypted information relative to the original data in the presence of corrupting noise. In the next round of FFT-based simulations we demonstrate the robustness of the approach to both amplitude and phase noise introduced both at the site of the input amplitude modulator and on top of the encrypted phase mask.

It should be noted that, for clarity, all phase values presented in Figs. 11.17 to 11.20 have been plotted as absolute values thereby flipping all negative phase values into the positive phase domain from 0 to π. This enables a much easier comparison with the corresponding noise free decryption plots in Figs. 11.14 to 11.16 but has the obvious drawback that the phase noise gives the visual impression of being limited to only half the stroke actually used in the simulations.

Fig. 11.17(a) shows the input amplitude modulation arising from a superposition of the binary amplitude modulation with a uniformly distributed random phase and amplitude noise. Both noise factors have been set to 5% of the input amplitude and 5% of that of a full phase cycle. Fig. 11.17(b) shows the generated phase key and the corresponding "flat" amplitude profile.

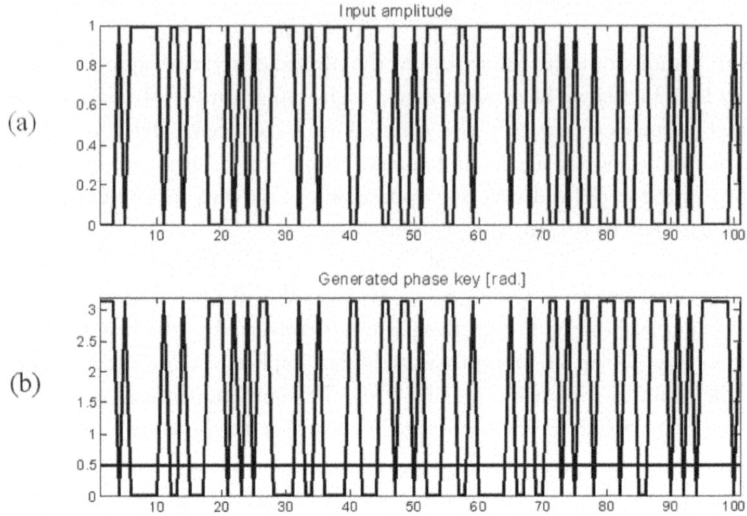

Fig. 11.14 (a) shows 100 pixels of a one-dimensional input binary amplitude pattern with a uniform random distribution providing for an average value of a half. Following the schematic of Fig. 11.10 this pattern is converted by the first RPC spatial filtering system (simulated by using a 16384 pixel FFT) into a binary phase key encoded onto an approximately flat amplitude value (=1/2) as shown in (b).

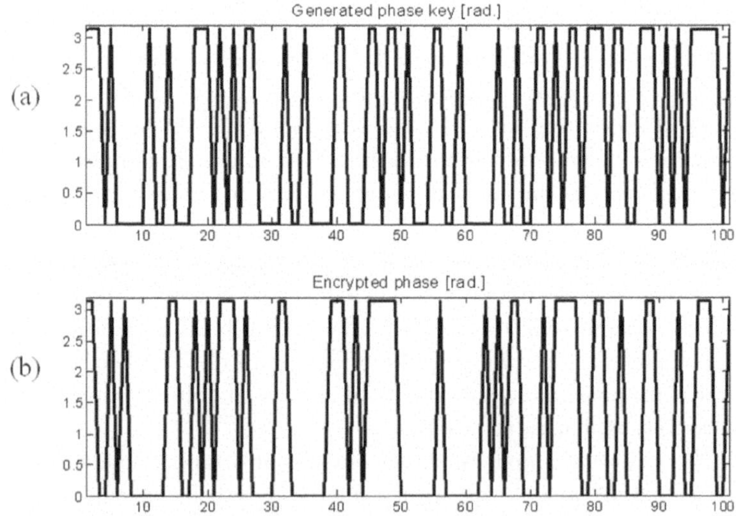

Fig. 11.15 These plots correspond to the left-most schematics in Fig. 11.9(b) and show (a) the generated phase key from Fig. 11.14(b) and the encrypted phase mask (b) side by side.

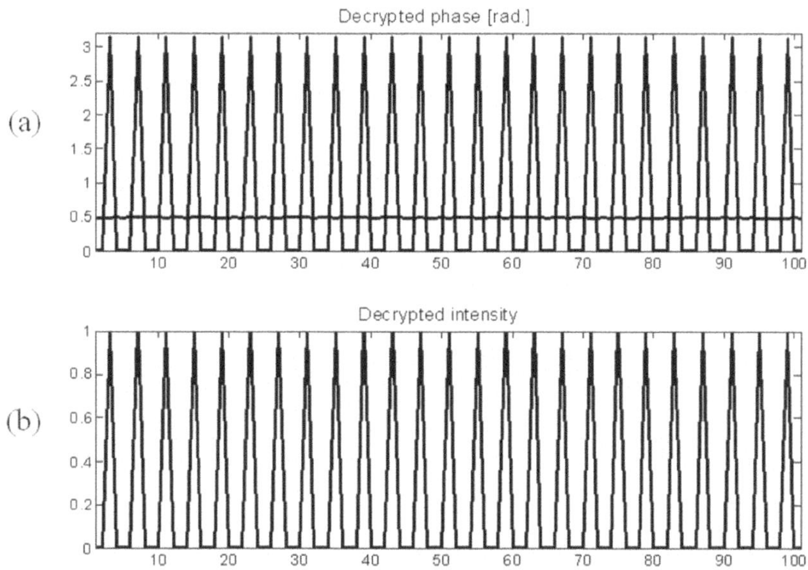

Fig. 11.16 (a) The decrypted phase shown with its approximately flat amplitude profile equal to a half. (b) The final phase-intensity conversion by the FFT-simulated second 4*f* system providing for the decrypted periodic spike array.

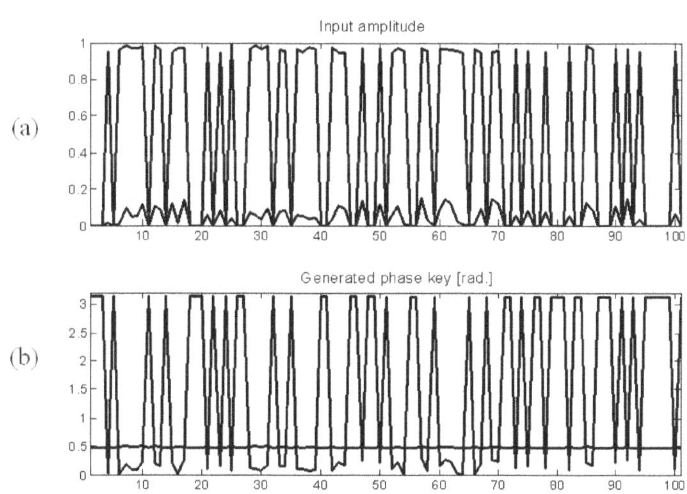

Fig. 11.17 (a) Input amplitude corresponding to Fig. 11.14(a) but superposed with uniformly distributed random amplitude and phase noise (the "invisible" phase noise is indicated in radians at the bottom of the plot). The plot in (b) shows the generated phase key and the corresponding "flat" amplitude profile as in 11.14(b). It should be noticed that for clarity all phase values have been plotted as absolute values leading to the effect that all negative phase values are mirrored into the positive phase domain from 0 to π.

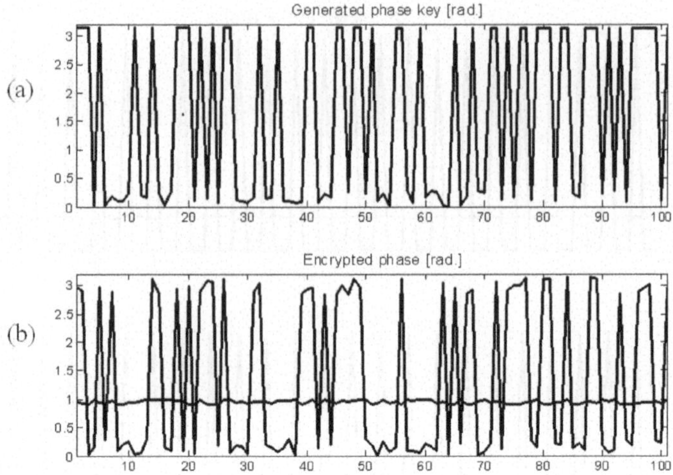

Fig. 11.18 The generated phase key (a);. the encrypted phase mask (b). In addition to the phase key generated from the noisy input amplitude of Fig. 11.17(a), the encrypted phase is also perturbed in the simulation by both amplitude and phase noise from a uniform random distribution. Note that all phase values are plotted as absolute values in the range from 0 to π and that the perturbed amplitude of the noisy encrypted phase is illustrated as the curve smaller than the value 1 in (b).

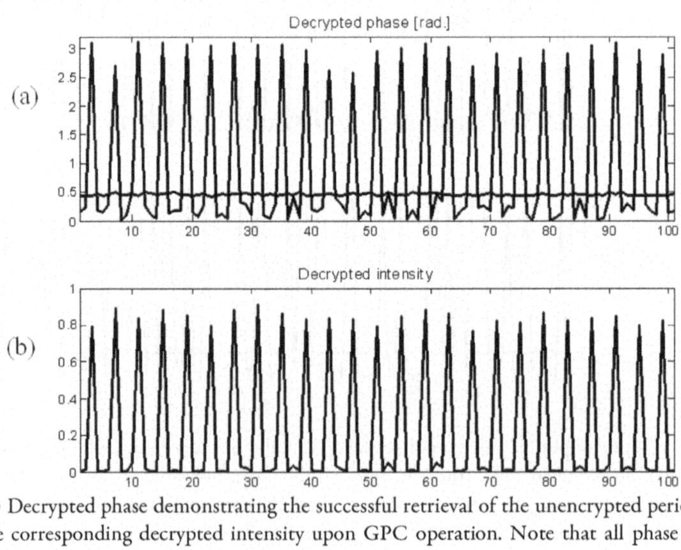

Fig. 11.19 (a) Decrypted phase demonstrating the successful retrieval of the unencrypted periodic phase array. (b) The corresponding decrypted intensity upon GPC operation. Note that all phase values are plotted as absolute values in the range from 0 to π.

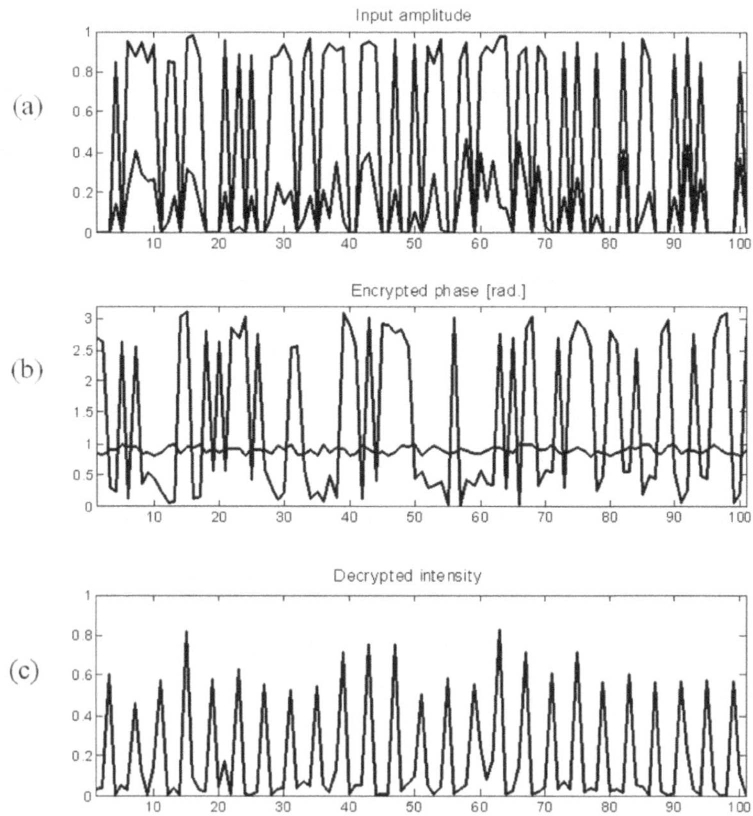

Fig. 11.20 The plots demonstrate the simulation results with uniformly distributed phase and amplitude noise tripled in magnitude at the input (a) in conjunction with a doubling in magnitude for the encrypted phase mask (b). The plot in (c) clearly shows that decryption of the same periodic phase array is still possible using a lower value for the threshold. Note that all phase values are plotted as absolute values in the range from 0 to π. The "invisible" phase noise at the input is again indicated in radians at the bottom of the plot in (a) and the perturbed amplitude of the noisy encrypted phase is illustrated as the curve smaller than the value 1 in (b).

Fig. 11.18 shows the phase key side by side with the encrypted phase mask. As with the input amplitude modulation, the simulation also perturbed the encrypted phase mask with both amplitude and phase noise of uniform random distribution. These noise factors have been set to 10% of a unit amplitude phase mask and 10% of a full phase cycle.

Figure 11.19(a) demonstrates the successful decryption of the encrypted periodic phase array and Fig. 11.19(b) shows the corresponding decrypted intensity. By implementing a thresholding operation with a threshold level equal to a half, a completely error-free decryption can be performed despite the introduction of amplitude and phase noise sources into the system.

To further examine the noise tolerance of the system, additional simulations were performed with increased noise levels. Figure 11.20 illustrates the simulation results in the presence of uniformly distributed phase and amplitude noise tripled in magnitude at the input in conjunction with a doubling in magnitude for the encrypted phase mask. The input amplitude shown in Fig. 11.20(a) is superposed noise factors set to 15% of the input amplitude and 15% of that of a full phase cycle. The encrypted phase mask in is shown in Fig. 11.20(b) with noise factors being set to 20% of a unit amplitude phase mask and 20% of a full phase cycle. Fig. 11.20(c) clearly shows that successful decryption of the periodic phase array can be made possible through an additional thresholding operation using a lower threshold level. This large tolerance to both phase and amplitude noise was maintained over a plurality of decryption simulation cycles each time performed with different random noise super-positions.

These results demonstrate the viability of the proposed method for decrypting a binary phase-only mask using an amplitude-only spatial light modulator. The system is based on two cascade-coupled common-path interferometers, with one implementing the RPC method and the other operating as a standard GPC system. Using an amplitude-only spatial light modulator, a binary amplitude key can be converted to phase for decrypting a binary phase-only encrypted pattern in the first spatial filtering system. The decrypted phase pattern is subsequently converted to an intensity pattern in a second spatial filtering system. The complete setup has been analysed analytically and subsequently using a highly accurate one-dimensional FFT-based modelling that includes all system aperture truncations for a realistic assessment. A key finding from the simulations is that the approach is found to be substantially robust to both amplitude and phase noise superposed the input amplitude and the encrypted phase mask simultaneously. Finally, a compact cascaded system can be implemented as a folded dual-path scheme operating the encrypted phase mask in reflection geometry or, for a robust implementation in real world applications, a truly miniaturized version of the unfolded system may be considered.

11.7 Summary and Links

In this chapter we presented yet another novel application of phase contrast – optical cryptography – which was made possible by our generalized analysis. As opposed to the weak phase modulation regime of traditional phase contrast, the expanded phase modulation range in GPC supports encoding encrypted information with higher security since the former can leave cues upon encryption that can compromise information security. The miniaturized device, discussed in Chapter 9 and applied here, is an attractive GPC implementation for supporting robust, compact and portable cryptographic systems. Furthermore, the discussion on the proposed system for decrypting binary phase patterns by amplitude, although applied for optical cryptography, illustrates a promising potential for creatively combining GPC, RPC and a suitable implementation scheme (e.g. a miniaturized device). This serves as an illustrative example of how the different aspects of GPC (and RPC), discussed throughout the book, can be used as potential building blocks for creating other novel optical systems.

References

1. B. Javidi, "Securing information with optical technologies," Phys. Today 50 (3), 27-32 (1997)
2. B. Javidi and J. Horner, "Optical pattern recognition for validation and security verification," Opt. Eng. 33, 1752-1756 (1994)
3. P. Réfrégier and B. Javidi, "Optical encryption based on input plane and Fourier plane random encoding," Opt. Lett 20, 767-769 (1995)
4. N. Towghi, B. Javidi and Z. Lou, "Fully phase encrypted image processor," J. Opt. Soc Am. A 16, 1915-1927 (1999).
5. B. Javidi and T. Nomura, "Securing information by digital holography," Opt. Lett. 25, 28-30 (2000)
6. J. Glückstad, "Phase contrast scrambling," International PCT patent WO 002339A1 (3 July 1998)
7. P. C. Mogensen and J. Glückstad, "A phase-based optical encryption system with polarisation encoding," Opt. Commun. 173, 177-183 (2000)
8. P.C. Mogensen, R. L. Eriksen and J. Glückstad, "High capacity optical encryption system using ferro-electric spatial light modulators," J. Optics A. 3, 10-15 (2001)
9. R. L. Eriksen, P. C. Mogensen and J. Glückstad, "Elliptical polarisation encoding in two dimensions using phase-only spatial light modulators," Opt. Commun. 187, 325-336 (2001)
10. J. Glückstad, "Image decrypting common path interferometer," in Optical Pattern recognition X, D. P. Casasent and T. Chao, eds, Proc. SPIE 3715, 152-159 (1999)
11. P. C. Mogensen and J. Glückstad, "Phase-only optical encryption," Opt. Lett. 25, 566-568 (2000)
12. P. C. Mogensen and J. Glückstad, "Phase-only optical decryption of a fixed mask," Appl. Opt 40, 1226-1235 (2001).
13. V. R. Daria, P. J. Rodrigo, S. Sinzinger, and J. Glückstad, "Phase-only optical decryption in a planar-integrated micro-optics system," Opt. Eng. **43**, 2223-2227(2004).
14. J. Glückstad, V. R. Daria, and P. J. Rodrigo, "Decrypting binary phase patterns by amplitude," Opt. Eng. **43**, 2250-2258 (2004).
15. F. Zernike, "How I discovered phase contrast," *Science* 121, 345-349 (1955)
16. J. Glückstad, "Phase contrast image synthesis," Opt. Commun. 130, 225-230 (1996)
17. J. Glückstad, "Phase contrast imaging," U.S. patent 6,011,874 (January 4, 2000)
18. J. Glückstad and P. C. Mogensen, "Optimal phase contrast in common-path interferometry," Appl. Opt 40, 268-282 (2001)

19. V. Daria, J. Glückstad, P. C. Mogensen, R. L. Eriksen and S. Sinzinger, "Implementing the generalized phase-contrast method in a planar-integrated micro-optics platform," Opt. Lett. 27, 945-947 (2002)
20. V. Daria, R. L. Eriksen and S. Sinzinger and J. Glückstad, "Optimizing the generalised generalized phase-contrast method for a planar optical device" J. Opt. A: Pure Appl Opt. 5, s211-s215 (2003).
21. J. Glückstad, V. R. Daria and P. J. Rodrigo, "Comment on: Interferometric phase-only optical encryption system that uses a reference wave," Opt. Lett 28, 1075-1076 (2003).
22. S. Sinzinger, "Microoptically integrated correlators for security applications", Opt. Comm. 209 (2002), 69-74.
23. S. Sinzinger and J. Jahns, Microoptics (2nd edition Wiley-VCH, Weinheim 2003)
24. J. Glückstad and P. C. Mogensen, "Reverse phase contrast for the generation of phase-only spatial light modulation", Opt. Comm. **197**, (2001) 268-282.
25. J. Glückstad, P. C. Mogensen and R. L. Eriksen, "Phase-only spatial light modulation by the reverse phase contrast method", Mol. Cryst. Liquid Cryst. **375** (2002), 679-688.

Chapter 12
Concluding Remarks and Outlook

Reflecting on the circumstances surrounding his discovery of the phase contrast phenomenon, Frits Zernike marvelled on the limitations of the human mind during his Nobel Prize lecture, remarking:

> *"How quick are we to learn – that is, to imitate what others have done or thought before – and how slow to understand – that is, to see the deeper connections. Slowest of all, however, are we in inventing new connections or even in applying old ideas in a new field."*

Now, more than three quarters of a century since Zernike's discovery of the phase-contrast phenomenon in 1930, most contemporary expositions of phase contrast, sadly, continue to *imitate* the simplifying approximations in the original formulation. The generalized phase contrast (GPC) method, in extending beyond the restrictive assumptions, is able to open up a wider range of applications. This monograph presents the theoretical foundations of the GPC method in an attempt to assist readers leapfrog over the conventional phase contrast treatments. We have also illustrated a repertoire of the broader applications that is enabled by the generalized formulation to provide motivation. These creative adaptations can provide a conceptual switchboard so that, by way of examples, readers can be inspired to begin "inventing new connections" and, hopefully, carry these generalized ideas further into other novel contexts.

12.1 Formulating Generalized Phase Contrast in a Common-Path Interferometer

In the early chapters, we developed an analytical framework for the design and optimization of common path interferometers (CPI) based on phase-only inputs and spatial filtering around the optical axis in the spatial Fourier domain. The GPC method was derived and introduced as the common denominator for these systems basically extending Zernike's original phase contrast scheme into a much wider range of operation and application. We have shown that the GPC method can be successfully applied to the interpretation and subsequent optimization of a number of different commonly employed spatial filtering CPI architectures. We have proposed that a considerable improvement in CPI design and the interpretation of experimental results should arise from the fact that we have a detailed treatment of the profile of the synthetic reference wave (SRW). We have derived optimal conditions for interferometric accuracy, taking account of the synthetic reference wave, and have shown that our results agree well with empirical results from the literature.

Our analytical approach makes it possible to characterise any CPI in terms of a combined filter parameter which places all CPI filters in the same phase space domain, drastically simplifying the comparative analysis of different CPI types. The basic GPC model presented in Chapter 3 is refined and adapted throughout the book, depending on the specific context, to exploit available design freedoms and match the practical constraints in the chosen applications.

12.2 Sensing and Visualization of Unknown Optical Phase

Conventional phase contrast is geared towards microscopic imaging of thin biological samples and, hence, its range of validity is limited to weak phase perturbations. The GPC framework breaks away from the weak phase limitations and can be applied to optimize more general phase visualization and other wavefront sensing applications, as discussed in Chapter 5. We have compared a range of well-known CPI types using the complex filter space plots we have developed, which indicate how their performance might be improved. Using the criteria of high fringe accuracy, high visibility and peak irradiance we have shown that it is possible to optimise a CPI system for operation with a given dynamic range of phase distribution at the input. The complex filter space plots show that the lossless operating curve (i.e., when using an absorption-free filter) provides an extremely good first choice for a variety of filtering applications. The operation of a CPI in a photon-limited regime should always seek to remain on the lossless operating curve. However, the inclusion of a certain degree of field absorption becomes increasingly necessary for large-scale input phase perturbations, if the visibility is to be maximised.

We have applied our analysis to extend the linearity of the phase-to-intensity mapping in a CPI and have shown that it is possible to improve the linearity of some currently applied systems. For instance, we illustrated that the generalization of the Henning method offers considerable practical improvements. We have also discussed the extension of linear, unambiguous phase-intensity mapping to the full phase circle and demonstrate through the use of originally designed phasor charts that this can be achieved by the operation of two CPI systems in parallel. We have shown that for high fringe accuracy and linear phase-intensity mapping conditions, the on-axis amplitude damping term is superfluous thus simplifying filter design. Moreover, we have demonstrated that the advantages offered by the generalized treatment extend to CPI-based phase-shifting interferometry and can be adapted for accurate quantitative phase imaging.

12.3 Synthesizing Customized Intensity Landscapes

Having provided an overview of the GPC framework and illustrating its functionality in a *camera mode* for sensing unknown phase inputs, we turned in another direction to show GPC functionality in a *display mode* where user-controlled phase inputs are used for synthesizing user-defined intensity landscapes, as discussed in chapters 6 and 7. Using appropriate phase-only, spatial light moadulation technologies to condition the CPI input, we are able to create desired intensity distributions at the CPI output. We derived equations for optimising light efficiency and visibility of arbitrary analog phase-encoded wavefronts. We exploited this control over the input modulation in wavefront engineering applications as additional design freedom. The availability of design freedoms for both the filter and input parameters enabled a more versatile optimization approach. We have also demonstrated how the user-adjustable phase input can be exploited to devise compensation schemes for improving output homogeneity. The developed design criteria were illustrated using practical design examples such as Gaussian laser beam shaping and laser projections of periodic and arbitrary light patterns not only for binary but also greyscale intensity levels. We tackled practical issues such as the influence of using the rectangular aspect ratios of contemporary devices and performance under broadband illumination. Looking at binary intensity projections, we showed the possibility of adopting more experimentally expedient approaches such as ternary-phase, and even binary-only wavefront encoding. When benchmarking GPC relative to phase-only computer generated holography, we showed that GPC is able to get the most out of the available information capacity in the modulation devices and, hence, highly suitable for projecting efficient dynamic greyscale intensity landscapes at high refresh rates.

12.4 Projecting Dynamic Light for Programmable Optical Trapping and Micromanipulation

One of the most successful applications of the dynamic, real-time reconfigurable light patterns created by the GPC method is in programmable optical trapping and manipulation of microscopic specimens, which is discussed in Chapter 8. Coupled through microscope objectives, the user-controllable light fields can be used as optical traps for exerting controllable optical forces onto a plurality of microparticles in conjunction.

The multiple traps can be simultaneously controlled in real-time, allowing for independent control of each particle. When dealing with extended micro-objects, such as a microtool, several traps can be operated in concert to grab different parts of the microtool so that it can be manoeuvred translationally and rotationally with a high degree of control. The optical force depends on the geometric and optical properties of the object being manipulated. Hence, aside from providing handy control over particle motion, the ease with which the optical fields can be reconfigured lends itself to the unique feature of the GPC-powered traps being able to match the particle properties. In practice, GPC-based trapping is implemented using counterpropagating beam traps, where the created light fields are introduced from opposite sides of the sample chamber. This counterpropagating geometry allows for using long working distance microscope objectives with relatively lower NA, which leaves ample space for incorporating accessory systems orthogonally, such as a secondary microscope for side-view imaging or even nonlinear imaging schemes for enhanced probing capacity. Chapter 8 presented optical trapping experiments utilizing assorted micro-objects ranging from dielectric microspheres and living cells with diverse properties, to extended silicon and photopolymerized microstructures. We also described a GPC-based biological experiment, where a GPC trapping system was used to gain novel insights on the cell growth dynamics in a mixed yeast culture. We have likewise provided supporting theoretical analysis of the optical forces in counterpropagating beam trapping that are relevant to optimizing GPC-based optical trapping systems.

12.5 Exploring Alternative Implementations

In most GPC demonstrations, the optical system is typically illustrated as a *4f* optical processing system consisting of two Fourier lenses with a static, prefabricated phase contrast filter (PCF) in between that shifts the phase of spatial frequency components around the zero-order. There are, however, many alternative experimental layouts that users can choose from to suit their requirements, and some of these are discussed in chapter 9. For example, we considered a PCF capable of introducing an adjustable phase shift. This can be used to implement dynamic optimization where the phase shift adapts to varying input conditions. Using a nonlinear material to automatically create a light-induced PCF offers practical advantages for optical alignment and can also dynamically adjust its dimensions to adapt to input conditions. An application calling on a compact, portable and robust GPC system can use an integrated implementation on a planar microoptics platform. A multi-beam implementation can be used in high power applications to redistribute energy in the Fourier plane and increase power tolerance. From linear systems theory, one can integrate the functionality of a succeeding optical processing system into the PCF to create a compact hybrid system. This was done in the mGPC, which integrates a matched filtering functionality into the PCF to enable synthesis of sharper light peaks, among others. A generalized hybrid system that integrates elements from the different alternative implementations can also be considered when customizing a GPC system for specific applications. While the experimental illustrations were focused on GPC, the same alternative schemes may also be considered for implementing the reverse phase contrast method discussed in Chapter 10.

12.6 Creating Customized Phase Landscapes: Reversed Phase Contrast Effect

The GPC method is a powerful tool for efficiently synthesizing dynamic greyscale intensity landscapes at high refresh rates. On the other hand, phase-only holograms are classic illustrations that, in Fourier synthesis, the phase can play a dominant role over amplitude in preserving signal information. Thus, in Chapter 10, we discussed a method for creating customized phase landscapes. This technique is termed reverse phase contrast method, seeing as generating an output with uniform amplitude intensity whose underlying phase modulation mimics the input amplitude modulation – effectively a reversed phase contrast effect. We adapted the general framework for CPI analysis, earlier used for the GPC method, to examine the CPI with amplitude-modulated inputs and found optimal design criteria for producing the reversed phase contrast effect. This method enables a given spatial binary intensity distribution to be converted into a binary phase distribution with a spatially uniform intensity profile, the phase step of which is determined by a Fourier plane filtering operation. The spatial performance of such a system is only limited by the input amplitude modulation source, which may be either a static fixed-mask or dynamic amplitude based spatial light modulator. The RPC method remains relevant given available phase modulating devices since high-quality amplitude masks are easier to produce and high-speed amplitude modulators are widely available. The RPC method offers possibility of achieving high performance phase-only spatial light modulation. As discussed elsewhere in the book, one application of the RPC method is in optical cryptography.

12.7 Utilizing GPC and RPC in Optical Cryptography

Optical technologies are becoming ubiquitous elements of modern information systems, performing roles that include encoding, processing, transmission, storage, among others. In this era of massive information traffic, cryptographic technologies are highly needed for securing information to ensure authenticity and prevent fraud. Here, optical cryptographic techniques can offer compatibility with other optical aspects of the information system. In chapter 11, we described various possibilities for applying the GPC and RPC methods in optical cryptography. We experimentally applied GPC in phase cryptography, showing macrooptical demonstration of optical decryption. With the vision of a fully integrated miniaturized future system, we revisited system miniaturization and described an experimental demonstration of optical decryption in a micro-optical system. Furthermore, we described a cryptographic system combining GPC and RPC functionality through dual passage through the same CPI system for the optical decryption of binary phase patterns by amplitude.

12.8 Gazing at the Horizon Through a Wider Window

Scientific and technical innovation is a work in progress and, as in practical examples, having the proper tools can make a big difference. In the case of phase contrast, we did not invent a new tool but sharpened an old, rather blunt one. In doing so, it was revitalized and imbibed with a wider set of functionalities and enables the creation of other tools, as we have seen throughout the book. This book attempts to make these tools available and useful to others.

Appendix
Jones Calculus in Phase-Only Liquid Crystal Spatial Light Modulators

In Jones calculus, the polarization state of light is represented by a Jones vector,

$$\boldsymbol{E} \Rightarrow \begin{bmatrix} E_x \\ E_y \end{bmatrix} \tag{A.1}$$

where E_x and E_y are normalized complex amplitude components along x- and y-axes, respectively. An optical element is represented by a Jones matrix,

$$\mathbf{T} \Rightarrow \begin{bmatrix} T_{11} & T_{12} \\ T_{21} & T_{22} \end{bmatrix}, \tag{A.2}$$

and light that impinges on it undergoes the transformation

$$\mathbf{E}^{(\text{out})} = \mathbf{T}\mathbf{E}^{(\text{in})} \Rightarrow \begin{bmatrix} E_x^{(\text{out})} \\ E_y^{(\text{out})} \end{bmatrix} = \begin{bmatrix} T_{11} & T_{12} \\ T_{21} & T_{22} \end{bmatrix} \begin{bmatrix} E_x^{(\text{in})} \\ E_y^{(\text{in})} \end{bmatrix}. \tag{A.3}$$

A phase-only liquid crystal spatial light modulator (SLM) is a two-dimensional array of independently addressable uniaxial birefringent pixels. The pixels are characterized by a common ordinary refractive index, n_o, along one axis and individually voltage-controllable extraordinary refractive index, $n_e(V)$, along an orthogonal axis. To understand the various operation modes of an SLM, consider the optical setup shown in Fig. A.1. The SLM orientation is specified by the angle θ between the extraordinary axis and the y-axis.

The Jones matrix for an SLM oriented with its extraordinary axis parallel to the y-axis ($\theta=0$) is

$$T_{SLM}^{(0)} = \begin{bmatrix} \exp\left(j\frac{2\pi}{\lambda}dn_o\right) & 0 \\ 0 & \exp\left(\frac{2\pi}{\lambda}dn_e(x,y)\right) \end{bmatrix}. \tag{A.4}$$

where d is the thickness of the SLM cell and the index modulation $n_e(x,y)$ is controlled by the applied pixel voltages $V=V(x,y)$. To study the output polarization states, it is convenient to write the unrotated Jones matrix as

$$\boldsymbol{T}_{\text{SLM}}=\exp\left(j\frac{2\pi}{\lambda}dn_o\right)\begin{bmatrix}1 & 0\\ 0 & \exp\left[\delta(x,y)\right]\end{bmatrix},$$

where $\delta(x,y)=\frac{2\pi}{\lambda}d\left[n_e(x,y)-n_o\right]$ describes a spatially varying phase retardation. Each SLM pixel acts as a programmable phase retarder, enabling various polarization states to be encoded on an incident light. Using this alternate expression with the rotation matrix, R_θ, we can apply a rotational transformation to get the new SLM Jones matrix for an arbitrary θ

$$\begin{aligned}\boldsymbol{T}_{\text{SLM}}^{(\theta)}&=\boldsymbol{R}_\theta^{-1}\boldsymbol{T}_{\text{SLM}}^{(0)}\boldsymbol{R}_\theta\\ &=\exp(j\phi_o)\begin{bmatrix}\cos^2\theta+\sin^2\theta\exp(j\delta) & \sin\theta\cos\theta\left[-1+\exp(j\delta)\right]\\ \sin\theta\cos\theta\left[-1+\exp(j\delta)\right] & \sin^2\theta+\cos^2\theta\exp(j\delta)\end{bmatrix}.\end{aligned}\qquad\text{(A.5)}$$

Note that the uniform phase offset, $\phi_0=(2\pi/\lambda)dn_o$, which does not affect the polarization, is commonly neglected in Jones calculus.

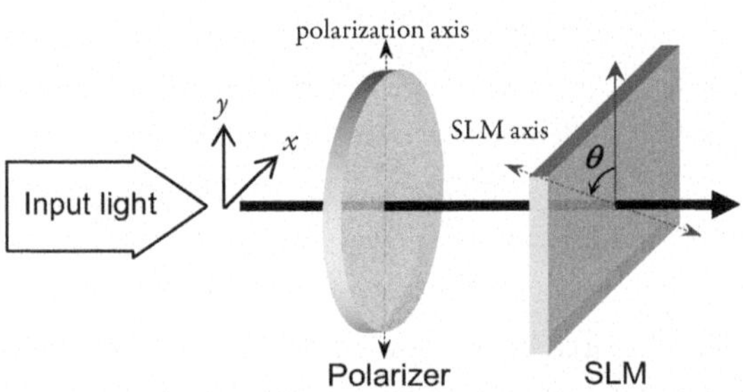

Fig A.1 Optical setup for realizing different SLM operation modes. A linearly polarized (along the y-axis) light is incident on a spatial light modulator (SLM). The director axis of the liquid crystal molecules in the SLM is oriented at an angle, θ, with the incident polarization.

A.1 Spatial Phase Modulation

Consider a vertically polarized input light

$$E^{(in)}=\begin{bmatrix}0\\1\end{bmatrix},\qquad\text{(A.6)}$$

incident on an SLM oriented at θ=0. Using the SLM Jones matrix in Eq. (A.4), the output after SLM is a new vector $\boldsymbol{E}^{(out)}$ given by

$$\boldsymbol{E}^{(out)} = \boldsymbol{T}_{\text{SLM}}^{(0)} \boldsymbol{E}^{(in)} = \exp\left[\mathrm{j}\phi(x,y) \right] \begin{bmatrix} 0 \\ 1 \end{bmatrix}. \tag{A.7}$$

This is a vertically polarized output that contains a spatial phase modulation,

$$\phi(x,y) = \frac{2\pi}{\lambda} dn_e(x,y). \tag{A.8}$$

Thus, the SLM operates as a *spatial phase-only modulator* when its extraordinary axis is aligned with the incident polarization.

A.2 Spatial Polarization Modulation

For an SLM oriented at $\theta = 45°$, the rotationally transformed T_{SLM} in Eq. (A.4) becomes

$$\boldsymbol{R}^{-1}(\theta)\boldsymbol{T}_{\text{SLM}}\boldsymbol{R}(\theta) = \frac{1}{2}\exp(j\phi_o) \begin{bmatrix} 1+\exp(j\delta) & 1-\exp(j\delta) \\ 1-\exp(j\delta) & 1+\exp(j\delta) \end{bmatrix}. \tag{A.9}$$

Using the same vertically polarized input $\boldsymbol{E}^{(in)}$, the output field $\boldsymbol{E}^{(out)}$ is

$$\begin{aligned} \boldsymbol{E}^{(out)} &= \boldsymbol{T}_{\text{SLM}}^{(0)} \begin{bmatrix} 0 \\ 1 \end{bmatrix} = \frac{1}{2}\exp(j\phi_o) \begin{bmatrix} 1-\exp(j\delta) \\ 1+\exp(j\delta) \end{bmatrix} \\ &= \exp\left(\mathrm{j}\frac{2\phi_o + \delta - \pi}{2} \right) \begin{bmatrix} \sin(\delta(x,y)/2) \\ \mathrm{j}\cos(\delta(x,y)/2) \end{bmatrix}. \end{aligned} \tag{A.10}$$

The uniform phase offset, which is not relevant to the polarization state, can be neglected and we can describe the output polarization simply as

$$\boldsymbol{E}^{(out)} = \begin{bmatrix} \sin(\delta(x,y)/2) \\ \mathrm{j}\cos(\delta(x,y)/2) \end{bmatrix} \tag{A.11}$$

In spatial amplitude modulation, an output polarizer isolates either the vertical or horizontal component of Eq. (A.11). Spatial polarization modulation utilizes the full output with both components. Spatially programmed polarization is achieved by controlling the spatial phase retardation of the SLM. Output polarization vectors for some values of δ are illustrated in Table A.1. Other δ-values yield elliptic polarization states with major axis aligned either along the horizontal or the vertical.

Table A.1 Polarization vector in Eq. (A.11) for various retardation, δ.

$\delta(x,y)$	$E^{(out)}$
0	$j\begin{bmatrix}0\\1\end{bmatrix}$; ↑
π/2	$\frac{1}{\sqrt{2}}\begin{bmatrix}1\\j\end{bmatrix}$; ○
π	$\begin{bmatrix}1\\0\end{bmatrix}$; →
3π/2	$\frac{1}{\sqrt{2}}\begin{bmatrix}1\\-j\end{bmatrix}$; ↻
2π	$-j\begin{bmatrix}0\\1\end{bmatrix}$; ↓

Table A.2 A graphical representation of the output polarization ellipticity and orientation as a function of the phase shift implemented in corresponding regions of SLM-1 and SLM-2

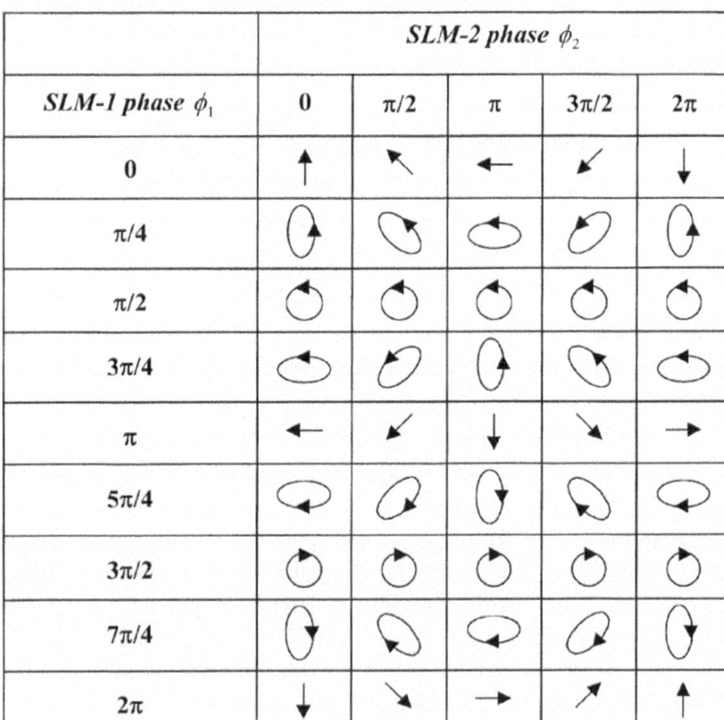

A.3 Spatial Polarization Modulation with Arbitrary Axis

A polarizer and SLM can generate elliptically polarized light with major axis that can be vertical or horizontal. It is possible to generate an arbitrary state of polarization by taking this elliptically polarized light and rotating its major axis to a desired arbitrary orientation. This rotation can be realized using SLMs and two quarter wave plates, and the full polarization encoding system is schematically illustrated in Fig. A.2. Both SLMs are oriented at $\theta=45°$.

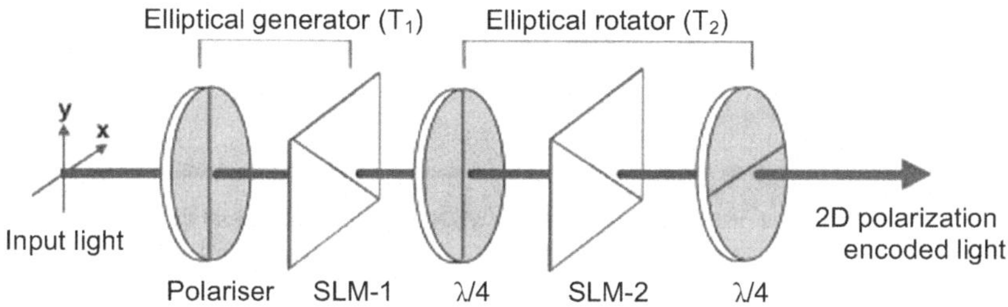

Fig. A.2 An optical system for converting incident polarized light into an arbitrary state of elliptically polarized light with the major axis of the elliptically polarized light rotated an arbitrary angle. The lines denote the extraordinary axis of the SLMs, the quarter wave plates (λ/4) and the polarization direction of the linear polarizer.

To verify the axis rotation, let us write the transformation matrix for the two quarter wave plates and SLM:

$$\begin{aligned}T_{\text{rotator}} &= \begin{bmatrix}1 & 0\\ 0 & -j\end{bmatrix}\begin{bmatrix}\cos\left(\delta_2(x,y)/2\right) & -j\sin\left(\delta_2(x,y)/2\right)\\ -j\sin\left(\delta_2(x,y)/2\right) & \cos\left(\delta_2(x,y)/2\right)\end{bmatrix}\begin{bmatrix}1 & 0\\ 0 & j\end{bmatrix}\\ &= \begin{bmatrix}\cos\left(\delta_2(x,y)/2\right) & \sin\left(\delta(x,y)/2\right)\\ -\sin\left(\delta(x,y)/2\right) & \cos\left(\delta_2(x,y)/2\right)\end{bmatrix}.\end{aligned} \tag{A.12}$$

This result shows that $\mathbf{T}_{\text{rotator}}$ is, indeed, a rotation matrix with a rotation angle of $\phi_2(x,y)/2$. The net transformation matrix for the optical system in Fig. A.2 is

$$\begin{aligned}&R_{\delta_2/2}T_{SLM}^{(\pi/4)}\\ &= \begin{bmatrix}\cos\left(\delta_2(x,y)/2\right) & \sin\left(\delta_2(x,y)/2\right)\\ -\sin\left(\delta_2(x,y)/2\right) & \cos\left(\delta_2(x,y)/2\right)\end{bmatrix}\begin{bmatrix}\cos\left(\delta_1(x,y)/2\right) & -j\sin\left(\delta_1(x,y)/2\right)\\ -j\sin\left(\delta_1(x,y)/2\right) & \cos\left(\delta_1(x,y)/2\right)\end{bmatrix}\\ &= \begin{bmatrix}\cos\left(\delta_1(x,y)/2\right)\cos\left(\delta_2(x,y)/2\right) & -j\sin\left(\delta_1(x,y)/2\right)\sin\left(\delta_2(x,y)/2\right)\\ j\sin\left(\delta_1(x,y)/2\right)\sin\left(\delta(x,y)/2\right) & \cos\left(\delta_1(x,y)/2\right)\cos\left(\delta_2(x,y)/2\right)\end{bmatrix}.\end{aligned} \tag{A.13}$$

Output polarization states from this arbitrary polarization encoding system are graphically illustrated in Table A.2. The phase retardations of SLM-1 and SLM-2 determine the type and direction of the output polarization. It is interesting to note that the direction of the elliptically polarized light changes from left handed to right handed when the phase modulation of SLM-1 is above π. This table shows that it is possible to generate an arbitrary state of elliptical polarization if both SLM-1 and SLM-2 can produce a phase modulation of at least 2π. Experimental demonstrations of arbitrary polarization encoding are available in Ref. [1].

Reference

1. R. L. Eriksen, P. C. Mogensen, and J. Glückstad, "Elliptical polarisation encoding in two dimensions using phase-only spatial light modulators," Opt. Commun. **187**, 325-336 (2001).

Index

GPSR Compliance

The European Union's (EU) General Product Safety Regulation (GPSR) is a set of rules that requires consumer products to be safe and our obligations to ensure this.

If you have any concerns about our products, you can contact us on ProductSafety@springernature.com

In case Publisher is established outside the EU, the EU authorized representative is:

Springer Nature Customer Service Center GmbH
Europaplatz 3
69115 Heidelberg, Germany

The manufacturer's authorised representative in the EU is Springer
Nature Customer Service Centre GmbH, Europaplatz 3, 69115 Heidelberg,
Germany. If you have any concerns regarding our products, please
contact ProductSafety@springernature.com

Printed and bound by CPI Group (UK) Ltd, Croydon, CR0 4YY
13/08/2026
02195493-0003